PROPOSAL PREPARATION

NEW DIMENSIONS IN ENGINEERING

Editor
RODNEY D. STEWART

SYSTEM ENGINEERING MANAGEMENT
Benjamin S. Blanchard

NONDESTRUCTIVE TESTING TECHNIQUES
Don E. Bray (Editor)
Don McBride (Editor)

LOGISTICS ENGINEERING
Linda L. Green

NEW PRODUCT DEVELOPMENT: DESIGN AND ANALYSIS
Ronald E. Kmetovicz

INDEPENDENT VERIFICATION AND VALIDATION: A LIFE CYCLE ENGINEERING PROCESS FOR QUALITY SOFTWARE
Robert O. Lewis

DESIGN TO COST
Jack V. Michaels
William P. Wood

OPTIMAL INVENTORY MODELING OF SYSTEMS: MULTI-ECHELON TECHNIQUES
Craig C. Sherbrooke

COST ESTIMATING, SECOND EDITION
Rodney D. Stewart

PROPOSAL PREPARATION, SECOND EDITION
Rodney D. Stewart
Ann L. Stewart

PROPOSAL PREPARATION

Second Edition

RODNEY D. STEWART
ANN L. STEWART

A Wiley–Interscience Publication
JOHN WILEY & SONS, INC.
New York • Chichester • Brisbane • Toronto • Singapore

In recognition of the importance of preserving what has been written, it is a policy of John Wiley & Sons, Inc., to have books of enduring value published in the United States printed on acid-free paper, and we exert our best efforts to that end.

Library of Congress Cataloging in Publication Data:

Stewart, Rodney D.

Proposal preparation / Rodney D. Stewart, Ann L. Stewart.—2nd ed.

p. cm.—(New dimensions in engineering)

"A Wiley Interscience publication."

Includes bibliographical references and index.

1. Proposal writing in business. I. Stewart, Ann. L. II. Title. III. Series.

HF5718.S85 1992

808'.066658—dc20 91-41581

ISBN 0-471-55269-0 CIP

Printed in the United States of America

10 9 8 7 6 5 4 3 2 1

To Our Marriage
The proposal that got it all started

CONTENTS

FOREWORD

AN INDUSTRY PERSPECTIVE

I first met Rod when he was working as a Manager of Cost Analysis for NASA. I was impressed by his professionalism and detailed knowledge of estimating, pricing, and the entire proposal process. It was obvious that he had a higher order of interest in the process and phenomena of these subjects and was destined for a role which would influence, teach, and lead others. Our periodic meetings during the ensuing years and his extensive accomplishments and publications have reinforced these original opinions. It has always been the free-flowing spirit, optimism and relentless charge toward innovation and change that has bonded us to this profession of developing proposals for state-of-the art products.

After more than 35 years of riding the waves of public perception, legislation, and changing world events impacting the funding of our process, we still produce proposals today which ultimately create growth in both economic and technological terms. The end of the cold war, the rapid change in world events, and the shift to domestic priorities, translates into an era where both technical and affordability issues are now paramount. From the early 1960s through the 1980s, technology was either the singular or the primary driver for systems procurement activities. Unsolicited proposal awards, technology for its own sake, and a cost type environment were common. Request for Proposal (RFP) responses were generally not focused on cost containment and cost performance. Cost credibility was rarely a major consideration in source selection. Today, in this new era of affordability and cost containment, both the offerer and the buyer are driven by cost concerns at the beginning of the procurement cycle. The current driver therefore is for smaller proposals, containing crisp, clear, verifiable data. This meets current goals to accelerate the source selection and award process through "awards without discussion" and, in the case of negotiated procurements, awards based on overall "best value". Both cases dictate that the proposal must be wholly self-contained.

On the selling side, proposal preparation funds derived from bid and proposal expense and profit are severely limited; therefore it is imperative that we incorporate Total Quality Management (TQM) throughout the procurement process.

On the buying side, customers have increased their focus on cost credibility and past performance as a way to measure the balance between optimism and realism. Although every chapter in this book has relevancy to the proposal process, the content of Chapter 12 relating to source selection especially sharpens this focus. The many facets of evaluation provide unique and valuable insights for both offerers and buyers.

Aside from the many technological advancements, a major change in the proposal preparation profession has been the trend towards the open sharing of information. Compilation of data bases that relate to every facet of our business, together with generic tables and processes are now shared with fellow professionals. While spirited competition is the springboard for improved products and overall lower cost, sharing information improves the proposal preparation and acquisition process contributing to the foundation of a strong economy and technology base for our nation.

As one of many people dedicated to this profession, I am pleased that Rod and Annie Stewart are part of the sharing and improvement of this process. *Proposal Preparation—Second Edition* should be required text not only for those who want education in this field but also for those engaged in oversight and legislative activities. The knowledge gained would be beneficial in each environment. It also pleases me that Rod and Annie have followed in the steps of John and Lillian Gilbreth in furthering insights into this esoteric profession.

JAY W. CHABROW
Director, Contracts, Pricing & Cost Data Systems
TRW Space and Defense

AN ACADEMIC PERSPECTIVE

One of the biggest challenges facing us today is the development and production of systems, products, and/or services that will meet new consumer needs in a cost-effective manner. There is a requirement to completely define current areas of deficiency, identify alternative approaches to meeting a given need, and select a preferred configuration that will fulfill *all* consumer expectations expeditiously and with a minimum of overall cost. This is happening at a time when our natural resources are dwindling and competition is increasing worldwide.

From past experience, it has become evident that much can be gained toward meeting this objective by improving the communications between the consumer and the producer during the early phases of a given program. It is during the initial stages of planning and design when decisions are made that have the greatest impact on the overall life-cycle cost of a system, product, or service. Further, many of these critical decisions evolve during the "proposal phase" when prospective suppliers are attempting to respond to specific consumer needs. In the past, *real* consumer needs have not been well defined, suppliers have developed elaborate proposals and marketing efforts in response to the *perceived* requirements, contracts have been initiated, and systems/products/services have been developed, only to learn at a later point in time that the real needs have not been met! This, of course, generally results in numerous modifications that are costly in terms of unnecessary resource consumption and waste.

With the initial proposal phase of activity being critical for any program, it is essential that (1) the *real* need(s) of the consumer be interpreted and defined completely, accurately, and with a minimum response time; (2) as many different feasible alternative approaches as possible be identified and evaluated in terms of meeting the defined need; and (3) that the selected configuration of the system, product, or service being proposed is presented in a clear, comprehensive, and concise manner. The objective is to promote the necessary communications between the prospective producer and the consumer at a time in the life cycle when there is still some degree of flexibility. Through a more indepth front-end planning and analysis effort (as part of the producer's proposal activity), there is greater assurance that the ultimate system/product/service will be more cost effective and that the risks of customer dissatisfaction will be less. It is during this proposal phase of the life cycle when the decision-making process is critical!

This text has been prepared to assist the prospective producer in the development and implementation of a formal proposal activity. Not only should excellent producer–consumer communications be established by virtue of the comprehensiveness of the text material, but the proposer's effort should be enhanced through the numerous real-world illustrations, checklists, and examples employing the utilization of various analytical and graphical aids. Through the application of many different computer-aided methods, the authors show how the objectives stated above can be realized.

Ann and Rod Stewart have done an *excellent* job in developing this very comprehensive "how to" guide covering the overall proposal preparation process. The entire process is presented commencing with the initial identification of a consumer need and extending through the steps of proposal preparation, marketing the proposal, and the final evaluation and final selection of the proposal by the consumer. The six basic steps of proposal preparation activity (marketing, analyzing, planning, designing, estimating, and publishing) are described in detail. The authors have included an organizational approach for proposal preparation, a case study entitled "A Winning Proposal", and a comprehensive bibliography. The reader of this book should gain significantly from its content. I congratuate Ann and Rod for their excellent product!

BENJAMIN S. BLANCHARD
College of Engineering
Virginia Tech

A GOVERNMENT PERSPECTIVE

Government proposal evaluators will be pleased with this revised edition by the Stewarts. Government always seeks to elicit, and industry always wants to provide, proposals that are fully responsive to the Government's request in a clear, concise, and orderly manner. This priority goal has been long sought after but infrequently met. The Stewarts have captured, organized and presented in this second edition of *Proposal Preparation* valuable information for both industry and government for preparing requests for proposals (RFP's). It will also enhance their appreciation of the process necessary for the private sector to prepare credible, responsive proposals.

The complete iterative process of reaching a meeting of minds between the government and its offerors has been a source of frustration and misunderstanding for many years. This edition of *Proposal Preparation* and the guiding principles offered to the reader are a major contribution to the printed body of knowledge available to all those involved in the procurement process.

The effective use of the graphics approach for the representation of work, and other simplifying techniques described in this book, should make proposals more readable and easier to evaluate.

The treatment of government trends designed to increase and improve use of automation and advanced communication methods in the procurement process adds realism and usefulness to the value of this edition.

Credible and clearly written proposals that are completely responsive to RFP's are the most straightforward way to fulfill government policies designed to enhance competition and to improve the efficiency, economy and effectiveness of the procurement process.

The Stewarts are to be commended for the creation and compilation of this revised edition and their success in presenting an approach to successful proposal preparation.

DONALD E. SOWLE

Former Administrator, OMB
Office of Federal Procurement Policy
Executive Office of the President

Editor's Note: Don Sowle, who is now a consultant in government procurement matters, is past Director of the Office of Federal Procurement Policy and was Executive Director of the Staff of the Congressionally Appointed Commission on Government Procurement. The Commission was chartered in the late 1960s and early 1970s to study the Federal procurement process and to make recommendations to the Congress and the Administration for legislation and policies to improve the efficiency, economy, and effectiveness of the Federal procurement process. He was the 1990 recipient of the Herbert Roback Memorial Award; the highest award given by the NCMA.

PREFACE

As we move rapidly toward and into the 21st century, we are experiencing an explosion of ideas, technologies, products, and services. These work activities and work outputs extend far beyond present concepts and are more numerous than we had ever envisioned. Spawned by ever-deepening knowledge in science, engineering, physics, mathematics, and computers, thousands of practical commercial, industrial, military, and humanitarian products, processes, projects, and services are germinating and growing into workable and cost-effective proposals for the protection, improvement, and fulfillment of human life. Technologies are emerging from unexpected sources, including military research on intelligent target-seeking sensors, space exploration initiatives, and biotechnical and medical research, which have, concurrently, broad and specific application in both wider and more focused marketplaces. Large investments are being made by industry, foundations, and government to exploit and expand the practical applications of these emerging technologies. How rapidly and effectively these emerging technologies can be adapted to new work outputs and work activities now and in the new century will depend on how well the ideas, concepts, designs, and approaches can be articulated, expanded upon, and supported in effective winning proposals that will attract the financial support of sponsoring industries, institutions, and investors.

The second edition of *Proposal Preparation* has been both broadened in concept and deepened in specific areas to permit the researcher, entrepreneur, applications technologist, and marketer to take advantage of expanding technology itself in effectively and successfully selling the myriad ideas, products, projects, and services that are now emerging and will be forthcoming in the next decade and well into the new century. From the identification and tracking of fund sources, through strategic market planning, to the final desktop publishing of the winning proposal, this book has been expanded and integrated into a powerful package that provides the concepts, information, tools, techniques, and procedures needed to transform ideas into reality.

New to this book is the concept of the Graphic Representation of Work (GROW) procedure for preparing proposals as implemented by storyboarding, a technique long used by high-technology industries in preparing large proposals. The storyboarding technique can be used in a proposal of any size to speed up the proposal process, to improve client and customer understanding, and to provide more effective communication of the ideas presented and their benefits. Also new to this book is the concept of straight-line control, a sales method developed for product marketing and applied directly to the preparation of written unsolicited proposals or to written responses to invitations to bid. Straight-line control is defined as the "shortest distance between the supplier's abilities and the customer's needs."

State-of-the-art proposal preparation software and new computer techniques for rapid and responsive bidding are included in this volume for the first time. Since the first edition was published in 1984, monumental advances have occurred in rapid retrieval of digitized drawings, text, databases, and graphics and in multiple font typeset-quality desktop publishing. Proposal managers now have available to them a whole new set of high-technology tools that will provide professional-looking documents in record time that will rival that of most professional publishing houses. Computer-aided purchasing and proposal preparation are becoming a reality in many organizations. New software that will prepare critical path networks and bar charts for small to mammoth-size jobs is available, and new proposal pricing computer programs are being introduced. Small proposals can take advantage of the same winning techniques as large ones. How to keep within page limitations, produce small and large proposals within often extremely short submission times (30 to 45 days), and how to engender synergy between corporate objectives and market objectives, as well as proposal strategies for acquisitions of all sizes are explained in the book.

For large proposals, a rainbow of team reviews are discussed in the book. Red teams, blue teams, green teams, and pink teams are described, as well as how to use these teams most effectively and how to use the same techniques for smaller proposals. Other newly included subjects are customized résumé writing for proposals, effective use of concurrent engineering principles, business development databases, cross-referencing of proposals to work statements, quad-chart planning for new business, and many others.

Last but not least, an up-to-date, winning proposal for a high-technology prototype product is included as a case study, which uses all of the techniques described in this updated and expanded volume.

In the environment of the 1990s, leading to the year 2000 and beyond, proposal preparation has become a necessity for anyone wanting to advance an idea, product, process, or service to gain financial support. Proposal preparation procedures have become sophisticated, organized, methodical, and structured to a point where those *not* employing the latest technologies in proposal preparation may lose the opportunity to advance their idea, to gain financial support for their project, and to win the contract. It is our conviction that adherence to the fundamentals advanced in this book, coupled with a conscientious adoption of even newer tools and technologies as they emerge, will result in more contracts, greater financial support, and in expanded opportunities to market ideas for products, processes, projects, or services.

RODNEY D. STEWART
ANN L. STEWART

PROPOSAL PREPARATION

1 WHY STUDY PROPOSAL PREPARATION?

Do you not know that those who run in a race will run, but only one receives the prize? Run in a way that you may win!

—I Cor. 9:24

This book is about proposals and their preparation. What is a proposal? A proposal is a plan of action for fulfilling a need. It is a sales document that is honest, factual, and responsive to the needs of others. It is a written description of work to be performed that provides enough information for a customer to make a purchase decision.

There are three principal factors causing proposal preparation to become a more important skill in the 1990s and, we predict, into the new century:

1. Greater complexity and technical content of work activities and work outputs, both in the public and private sectors
2. Increasing sophistication of source evaluation and selection techniques
3. Growth of the number and complexity of requirements imposed by customers or clients in the procurement and purchasing processes—with no relaxation of the short time intervals allocated for proposal preparation and submission.

This chapter discusses the effects of these factors on the business and technical community and describes the benefits of developing proposal preparation skills. It outlines the basic steps and flow processes required to meet the proposal preparation needs brought about by the impact of current requirements.

THE AGE OF HIGH TECHNOLOGY

As John Naisbitt pointed out in his book *Megatrends*, high-technology products and services abound in the commercial marketplace as well as in business and industry. The purchaser of work outputs and work activities encounters multiple options more and more frequently, rather than either–or decisions. The number of variables to be considered in making a purchase decision is increasing because of the growing number and complexity of products and services offered. These trends have resulted in the need for greater reliance on more objective and analytical approaches to expenditure decision-making. The use of analytical approaches to source selection requires that the suppliers of goods, services, or projects provide formal, organized, and detailed information on the characteristics and price of the proposed product or service. Consumers—whether individuals, businesses, industries, or government agencies—are demanding good value for what they can afford to pay. Documented evidence must be provided along with sales information to assure these consumers that they are getting the highest quality available for their money. The written proposal is the document that is designed to perform this function.

SOPHISTICATION IN SOURCE EVALUATION

Today, even the more unsophisticated customer is likely to have some rather sophisticated analytical tools and methods at his or her disposal. Personal, professional, engineering, and business computers and their related computer software permit a rapid and economical comparison and analysis of multiple options with multiple features. The growing availability and capability of the computer, greater knowledge about its use for analytical comparisons, and the seemingly constant decrease in cost per unit of computing power have resulted in increased automation, speed, and accuracy in making the economic and technical decisions required to identify the most attractive choice among the many available alternatives. Using the computer and other high-technology tools and methods of management for structured analysis, organized procedures for source evaluation are evolving into increasingly systematic and objective processes. Further, some evaluation teams have access to large numbers of skilled technical and management personnel and occasionally have more time and resources allocated for their evaluation than the bidder had to prepare the proposal. Availability of advanced tools and techniques for the buyer requires the seller to use equally advanced tools, techniques, and methods in generating the sales approach. The written proposal presents the resulting sales information in a way that will favor the offered work activity over that of competitors, when it is subjected to a searching comparative analysis.

THE EFFECT OF GOVERNMENT LAWS AND REGULATIONS

Government procurement regulations, laws, and socioeconomic objectives have far-reaching effects on the conduct of business in the United States. Growth in the number of these regulations, laws, and policies in the past several decades has had a pronounced effect on the amount of information that must be accumulated (and usually reported) prior to entering into a new venture. In a 1972 study-group report published by the Commission on Government Procurement, 101 laws were cited as affecting the procurement process for major systems alone; 80 of these were associated with purely socioeconomic objectives. Recently the Department of Defense (DOD) issued an instruction referencing 35 directives covering 39 acquisition management and system design principles required in developing and producing major systems (U.S. Department of Defense, 1983). These laws and regulations affect the procurement process—and they have increased the amount of detailed information required for submitting responsive proposals both in the private and public sectors.

THE PROPOSAL: TRAINING GROUND FOR PERFORMANCE

The proposal preparation activity itself is an excellent training ground for actual performance of the work. The proposal preparation activity organizes the team, establishes the conceptual design, initiates planning, identifies needed equipment or facilities, formulates needed work methods and procedures, creates employee interest and involvement, and "debugs" the work activity or work output. The proposal preparation activity not only serves to fully inform the customer of what he or she can expect in the way of job performance, but, equally important, it familiarizes the proposer with the scope, intent, and content of the work. A flaw in design, planning, or estimating found during proposal preparation creates less impact than if it is found during work progress, when it is too late to make a change in the contract.

Experience has shown that there is a common thread of methodology that exists in virtually all proposal preparation situations. There are certain principles, practices, and procedures that hold true among the various categories of proposals, whether it is an industrial process, a manufactured product, a multimillion dollar construction project, or a business service that is being proposed. The several categories of proposals are described below.

CATEGORIES OF PROPOSALS

Fixed-Price "Advertised" Bids. Fixed-price bids and quotes in response to advertised procurements are common methods of providing information to potential customers for work activities and work outputs that are relatively straightforward in content. The bid may be in the form of a sealed bid referencing a specification or scope of work, or it may be accompanied by some form of proposal. In either case, it is to the advantage of the bidder to have completed the steps outlined in this book, even though all of the information produced is not supplied to the potential customer. The internal marketing analysis, planning, preliminary design, estimating, and publication steps will result in a more realistic bid and one which will more nearly represent a task that can actually be accomplished with the projected resources.

Negotiated Procurements. It is in preparation for solicited or unsolicited bidding on fixed price or cost-reimbursable negotiated procurements that improved skills in proposal preparation will yield the greatest rewards. Proposals for large procurements, particularly in the field of federal, state, and local government projects, involve extensive substantiation and supporting data. The larger the procurement, the more likely the procuring organization is to establish formal evaluation and competitive negotiation procedures. When the proposal is being reviewed by a team of experts rather than by a single purchasing agent, the full spectrum of proposal preparation skills must come into play. Figure 1.1, a typical proposal flow chart for a negotiated procurement, will be used throughout this book to describe the proposal preparation process.

In-Company Proposals. In-company proposals (those that are not submitted to a potential external customer or buyer) are solicited or unsolicited proposals that are prepared within a company to convince management, the stockholders, or a department head that the company should proceed with a work activity or work output that will improve the company's efficiency, economy, or effectiveness or enhance the potential of capturing future work. Typical of in-company proposals are requests for: (1) the acquisition of new equipment, facilities, methods, or techniques that will improve company profitability; and (2) independent research and development that the company sponsors with its own funds. A proposal for the location of a new plant or office facility is a typical in-company proposal that frequently employs all of the skills, methods, and techniques of proposal preparation.

Figure 1.2 shows some typical internal and external proposal types that are commonly encountered in the business and industrial community.

SIX BASIC STEPS IN PROPOSAL PREPARATION

There are six basic steps that are taken in proposal preparation, not necessarily in the order listed. They are: (1) marketing; (2) analyzing and making a bid decision; (3) planning; (4) designing; (5) estimating and (6) publishing the proposal. Although these steps generally are taken in the order listed, the marketing, analysis, and planning steps continue throughout the process.

Marketing. The principal step required to start out in a business venture is to find or identify a need and fill it. As a subset to this step, one can identify and generate new needs that are waiting to be filled by someone who makes available a work activity or work output on which the need is dependent. Fashion designers, for example, know that they can generate a need for a new style of clothing by widely advertising and publicizing the new style as being the latest and most chic. Successful businessmen realize that the idea of a need must be planted in the minds of their customers long before the proposal cycle is initiated. Throughout the proposal preparation process, the marketing function plays an important role in shaping and directing the policies, ground rules, and procedures used. As will be mentioned later, care must be taken to counterbalance the marketing department's optimism and desire to win new work for the company with realistic independent planning, scheduling, and estimating of the project by the performing elements of the organization.

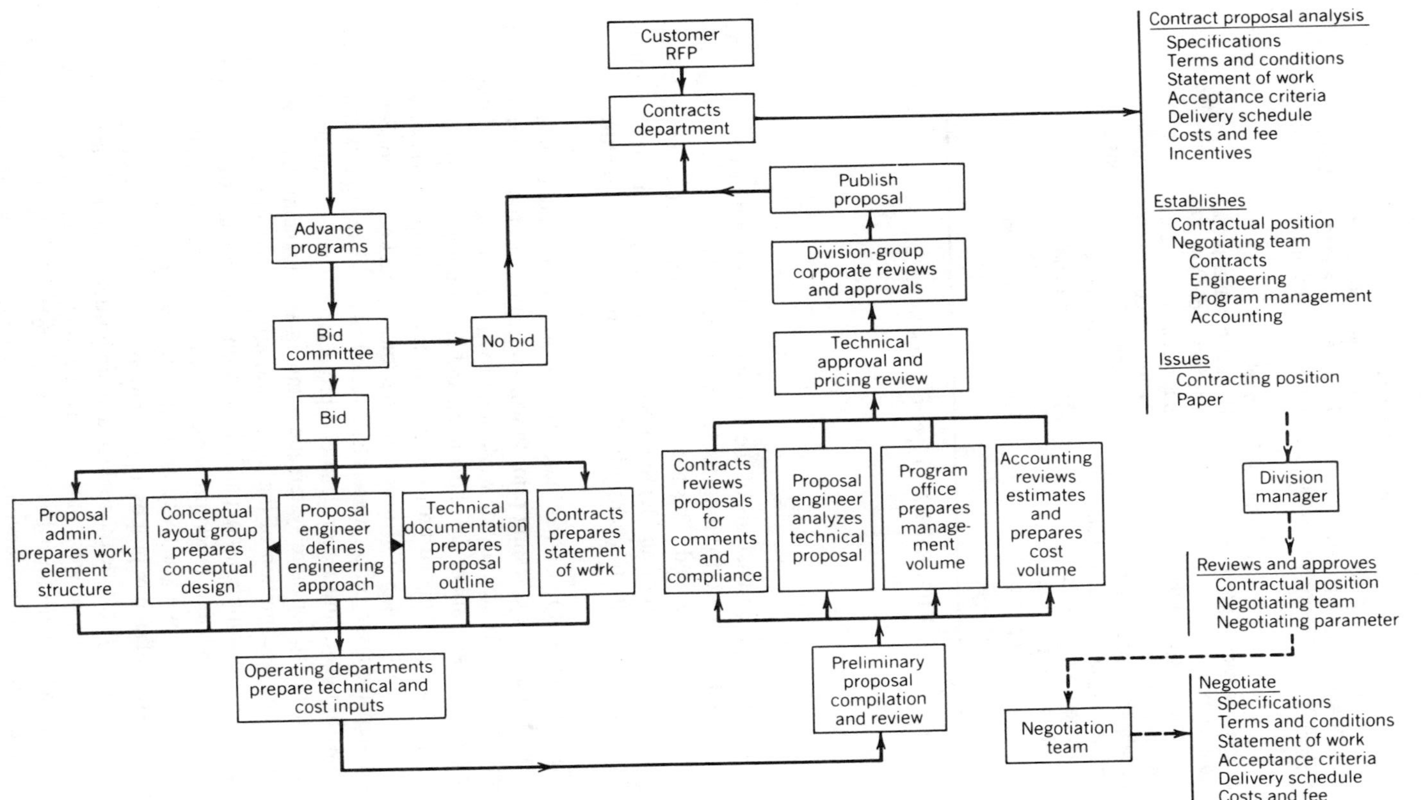

FIGURE 1.1 Proposal flow chart. RFP, request for proposal.

Internal Proposals

Independent research proposals	Plant relocation proposals
New product line proposals	Process engineering proposals
Operational procedures proposals	Product improvement proposals

External Proposals

Architect-engineer proposals	Research grants proposals
Book proposals	Services proposals
Construction proposals	Subsystems proposals
Design proposals	System production proposals
Development proposals	Systems development proposals
Engineering change proposals	Testing proposals
Maintenence proposals	"Time and materials" proposals
Manufacturing proposals	Tooling proposals
Materials supply proposals	Training and education proposals
Parts supply proposals	

FIGURE 1.2 Types of proposals.

Analyzing. It is through an analysis of the customer's needs and of the proposing company's capabilities of performing useful work that identifiable products and potential proposals are generated. This analysis will take many factors into account, but the criteria for selection of an identified need for further pursuit in the form of a specific proposal will most frequently be the profitability of the venture.

Planning. Experience in the United States as well as in other countries has shown that meticulous planning and testing prior to work initiation or product introduction helps assure that the work activity or work output will be profitable and beneficial to producer and consumer alike. Good planning and definition will help avoid dead-end projects, partially completed projects, and unnecessary duplication or overlap of work activities. A better job of proposal preparation, backed on conceptual or preliminary design and testing effort, means less wasted effort, more projects completed on time and within cost, and fewer dead end projects. Businesses are developing an awareness of the need for more systematic planning and financial analysis before initiating a venture. Recognition of the significant effect of economic factors and of the need for good planning has expanded the scope and content of the proposal from a simple document into a comprehensive technical, organizational business plan for accomplishing a work activity or producing a work output.

Planning, an essential step in proposal preparation, coupled with its close cousin, scheduling, is required to provide the realism and credibility needed in the proposal. In planning a work activity or work output, it is necessary to concentrate on as few alternatives as possible. One must beware of the professional planner who wants to be in a continuous planning mode with two or more alternatives being simultaneously analyzed in depth. Comparison of alternatives early in the need-identification phase is always helpful; but an early choice of a single alternative is usually very beneficial. The choice of a single alternative forces the resolution of key questions and assumes that the work can and will be performed in a selected manner. Planning includes all of the technical, organizational, and management aspects of the project and considers all marketing inputs, such as projected quantities, cost targets, and capture potential for the work.

Designing. Design work that is done as part of proposal preparation is usually conceptual design or preliminary design, although a final design of the work activity or work output will occasionally be required. When a proposal is to be submitted for performing final design work, conceptual or preliminary designs are all that is needed. Design work for proposal preparation includes preliminary sketches, plant layouts, flow diagrams, scale models, mockups, and prototypes. The degree of completion of design work, as evidenced by the number and types of drawings, models, mockups, components, or prototypes, is often a source evaluation factor. The design step of the proposal preparation process usually culminates in the preparation of the technical volume or technical section of the proposal.

Estimating. Estimating is one of the most important steps in the proposal preparation process. Estimating includes predicting or forecasting the amounts of materials, numbers of labor hours, and costs required to accomplish the job. Credible estimating cannot be done without adequate planning and preliminary design of the work activity or work output being proposed. Estimating requires unique skills, usually multidisciplinary in nature, that must either be acquired by experience or by training in a special mixture of technical and business disciplines. This unique mix is most nearly approached by the industrial engineering profession, but includes business skills that analyze and optimize profitability, which have not traditionally been a part of the industrial engineering discipline. The estimating step of the proposal preparation process culminates in the cost volume or cost section of a proposal.

Publishing. The publication step of proposal preparation includes the organization, writing, editing, art work, printing, and binding of the proposal document or documents. The publication capability and publication team should be an integral, responsive part of the proposing company's organization. Opportunities for using high technology in the proposal publication process must be taken if one is to submit a competitive proposal. The appearance and accuracy of a proposal, although not usually numerically scored by the evaluator or by the evaluation team, are important factors in the general impression made by the proposal. They can engender confidence and could be a basis for initial acceptance or rejection. Fancy or elaborate formats or displays are unnecessary and in many instances are even undesirable as they give an impression of lack of cost consciousness. A neat, accurate, easily read, easily referenced proposal is an aid to evaluators and is an indication to the customer of the type of work that he or she can expect to receive in reporting and documentation during the performance of the work.

URGENCY: A COMMON TRAIT OF PROPOSALS

Because there is usually a limited time available to respond to requests for proposals (RFPs) or requests for quotations (RFQs), urgency of publication is a common characteristic of proposals. The proposal preparation process involves the marshaling, managing, and utilizing of a broad range of disciplines, skills, and specialists in a very short time period and often under conditions of great stress: to (1) present the solution to the problem; and (2) compel action or adoption by the customer. This urgency of action and the stressful environment that exists in the days or weeks immediately preceding proposal submission is the reason for strategic planning and in-depth preproposal marketing and planning of the proposal activities themselves, as explained in chapters 2, 3, and 4. If preplanning is done properly, the urgency will not be a source of confusion or frustration to the proposal team, the proposal managers, or the company.

Below we list what one might call basic truths about proposal preparation and its treatment in this book:

- There are basic work elements and work activities that are common to all proposals. We emphasize the commonality rather than the differences because we would like to share this information with the widest possible audience and have it applied to the greatest variety of proposal situations.
- There are some highly successful instructions that will help produce a more credible, accurate, and attractive proposal.
- The first step in the proposal preparation process is introspection by the proposer to determine if he or she truly feels able to do the job better (by "better" we mean more advantageously to the customer) than anyone else.
- A really good proposal is often like a breath of fresh air, and many a completely unknown bidder has confounded the big name competition and won against a company that thought it had guaranteed success in having its bid accepted.
- In an almost-even contest, the scales might be tipped toward a favored bidder, but a poor proposal cannot win against a really good proposal, no matter who wrote it.
- One cannot win a contract just because one believes he or she can win—although that must come first. A person must put this belief into words and actions, which is the function of the proposal. These actions involve launching out in faith to do some of the preliminary laboratory work, design, and development.
- No need exists if the customer does not know that there is a work activity or work output to fulfill it. A proposal is the means of letting a customer know.
- The successful proposer is the one who can identify relevant information, process it, and present it more efficiently and effectively than his or her competitors.
- Survival in a tough, competitive environment is often thought to be possible only at the expense of someone else. This is true only in the sense of the losing

proposers. A successful proposal is a win–win situation for the winning supplier and the customer because they both gain. The successful proposer looks for and finds the synergistic effect between himself and his customer.

- The plain fact is that it is all but impossible to win with a poor proposal, no matter how effectively other marketing activities have been carried out, especially if there are better proposals submitted. On the other hand, many contracts are won by those submitting superior proposals and using no other marketing activity whatsoever.

THE CONCEPT OF STRAIGHT-LINE CONTROL

Preparing a winning proposal requires *straight-line control.* Straight-line control (see Figure 1.3) is the process of heading directly toward the objective of winning and *staying on* the most direct path until the objective is achieved. To do this, the proposal manager, every member of the proposal team, and corporate management must:

- *Address every point* covered in the request for proposal
- *Expand on every point* or issue covered in the request for proposal
- *Avoid including extraneous information not requested* or addressed in the request for proposal, unless the information serves to conclusively illustrate capability, competence, compliance, comprehension, risk, or credibility
- *Answer and expand on all questions* put to the proposer in other phases of the procurement process:
  1. When formal written or verbal questions are asked by the customer
  2. At the best and final offer

SEVEN STEPS TO STRAIGHT LINE CONTROL

1. Straight Line Control starts where you are now!
2. Straight Line Control puts first things first!
3. Straight Line Control takes the most efficient path!
4. Straight Line Control completes intermediate milestones on time!
5. Straight Line Control fulfills <u>all</u> requirements!
6. Straight Line Control takes the most efficient path!
7. Straight Line Control reaches the objective!

FIGURE 1.3 Seven steps to straight-line control.

3. When informal questions are asked to verify credibility of the proposal content.

- *Refuse to be diverted* from the written request by "perceived" requirements that are not requested
- *Respond immediately* to internal and external comments, criticisms, and inputs, either accepting them as positive inputs and encompassing corrective information in the proposal or by rejecting them if they are not in keeping with the request for proposal.

Several corporate and government proposal evaluators for large and small procurements alike told us the most common fault in proposals is the tendency of proposers to write all they know about the subject even if it is not requested or germane to the request for proposal. Straight-line control will avoid this common fault. Straight-line control will not only assure a winning proposal, but it will also conserve the precious resources required to prepare proposals. Taking paths that do not lead to fulfilling customer objectives as outlined in the request for proposal is wasteful and cost-inefficient. Straight-line control will minimize the disturbances and deviations from the most efficient and direct path to cost-effective proposal preparation.

Throughout this book we will be emphasizing the need for straight-line control and how to exercise it in the following areas, which are vital to proposal success:

1. Establishing corporate objectives, goals, and strategies
2. Developing teaming arrangements
3. Developing a comprehensive and targeted corporate experience description.
4. Preparing a comprehensive résumé database and rapidly preparing selectively targeted résumés for specific procurements.
5. Preparing for and carrying out computer-aided proposal preparation
6. Responding to the request for proposal
7. Proposal-writing
8. Proposal follow-up actions.

2

STRATEGIC MARKET PLANNING AND BUSINESS DEVELOPMENT: THEIR EFFECT ON PROPOSAL PREPARATION

Where there is no vision, the people perish.

—Proverbs 29:18

For multidisciplinary high-technology products, processes, projects, and services, the new-business acquisition cycle begins on the very day that corporate strategic planning is started. (This is not so for routine commercial products and services, as bids to sell existing goods can be submitted at the last minute before the bid receipt deadline.) For the high-technology marketplace addressed principally in this book, straight-line control starts as corporate management acquires a vision of the market to be addressed and of the goods or services to be sold. Through luck or dogged determination, some firms succeed without vision or a long-range plan; but those who have been most successful in capturing a market segment or a series of large contracts have done so through some initial concept, vision, or dream in the mind of the founder or top executives of the company.

Steven Jobs, for example, envisioned a user-friendly, people-oriented, affordable computer system when he founded the Apple Corporation. Ray Kroc, an early executive of McDonald's, wanted to make the best hamburgers available anywhere and serve them in an atmosphere of "quality, service, cleanliness, and value." A review of the history of excellent corporations invariably reveals key people in the company who had a strong and enduring vision. The ability to formulate a vision, to build goals, objectives, and a strategic plan around the vision, and to keep the activities of the company focused on the vision and its implementing strategic plan are keys to long-term business success.

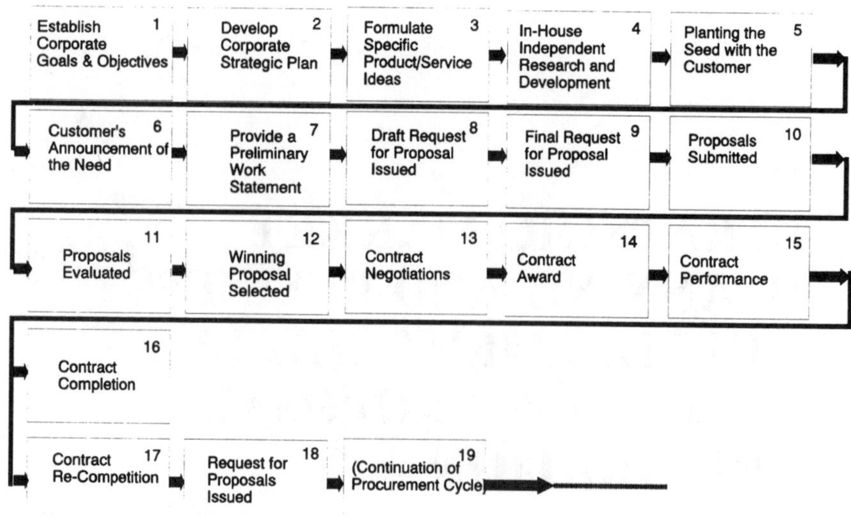

FIGURE 2.1 Strategic planning and procurement cycle.

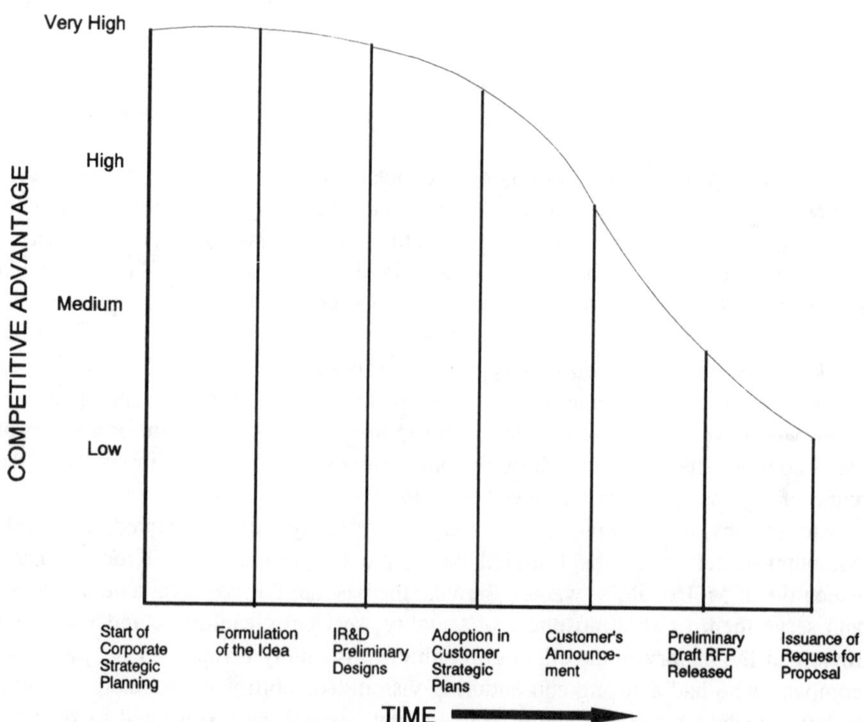

FIGURE 2.2 Competitive advantage versus relative time of injection into the procurement process. IR&D: in-house independent research and development.

As can be seen in Figure 2.1, the typical strategic planning and procurement cycle consists of 16 steps, beginning with the establishment of corporate goals and objectives and ending with contract completion or recompetition. The first seven of these steps are essential pre-RFP (request for proposal) marketing functions. The first four comprise strategic market planning and business development. Early entry into this 16-step process is vital to securing an advantageous competitive position in the marketplace. The firm that assists in the identification of product service or project requirements is clearly at an advantage in competing for the contract. Waiting until steps 7 or 8 to become involved in a new endeavor would severely limit the firm's ability to set or influence requirements. The firm that waits until step 9, issuance of the request for proposal, before taking the measures needed to form teaming arrangements or to start developing proposal information is clearly at a disadvantage in competing with firms that have entered the cycle at a much earlier stage. Figure 2.2 illustrates the relative competitive advantage of early entry into the strategic planning/procurement cycle. Notice how the relative competitive advantage falls off rapidly after the customer's announcement of opportunity. In general, if your firm can be instrumental in the adoption of the product or service in the *customer's* long-range plan and budget cycles, your firm will have an advantageous position in the procurement.

FIRST STEPS: VISION, GOALS, OBJECTIVES, AND STRATEGIC PLANS

Procedures

There is some disagreement in the planning community concerning which step comes first and the name to provide for each step. For this discussion, we will place the list in increasing order of specificity with regard to time (see Figure 2.3).
The corporate vision is a broad concept of where the company should be at some time in the distant future—say, the year 2020. A vision usually comes from a real or perceived need and usually provides long-term benefits to the customer as well as to the seller. A company with a 2020 vision would be way ahead of competitors without it. The important thing about vision is that it should be *focused*. Lack of focus or a blurred vision results in unclear, sporadic, and potentially incompatible goals, objectives, and strategy. The broad concept of a company with X employees and Y amount of profit or Z gross sales per year is not a focused vision. A focused vision would include the product or service lines the company would like to be selling in the future, the type and size of projects, and the geographic or demographic market areas. A focused vision can then be translated into sharply focused corporate goals.

An important word in establishing corporate goals, corporate objectives, and corporate strategies is the word *synergy*. Through the use of synergistic activities, a company can multiply its income many fold. Profits can be made by repackaging the same or modified information; complementary products or services can be added

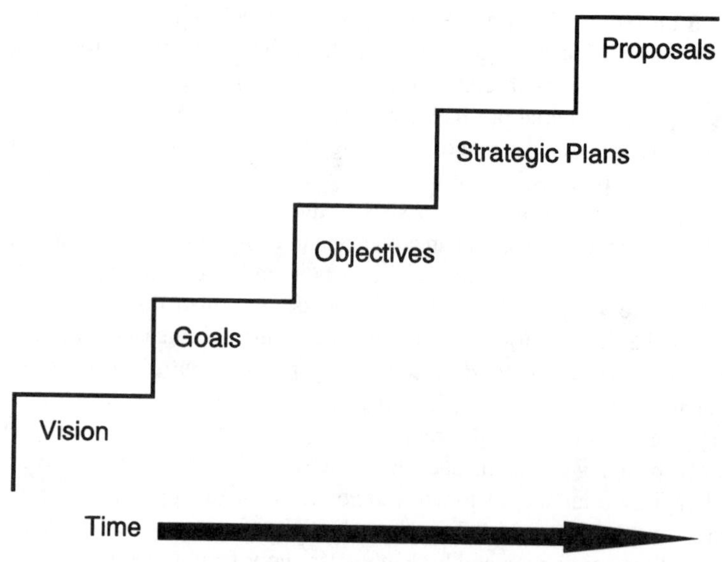

FIGURE 2.3 Steps leading to proposal preparation.

to enhance, supplement, or resupply the primary products or services; and multiple uses can be found for the same skills, equipment, or facilities. A classic example of synergy is the DisneyWorld complex near Orlando, Florida. Disney not only sells admission tickets to the major attractions, but also sells food, supplies (film, sunglasses, suntan oil, batteries, etc.), apparel, gifts, and souvenirs. Disney markets training and education to schools, and, outside the complex, offers a wide variety of tourist services such as lodging and camping, restaurants, entertainment, special tours, transportation, and many more. Many of the products sold by Disney are advertisements for Disney attractions and services. Disney has learned to interrelate the needs of vacationing families to turn a handsome profit.

A company that has its focus on synergy is bound to find complementary products, projects, or services that, with just a little added effort, can provide significant income increases.

As shown in Figure 2.4, two types of goals should be established: corporate goals and market goals. Corporate goals should be synergistic with each other, and market goals (product/service lines and customers) should also be synergistic with each other. *Corporate goals* are distilled into even more specific and sharply focused *corporate objectives*, and *market goals* are distilled into even more specific sharply focused *market objectives*. Corporate and market objectives are merged with each other as shown in the center of Figure 2.4, and the merging of corporate and market objectives is subdivided into implementation plans for each of the company's organizational segments. Organizational segment implementation plans should also be synergistic with each other to make maximum use of the personnel, skills, equipment, and facilities in the corporation.

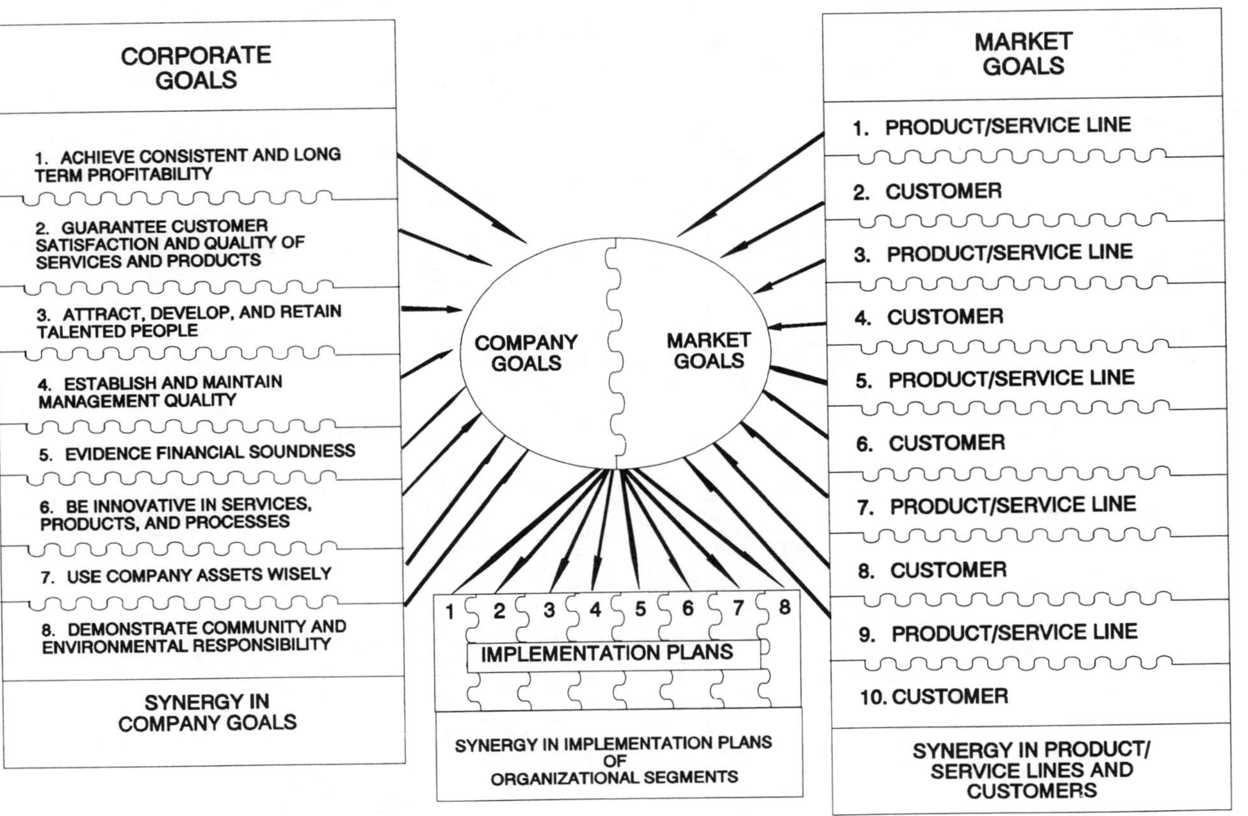

FIGURE 2.4 Synergy in business acquisition.

Synergistic product/service lines and customers can be generated in a stepwise approach as shown on the right-hand side of Figure 2.4, reading from the top to the bottom under "Market Goals." If you are providing a product/service line 1 to customer 2, perhaps this same customer 2 would be interested in another product/service line, 3. Having developed and successfully marketed this new product/service line 3 to customer 2, perhaps you can interest a *new* customer 4 in the same product line; and so on down to customer 10. Having gone through this progression, you now have five customers (2, 4, 6, 8, and 10) and four new product lines (3, 5, 7, and 9) as well as the original product line. This figure represents a linear progression. If translated into a hierarchical flow-down, with two or more product lines or customers flowing out of each preceding relationship, many more product/service lines and customers can be generated. This flow-down principle will be important in developing new product/service lines and new customers. It represents an orderly and deliberate plan of action that will help create more business and corporate growth, if growth is a corporate goal.

Corporate Growth

If corporate growth is a major goal of the company, and in many firms it is, the direction and intensity of growth can be both controlled and encouraged through the allocation of independent research and development (IR&D) funds and the disbursement of bid and proposal budgets to the areas where growth is desired. There is a way of providing an even more sharply focused emphasis to those customers and product lines targeted by the corporation for the near-term, mid-term, and long-term future. A matrix (see Figure 2.5) can be generated for near-

	PRESENT CUSTOMER	NEW CUSTOMER
CURRENT PRODUCT LINE	Quadrant I	Quadrant IIa
NEW PRODUCT LINE	Quadrant IIb	Quadrant III

FIGURE 2.5 The growth matrix.

term, mid-term, and long-term growth in which new or continuing product or service lines can be entered into one of four quadrants: Quadrant I (present customer, current product line); Quadrant IIa (new customer, current product line); Quadrant IIb (present customer, new product line); and Quadrant III (new customer, new product line). This is a proven way of categorizing business prospects. Using this system, company management can identify and control future business directions by allocating a percentage of independent research and development and bid and proposal funds to each quadrant, depending on the direction of desired growth in the near-term, mid-term, and long-term future. Assuming that sales efforts are successful, new customers for near-term business will become present customers for mid-term business, and so on. Because these growth matrix charts will be operating in a dynamic business environment, they should be updated constantly. This process is relatively easy if done by computer in conjunction with a computerized business development database.

A STEPWISE PROCESS FOR STRATEGIC PLANNING

Assuming that your company is an existing corporation and not a new-start venture, an inventory must first be made of existing capabilities (people, equipment, facilities, patents, products, and services) and existing or past goals and objectives. In straight-line control, as mentioned in Chapter 1, you start where you are now, so it is necessary to have a very clear picture of where you are now in order to establish a known beginning point. Assuming that you want to change directions or to establish more clearly defined and sharply focused goals, objectives, and implementing strategies, the next step is to compare the new vision and goals with the old vision and goals.

From this comparative analysis, new and exciting company goals are established. All levels of management as well as lower-level employees should participate in the goal-setting process. Corporate management sets the vision, but the employees should help develop this vision into goals and then help develop the implementing objectives. Once goals and objectives are established, priorities must be set through an iterative process between the employees and management. It takes many people to reach a goal; and, therefore, many should participate in its formulation and implementation.

Prioritized goals and their implementing objectives are then used as a basis for developing alternative strategies. Strategies are specific approaches, methods, and plans for carrying out corporate objectives. Evaluation and selection of the best of the alternative strategies is next, followed by the development of plans of action and implementing procedures. Throughout this planning cycle, groups in the company should evaluate the emerging results, critique the approaches, and monitor process. Since the strategic market planning and business development process is dynamic, synergistic, and iterative, it should continue to be updated, massaged, and redirected as required, as long as the company exists. Plans must continually be updated and details changed to meet changing market conditions, customer base, and degree of company success in each product or service line arena.

IMPORTANT CHARACTERISTICS TO DEMONSTRATE IN PROPOSAL PREPARATION

Does all of this planning directly affect the proposal preparation process? Yes! Definitely! The planning process forms a framework for all future proposals prepared by the company. Bid decisions are made based on these previously thought-out plans. Effective translation of company goals, objectives, and implementing strategies into the specific characteristics of specific proposals requires a conscientious effort to adhere to straight-line control to develop quality proposals.

In an interview with the Canadian government we obtained perhaps the best and most descriptive list of desirable proposal characteristics that we have ever seen. These are: *capability*, *competence*, *compliance*, *comprehension*, and *risk*. We have added another one in this book: *credibility*. In the evaluations in which we have participated, credibility has proven to be one of the most important cost proposal characteristics. You could call this the "Five *C*s and an *R*" concept: *capability*, *competence*, *compliance*, *comprehension*, *credibility* and *risk reduction*.

Capability. Capability is potential effectiveness in performing the work. How is capability evidenced in a proposal? Capability is evidenced by the qualifications of the team chosen to do the work; the equipment and facilities allocated to the job; the company organization, project organization, and key personnel; and the company's resources and financial backing. Potential effectiveness (capability) is conveyed by a clear and factual description of the attributes of the proposed personnel, equipment, facilities, and organizations.

Competence. Competence is the actual and projected record of reaching the very highest potential. How is competence measured and how is information on one's competence conveyed to the customer? It is conveyed through (1) documented evidence of past successes; (2) skill in estimating and allocating resources; (3) demonstration of a high quality of work; and (4) the display of a winning attitude.

The easiest of these to convey is the documented evidence of past successes. These are tangible accomplishments that can be stated by the proposer and verified by others who have dealt with the company.

Skill in estimating and allocating resources is more difficult. It takes a great deal of knowledge and experience to accurately time-phase skill categories and skill levels for a projected program. A good job done here will reveal much about one's competence in performing the work.

Then there is "demonstration of a high quality of work." One might ask, "How can I demonstrate a high quality of work when the work is yet to be performed?" We asked the following question of participants in a seminar on proposal preparation: "What characteristics in a proposal will most surely convince the company of your ability to perform high-quality work?" After several answers, which included responses regarding percent inspection, depth of quality assurance, program, personnel morale, etc., one student replied, "The quality of the proposal itself!" Indeed, this is the most important way in which competence can be displayed to a customer: the development of a high-quality proposal. Since the proposal itself is a product of the organization, one would expect it to reflect the quality to be expected in the work being proposed.

What about the "winning attitude"? A winning attitude can be detected easily in a proposal because the proposer is functioning in a way that ensures success. This winning attitude reflects itself in the confidence (another "C" word) of the proposer.

Compliance. Compliance to the requirements stated by the customer has four facets that must be covered in a proposal:

1. *Demonstration that specifications can be met.* How can this be done? It can be done through analytical studies or through actual hardware testing. Proof of compliance is provided as analysis results or as test results.
2. *Suggesting alternatives not included in the requirements.* These alternatives can involve compliance with the *intent* if not the *letter* of the requirement.
3. *Suggesting deviations from the requirements.* (Don't agree to unrealistic ones, however.)
4. *Demonstrating and emphasizing where requirements can be exceeded.*

Compliance is an attribute or characteristic of a proposal that can be *measured.* It can be measured by the proposer prior to proposal submittal. Compliance, therefore, should enter into the bid/no-bid decision.

Comprehension. Comprehension is the characteristic of recognizing and understanding the various facets of the total job. Comprehension of program requirements can be best conveyed and displayed by a sound program approach, a realistic treatment of potential cost and schedule deviations, and recognition and treatment of the risks involved. A true comprehension of the job prior to proposing can affect an early bid/no-bid decision.

Credibility. Credibility enhances the ability of the evaluator (1) to believe the information, data, and supporting statements in the proposal and (2) to have confidence in the resource (cost) estimates in the proposal. Credibility is developed and established in a proposal by supporting and substantiating all data with suitable rationales. Supporting rationales can be obtained from: (1) actual historical records from previous projects; (2) quotes from authoritative literature, handbooks, and catalogs; or (3) factual data from previously published reports.

Credibility is improved with the depth of rationale provided, since understanding the methods that have been used for synthesizing technical and cost data increases the reader's confidence and provides the thread of evidence needed to verify proposal contents.

Risk Reduction. Risk is the possibility of incurring failure and is usually a result of multiple smaller risks *within* a new work activity or work output. The objective of the proposal is to show how the risk to the customer is eliminated by: (1) realistically assessing each risk area; (2) addressing each risk area; (3) devising an approach to each risk area; and (4) building an airtight case for successful program completion.

You may be saying by now, "Yes, all of these generalities are true, but what specific things can I do to develop a winning proposal that reflects customer requirements as well as company long-range objectives? There are four specific factors that will help more than any other in developing a winning proposal.

First, the proposed effort has been selected from among many alternative activities the company can pursue. The bid decision has been made based on a careful evaluation of the company's long-range goals, objectives, and strategic plans. Capabilities already exist or are being built that are compatible with both the corporate goals and the customer's wants.

The second is adequate planning and organization of the total proposal effort in keeping with corporate goals, objectives, and strategic plans. We refer to the quality of this effort as well as to the quantity. Proposal development must be attacked with the same rigor and enthusiasm as any job in order to succeed. Admittedly, proposal development is not an end in itself. But to many, it is merely a troublesome step that has to be taken before one can get on with the real meat of the work. In this respect, proposal preparation has taken on a professional flavor, and those engaged in proposal preparation have become recognized as true professionals in their field.

Third, in-depth consideration, analysis, adjustment, and time-phasing of skills is key to a competitive proposal. Careful attention to the mix of skill categories (engineering, manufacturing, testing, quality assurance, etc.) and the mix of skill levels (senior, junior, apprentice, technician levels) as time progresses is vital. Many proposers fail to adjust this mix of skill categories and skill levels as the program progresses. Higher skill levels are needed at the start of an operation and lower skill levels (with accompanying lower labor rates) can be phased in as the work progresses. This practice has the threefold benefit of training new people as the work progresses, lowering labor costs during the program, and releasing people with higher skills for the next job.

The fourth specific factor is the ability to accurately tie resources to results in estimating job performance and in estimating costs. The only way to do this is to employ excellence in cost estimating. Excellent cost-estimating tools, methods, and techniques exist. It is the proposer's job to use these tools in accurately predicting the resources to do the job in the conventional estimating situation and in accurately predicting or forecasting the job that can be done with a given amount of resources.

STRATEGIC PLANNING MODELS AS SOURCES OF PROPOSALS

The proactive strategic planning and decision-making process is summarized on Figure 2.6. The process starts with objective goal-setting and review and continues through the submission of relevant proposals to the customer. Notice the iterative nature of the process, which is effected by continuous performance measurement.

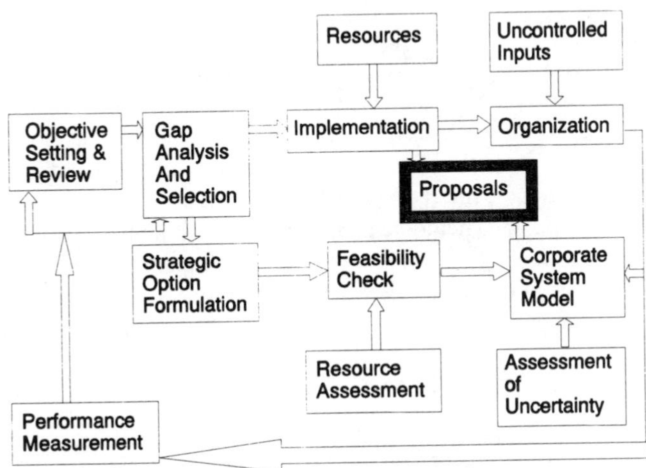

FIGURE 2.6 The proactive decision-making process. Adapted by permission from Robert G. Dyson, *Strategic Planning: Models and Analytical Techniques*, Chichester, U.K.: John Wiley & Sons, Ltd., 1990.

Planning as a Prudent Corporate Process

The recognition of planning as a prudent corporate process is essential for the successful implementation of strategic decisions. This prudent corporate process focuses on the corporate vision, but takes into account all relevant factors of a financial, personnel, geographic, demographic, and socioeconomic nature. Gap analysis and selection is a means of comparing future possible states of the organization with long-range goals and objectives to decide if existing strategies are acceptable or if new strategies should be implemented. Hence, the dynamic implementation process must continue to be updated as the firm moves toward its long term goals and objectives. Internal and external proposals created by the firm are the tools for implementation.

Classical Models

Classical models of the strategic planning process are invariably dynamic in nature and involve one or more feedback loops, which provide the internal checks and balances needed to ensure timely and responsive mid-course corrections to corporate implementation plans. A generic strategy should be developed for *each product line*, if a company has more than one product or service area. Three generic strategy options are available: cost leadership, differentiation, and targeting. Cost leadership depends on the firm's products having a lower cost than its competitors'; differentiation comes from the marketing of distinctive products or services; and targeting consists of a focused effort to develop a distinctive strength to fill a specific need.

Situational Analysis

Situational analysis, another tool used in strategic planning, is a two-dimensional matrix method of assessing the threats and opportunities of the internal and external environment against the weaknesses and strengths of the organization. A strength and weakness audit of internal and external threats to and opportunities for a corporation can be conducted using the matrix format shown in Figure 2.7. This matrix approach provides a systematic and methodical framework for simultaneously matching the external marketplace with the internal company's capabilities within the overall goals and objectives of the firm. Once the iterative process is completed, proposals that fit both criteria can be produced, avoiding weak areas. The matrix is dynamic,

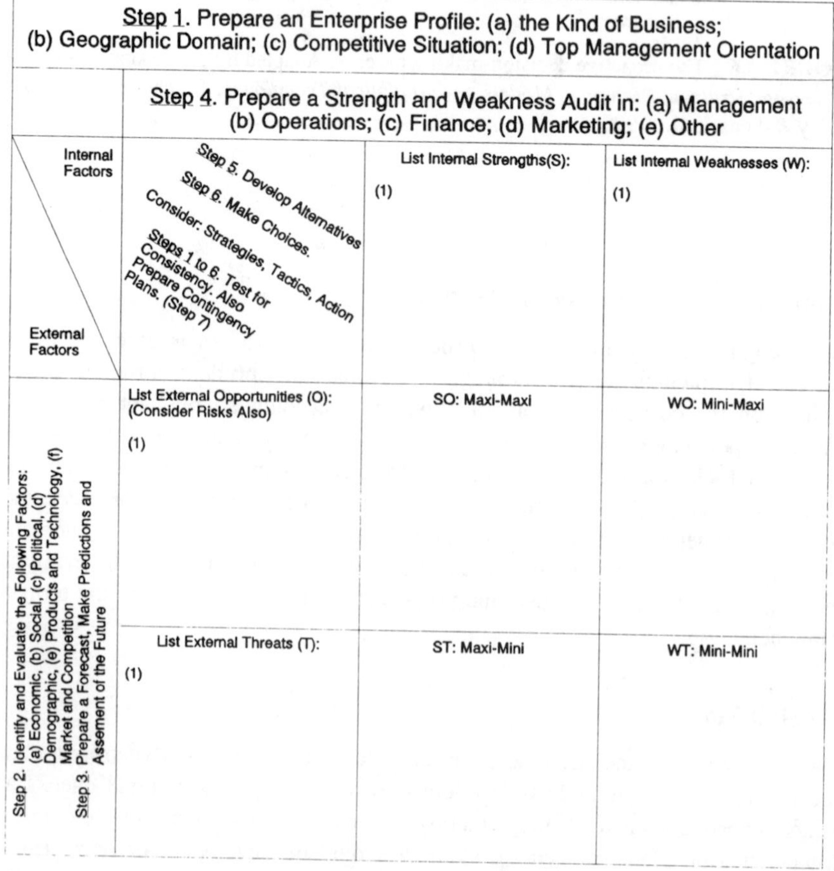

FIGURE 2.7 Strengths and weaknesses audit for internal and external factors. Adapted by permission from *Strategic Planning: Models and Analytical Techniques*, Chichester, U.K.: John Wiley & Sons, Ltd., 1990. SO: strengths of opportunity, WO: weaknesses of opportunity, ST: strengths of threats, WT: weaknesses of threats.

as are other forms of strategy development and implementation, so it must be updated as the complexion of the firm and its marketplace change.

Cognitive Maps

An important team-based strategic-planning concept is the use of cognitive maps as visionary tools. Most of the theory and practice of strategic planning fails to throw much light on the question with which most executives are concerned: *How to get something to happen*. The focus of any strategic planning activity should be that of the embedded vision of the key powerful actors in the organization. Cognitive mapping is a powerful way of linking embedded vision to action. Cognitive mapping is the team development of a diagram of the thought process that leads from nebulous ideas to firm actions. The cognitive-mapping team should consist of key corporate personnel. Team meetings should be held in a secluded and undisturbed environment. Like many of the other strategic-planning and implementation techniques, cognitive mapping is a systematic and methodical way of reducing ideas to flow diagrams in a way that will result in a clear and decisive output that creates results. Cognitive maps should lead from the vision to goals to objectives to implementing strategies to implementing plans to proposals. Figure 2.8 is a cognitive map for a hypothetical company, Applied Technology Corporation. The map leads from the imbedded corporate vision into the key areas of importance: capabilities, staff people, and customers. It defines what types of each are required and shows how these combine to reach specific markets with specific proposals. Each box in this master corporate cognitive map can be, likewise, expanded to show the flow of thinking and resulting flow of action required to move from the vision into the actions that will produce results. Thus, the resulting proposals are not only the way to get things to happen, but they incorporate the fundamental directions in which the firm wants to proceed.

The cognitive mapping process was developed and perfected in the United Kingdom to help project teams solve complex strategic issues. It will be found to be useful, we predict, in many other countries in this age of increasingly complex technologies. The aims of cognitive mapping are to provide the following:

- An instrument to help negotiation towards best solution
- A way of "hearing several people at once" by setting the views of one person in the context of the ideas of others
- A method for providing structure to multiple and conflicting aspects of argumentation
- A method that is designed to suggest actions to resolve issues
- A method for developing a consensus about a system of goals
- A method that does not violate the natural role of discussion
- An efficient way of avoiding "group-think" and "bounded vision"
- A designed scheme for attending to both the content of issues and to the need for a recognition that people change organizations
- A designed environment for ensuring effective decision-making.

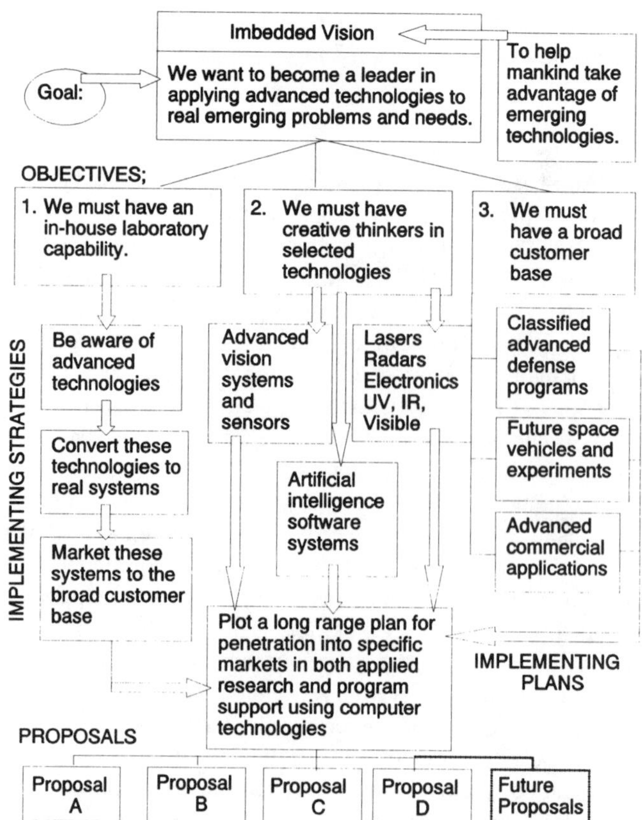

FIGURE 2.8 A cognitive top-level map for Applied Technology Corp. UV: ultraviolet; IR: infrared.

Cognitive mapping provides a means to determine goal interaction. Effective interaction and mutual support of goals creates synergy in a company's long term activities.

Growth and Profitability

Strategic and tactical corporate planning processes concentrate heavily on cash flows, return on investment (ROI), and corporate growth. These concepts are well-described in many economics texts. But some strategic planning researchers challenge many of the commonly held ideas about the interrelationships between these variables. Marakon Associates, in a presentation made to the Institute of Management Sciences on November 11, 1980, concluded the following:

- Growth and profitability are not generally tightly linked. In fact, they tend to compete or trade off.

- Good planning should not call for passing up profitable investment opportunities.
- Ideal business portfolios are not necessarily balanced in terms of internal cash flows.

Growth and profitability, in fact, do tend to compete. Careful analysis of corporate goals and objectives is needed, therefore, before deciding to go after the larger jobs or the larger markets in search of greater profitability. Proposals need to be tailored to the proper objectives, as guided by prudent consideration of growth versus profitability tradeoffs.

The analytic hierarchy process (the development of a composite scenario based on multiple actors, policies, and scenarios); robustness and optimality as criteria for strategic decisions; strategy support models; simulation as an aid to policy-making; and the assessment of uncertainty are other subjects explored exhaustively by strategic planning analysts. We would like to comment on only one of these at this point, the assessment of uncertainty.

Assessment of Risk

There has been substantial effort in the past devoted to risk assessment, risk analysis, and risk abatement. Volumes have been written on the subject, and dozens of statistical and mathematical models are available. In most cases, however, risk abatement (the reduction or elimination of risk) comes down to practical technical and business decisions that must be made based on corporate vision, goals, and objectives. In exercising straight-line control, risk is minimized by meticulous adherence to specific requirements. In proposals, the focus must be on the solution rather than on the risk. A healthy respect for the uncertainties in any project, process, product, or service, however, will result in a thoroughness in the plan of action to provide the work activity or work output. Unless the proposal itself is for risk analysis, the time has come for risk elimination rather than mere risk reduction, as the inherent risk of the procurement process itself is a sufficient deterrent to proposal acceptance.

The reason for meticulous strategic planning, held in a group environment with key corporate players present, is that no development occurs in isolation. In an efficient organization, organizational elements and people are interdependent. Synergy between work functions is enormously beneficial. The entire matrix of resulting proposals submitted by a firm should employ synergy with each other to the greatest extent feasible. It is through this interaction, and the interaction of the resulting work activities and work outputs, that the true serendipity can be developed to engender efficiency, economy, and effectiveness in products and services.

With the proper strategic planning accomplished, the firm's business activities can proceed into the first phase of proposal preparation: preproposal marketing. Preproposal marketing and the decision to bid are the subjects of the next chapter.

3

PREPROPOSAL MARKETING AND THE DECISION TO BID

And your ear will hear a word behind you, this is the way, walk in it, whenever you turn to the right or to the left.

—Isaiah 30:21

The success of the entire proposal preparation activity is heavily dependent on steps that are taken far in advance of actual proposal submission. Straight-line control starts before the request for proposal (RFP) is received—even before the RFP is requested or a bidder's list entry is made. Advanced steps that enhance straight-line control are associated with the marketing of the work output or work activity to one or more customers in keeping with corporate goals, objectives, and strategies. Marketing involves identifying a potential customer, letting the potential customer know that a work activity or work output exists to fill his or her needs, and getting the customer to commit to pay a stated price for performance of the work.

THE MARKET SURVEY: IDENTIFYING THE CUSTOMER

In identifying a potential customer and a specific need for work activities or work outputs, one has access to a wealth of readily available, sometimes free information. A principal source of information—often overlooked by marketing personnel—is the publications of state, federal, and local governments. They are available at little or no cost. Government publications are full of useful socioeconomic, geographic, and demographic data that reveal the buying habits of individuals, companies, and government agencies themselves. Labor markets where skilled personnel are available, along with complete wage histories and projections, can be obtained from the Bureau of Labor Statistics, one of the largest statistics-gathering, analysis, and reporting organizations in the free world. Literally thousands of other special reports ranging from agricultural to industrial and business subjects can be obtained at nominal costs from federal, state, and local agencies. These special reports and studies cover an almost unimaginable variety of subjects useful to the marketer. They cover expansion and growth trends, economic trends, and taxation and tariff effects on business, industry, and the economy.

Professional and industrial associations, trade shows, trade magazines, and business magazines and newspapers also contain sizeable amounts of useful marketing information. If used selectively and wisely, this information can result in the identification of market needs and even specific potential customers. The *Commerce Business Daily* alone lists hundreds of procurement opportunities every day. Publications of professional and technical societies such as the American Marketing Association have much useful information on techniques and methods as well as clues to potential open markets.

Wholesalers, retailers, and distributors of products possess real-life statistics on how various products are performing in the marketplace. They know which products are selling the fastest and know when needs or demands are building up in the marketplace. These organizations predict how products or services rate in comparison to competitors.

Market research organizations which specialize in gathering specific data for the introduction of a new product or line of products are becoming more sophisticated in their information gathering, analysis, and reporting techniques. These organizations can be hired at a nominal fee to do the primary research work to provide needed market data for entering into a new venture.

Stockbrokers and banks have, and are willing to disburse, a wealth of information about the volume of business, net worth, profit history, growth, and overall financial health of companies, which could be your customers or your competitors. Not only do they have financial data from publicly listed companies, but they are also quite willing to analyze and predict future growth, acquisition, and merger trends of these companies.

In-depth knowledge of the customer—a vital requirement during the preproposal marketing period—can be obtained from open literature, company-supplied information, and above all from personal contacts with management as well as with workers in the customer's organization. It is important to obtain as much information as possible about the customer's procurement or purchasing methods, techniques, organization, and personnel. If one is familiar with purchasing and procurement circles, work statements, contract provisions, and standard parts of the contract can be prepared easily.

Marketing policies and overall company policies as far as practicable should also be known. It should be remembered that the proposer's customer also is supplying products or services to an end customer. This end customer is the final recipient of the results of the bidder's work, and satisfaction of the end recipient is critical.

It is important to know the customer's other product lines to permit as much synergy as possible in the final work content. If the customer has a product or service similar to the proposer's, a mutually beneficial interdependence or combination of the product lines may be beneficial. The customer may also supply services or products needed or desired by the proposer. A willingness to use the customer's products will enhance the marketer's knowledge of the customer's company and could have a potential impact on decisions regarding the proposed work activity or output.

A friendly, cooperative relationship with the customer, rather than an adversative one, will open up the opportunities for greater synergism and for a greater two-way flow of information. Not only will the marketing representative learn more about the customer's organization, but the customer will learn more about the proposer's capabilities, aspirations, and competence. Knowing the customer's key personnel opens up the flow of information that is necessary for the proposing team to evaluate the requirements fully and responsively. The marketer should make it a point to personally contact and become acquainted with those who originate the need, those who implement the procurement, and those who make the final procurement selection decision. This personal contact and acquaintance period must come in the early preproposal phases, because communication is often severely limited or cut off entirely when the competitive process heightens in intensity.

With these sources of information available, the marketing representative is well-equipped to identify a market area, select one or more specific potential customers, plant the seed of the idea to identify or generate the need, and embark upon the preproposal marketing activities that will lead to a decision to bid or not to bid on the proposed work.

PLANTING AND CULTIVATING THE SEED

The secret of success is finding a need and filling it; and, as mentioned in Chapter 1, a proposal is a plan of action o fulfill a need. Needs can either be existing, latent, or generated. Regardless of the source of the need, the potential customer must be made aware of an available work activity or work output, and this awareness must be cultivated and nurtured until it becomes an interest, a desire, and an action on the part of a customer.

The AIDA Approach

Marketing professionals use what is known as the *AIDA approach* in making a sale. The steps are generally the same as those required in preproposal marketing. The AIDA approach involves gaining the customer's *A*ttention, creating *I*nterest, nurturing a *D*esire for the service or product, and stimulating *A*ction to request or purchase the work or product. The proposer should be prepared to establish relationships with potential customers *while the customer is formulating his or her needs*, and the proposer should be prepared to actually plant the notion in the customer's mind for the specific work *or* the specific work requirements.

Gaining the Customer's Attention. Gaining the customer's attention involves the transmission of information to the customer regarding a process, product, project, or service that is available to fulfill an actual, latent, or previously undeveloped need. The business customer's attention can best be gained by emphasizing that the work to be performed will improve overall profitability by increasing efficiency, economy, effectiveness, or overall growth of the company.

Many ways are available for gaining the customer's attention—advertising, direct mail, telephone, and so on—but by far the most effective, particularly for large procurements, is personal contact. Personal contact provides a two-way communication that is more effective than any other means in bringing together a need with a way to fulfill that need. Personal contact permits an early exploration by both parties into the potential for any work activity or work output and allows the supplier and customer to communicate in real time in an atmosphere of personal empathy and in a way that enhances the germination and evolution of fruitful work. A personal visit is by far more persuasive than a billboard, magazine advertisement, television commercial, or direct-mail brochure when it comes to interacting with a customer to develop or identify a means of fulfilling a need.

The personal-contact technique should be used extensively by the proposer in the early phases of any new work activity to develop a close and preferably indistinguishable interrelationship between the required task or project and the supplied task or project. To do this effectively, the seller must have considerable freedom and flexibility. He must be bound only by overall business principles rather than by detailed procedural constraints in order to retain negotiation and discussion flexibility. As the work activity is defined and scoped through the interactive customer–supplier interrelationship, fixed parameters, rules, and business arrangements can evolve.

Creating Customer Interest. Once the customer's attention is gained, the seller must create interest and enthusiasm for the service or product being offered. This is accomplished by fostering a familiarity with the potential work to be performed, the organization that will perform it, and the overall benefits of the completed work. Continued communication, including but not limited to additional personal contact, will create a more than passive interest in the customer and will serve to further define the work to be performed.

Nurturing Customer Desire. As preproposal marketing proceeds, the customer's desire should be nurtured, maintained, and increased through the supply of additional information that will reinforce the urgency of fulfilling the need. This is done by personal, telephone, and written contact to convey information that will accentuate the positive aspects of the project and diminish any negative aspects or objections that have surfaced during the discussion and evolution of the work's characteristics. The purpose of this step is to bring the customer's desire to a point where he or she is not only willing but is actively seeking an opportunity to take the action required to engage the services of the seller.

Stimulating Customer Action. In the proposal preparation process, the final customer action that must be taken is the signing of a contract with the proposer's firm to perform the work. There are often interim customer actions that must be stimulated, however, one of which is the issuance of a request for proposal (RFP) or a request for quotation (RFQ). It may be either verbal or written. The purpose of this step is to assure that the customer has everything required to act on the desire that has been developed. Customer action can be best stimulated by requiring a definite response within a given time period or by making the customer aware that there are other organizations, which may take action sooner, competing for the proposer's services. In addition to stimulating customer action, these activities improve the proposer's competitive position because of the proposer's intimate knowledge of the request for proposal or request for quotation.

Build in Acceptance Time

The acquisition process for any new work activity or work output must include sufficient acceptance time. Acceptance time is the time required for an organization and its key individuals to absorb, understand, adjust to, and become advocates of the proposed work. Acceptance time is usually proportional to the amount or degree of deviation of the work content from existing practice. Very large multiple-year projects may take months or even years to initiate, simply because of their size and complexity and the resources required to execute them. Overall marketing strategy should build in enough acceptance time; through patience, perseverance, and hard work, the project will materialize into a fruitful venture.

Activate Latent Needs

A latent need is one for which the customer has a vague idea of a need but can't fully describe it because the work activities or work outputs available to satisfy the need is not known. The key to the identification of a latent need is to make the customer aware of what is available. This step usually does not require a full-blown proposal, but can be accomplished by simply informing the customer that there is something that will (or might) fill the not yet fully defined need.

Help the Customer Decide

Not only can one plant the seed of desire in the mind of a potential customer, but one can develop more specific details about the characteristics of the item or service to be provided. This facet of proposal marketing requires that the salesperson be intimately familiar with the technical characteristics of the product or service being marketed. It is in this area of marketing that great care must be taken not to mislead the customer into believing that performance will be above that which can actually be provided. The most recent examples of shortcomings in this area are in the computer hardware and software industries. A number of computer users have been misled concerning the present or future capabilities of their computer equipment. A dissatisfied customer, who is disgruntled about the actual versus advertised performance of his or her recent acquisition, can do more harm than many satisfied customers can do good to credibility and future sales. Helping the customer decide can begin at the inception of the need and continue throughout the preproposal, negotiation, and contract performance periods. One must realize that in defining a need for a customer, it is necessary that the selling company be capable of fulfilling that need in order to have a successful business relationship. The person or company who can identify relevant marketing information and process it more efficiently and effectively than his or her counterpart in competing firms will be successful at this first but essential phase of marketing.

EVALUATING THE NEED

Since establishment of the need includes the continued assumption that the proposer's company can fulfill that need the next step is to determine that the proposing company can do the job better than anyone else. (If at any time it is determined that the customer's need cannot be fulfilled, preproposal discussions should be terminated.) By "doing the job better," we mean doing it in a way and at a cost that is most advantageous to the customer. The reason the customer's advantage is paramount is that the customer is the one who is doing the evaluating. The selection decision will not be based on what yields the greatest advantages to the supplier. On the other hand, the proposer must keep in mind his company's overall goals and must select only those pursuits that will ultimately accomplish these goals.

Keeping the Company's Vision, Goals, and Objectives in Focus

In the preproposal time frame, it is essential for a proposer to keep the overall company objectives in mind as well as those objectives that have been established for this particular procurement. Is the objective to make a short-term profit to improve cash flow or cash on hand? Or is it the primary objective of the company to improve its net worth and/or long-term business potential over a number of years? It is a proven fact that in business the second objective is the one having the greater potential of creating sizeable wealth in the long run. But if the company's cash flow is in trouble, a decision to market a dead-end job (one that doesn't lead anywhere) may be necessary to solve the short-term cash-flow problem. In general, it is best to search for work that will accomplish the long-term goals and objectives of the company. If this can be done while at the same time improving the short-term cash flow, then the best of both worlds has been attained.

Matching the Company's Unique Capabilities to the Customer's Need

In matching the company's capability to the customer's need, consideration should be given to specialized team skills, equipment, facilities, personnel, and experience, as well as to the unique capabilities available to perform the proposed work. Specialization is important when looking for a specific job acquisition. Competitive advantage over other firms bidding for the same work will lie principally in one or more of the six areas listed above. Even though the project may be open for competition, if a particular company has on its payroll a qualified expert in a skill required to do the job and the competitor does not, that particular company has the greatest potential of winning. The proposer company's pattern of acquisition of new equipment, skills, and capabilities can thus influence the type and magnitude of its new work.

Watch for Opportunities for Synergy

Often new work acquisitions can have a synergistic effect with current or other new company products or services. When a college professor combines teaching with writing, when a homemaker combines homemaking with a catering service, or when a physician combines research with medical practice, they are engaged in synergistic activities, which are mutually supportive and mutually beneficial. Synergistic activities are those that can use the outputs, resources, facilities, or knowledge acquired to help each other. Some companies have found that a by-product from one product line can give rise to a whole new product line. Other companies have found that services related to a product are often more profitable than sales of the original product itself. Synergism has a way of giving more to the customer for his or her dollar while at the same time providing a more profitable venture for the performer.

Beware of the Cuspetitor

The "cuspetitor" is the company that is potentially both a customer and a competitor. The importance of knowing the organization one is dealing with is emphasized when working with this type of company. Many companies have worked long and hard on detailed proposals only to have the supposed customer absorb, modify, and sell the proposal ideas and approaches as his or her own. Then there is also the customer who lets the contractor do the more difficult design and development work and then takes over in the more profitable production phase. It is best to establish at the outset in dealing with potential customers that certain product lines or service areas are within the customer's purview or area of competence and that others are within the proposer's area of expertise. This early informed but important agreement will provide smoother sailing during negotiations with a new customer and will eliminate possible areas of suspicion or distrust relative to overlapping target markets or possibilities of competition between customer and supplier.

DETERMINING LONG-TERM EFFECTS

In marketing a new work activity or work output it is essential to be certain that this new activity will be profitable in the long run to the company. It is also desirable to know that the work will be more profitable than other projects or activities that could be pursued with the same investment of resources. For these reasons, it is not only desirable but essential that some form of long-range analysis be performed. This should determine the profitability of the venture in relation to other alternatives and whether the new activity will result in a net worth growth equal to or better than that experienced in the past or desired by the company. It will also identify and evaluate any other long-term benefits to the company in acquiring the new work. Since companies are in business to be profitable, a profitability analysis and profitability optimization study should be conducted before investing in the efforts required to submit a formal proposal for the work and to implement the subsequent expenditures that must be undertaken to successfully compete in the chosen marketplace with the chosen customer.

Profitability Optimization

A profitability optimization study includes consideration of past and desired company profitability; the profitability of the new acquisition and its alternatives; the company's past and desired growth rate; the increase or decrease in growth rate brought about by the new work; and other factors such as changes in company capacity or capability, diversification, or follow-on work brought about by the new work acquisition. The ten steps required to do a profitability optimization study are shown on Figure 3.1. The profitability of a venture is the return on the investment (ROI) for that venture. It is the quality of possessing an operating profit while contributing to the growth of the company. If growth is measured by an increase in net worth, then a positive growth can be achieved when the net worth increases.

Businesses should acquire new work with a high profitability index (income divided by investment) rather than merely a high profit percentage (Fee divided by all other costs). This is emphasized by the fact that growth is necessary (1) in order to attract capital to provide continued merit advancement of a work force and (2) to provide continued modernization of plant equipment and facilities. Questions to ask in evaluating the profitability of a venture are:

1. Does the work include the expenditure of some nonrecurring resources that will result in something usable in the performance of the next job?
2. Does the work improve the company's overall competitive position in the industry?
3. Is the new work challenging, and does it represent a significant advance above the current work activities, which would permit acquisition of new capabilities?
4. Is the work synergistic (interrelated or mutually supporting) with the company's other activities?
5. Does the work improve the company's future growth potential, sales potential, profit potential, and diversification potential?
6. Will the work improve the company's net worth?

Profitability Index

The profitability index is a means of testing a work activity or work output to determine if its profitability is greater or less than past work or other work that is

Step 1

Establish long term company growth goals, profitability goals, and goals for increasing capability and capacity. Establish excellent cost estimating systems and methods.

Step 2

Determine *present* and *past* company profit category, profitability index, and net worth growth from past financial reports. If divisional financial reports are available, do this on a divisional or departmental basis.

Step 3

Determine cash flows of expenditures and incomes for the project, product, acquisition, or service under study. Determine which are nonrecurring investments and which are recurring incomes and costs.

a. Establish one or more assumed payment modes for expenditures (cash payment, lease, or mortgage loan).
b. Determine all federal and state tax benefits based on the *latest* tax laws.
c. Determine depreciation allowances for equipment and salvage values.
d. Determine increase in company capability and capacity in workforce, equipment, and market potential due to the new activity (nonmonetary value but very important).

Step 4

Bring *all* expenditures and incomes back to present value using a *discount rate* computed based on the most recently available economic projections of *interest rates* and *inflation rates*.

Step 5

Determine profitability index using the *present value* of incomes, expenditures, and nonrecurring investments.

Step 6

Determine net worth growth effects of the new project on overall company net worth growth.

Step 7

Compare the profitability of the new work activity or acquisition with past history by comparing its:

a. Profitability index with overall past company profitability index.
b. Net worth growth effects with overall past company net worth growth.
c. Increase in capacity and capability compared with past company capacity and capability.

Step 8

Compare the effect of profitability index, net worth, and capability/capacity increases for each alternative new project or method of acquisition of a new project.

Step 9

Select for implementation, propose, negotiate, and carry out the alternative that best fits company growth and profitability goals.

Step 10

Keep excellent financial records on progress of implementation of the new work activity or acquisition to see—*on a real time basis*—if it is living up to its expectations. If it is not, be prepared to cancel the project provided you have thoroughly evaluated the consequences of cancellation.

FIGURE 3.1 Steps in optimizing profitability.

currently under consideration for acquisition. The following is the equation for the profitability factor:

$$\text{Profitability Index} = \frac{\text{Volume (Units)} \times (\text{Unit Price} - \text{Unit Cost})}{\text{Investment(\$)} \times \text{Time}}$$

For example, if a company had a present net worth of $1.5 million and a past overall profitability (or return on investment) of 2% per year, and the company had a prospect for selling 1000 transformers per year for three years at $25.00 each, the profitability index (PI) for a cost of $17.00 each and an investment of $100,000 per year would be:

$$\text{PI} = \frac{3{,}000 \times (\$25 - \$17)}{\$100{,}000 \times 3} = .08 \text{ or } 8\%$$

The company would accept the job because the profitability index of 8% is four times the past profitability of 2%.

If the actual unit cost turns out to be $24.00 per unit and the equipment budget is overrun by $25,000/year, the profitability equation is changed as follows:

$$\text{PI} = \frac{3{,}000 \times (\$25 - \$24)}{\$375{,}000} = .008 \text{ or } .8\%$$

The profit of $3,000 is offset by the 2% normally expected gain on $125,000, which would have been $2,500 per year, or $7,500 for three years. Subtracting the $3,000 profit from the normally expected growth ($7,500 − $3,000) results in a $4,500 net loss for the venture. This example illustrates the importance of good forecasting and estimating of manufacturing costs and production equipment costs and shows how profitability can be decreased dramatically by faulty estimating.

Growth-Rate Assessment and Goals

The growth rate of the company can be measured best by examining the balance sheets of the company. Balance sheets tabulate both the assets and liabilities. Since they are "balance" sheets, liabilities are made equal to the assets exactly by including shareholders' equity as part of the liabilities. To determine growth rate, net worth must be computed by subtracting liabilities (not including shareholders' equities) from assets.

Let us use as an example Scientific Software Services (SSS), a hypothetical professional service specializing in providing computer programs (software) for a wide range of computer-based technical and business-monitoring and estimating systems. The SSS annual financial report for 1993 contained balance sheets for assets and liabilities (Figures 3.2 and 3.3) and an income statement, as shown in Figure 3.4. Since financial reports usually show the previous year's financial figures, growth in net worth can be calculated as well as the increase or decrease in the

BALANCE SHEETS

ASSETS	December 31 1993	1992
CURRENT ASSETS		
Cash	$ 13,840	$ 25,303
Time deposits and certificates of deposit	245,001	25,001
Short-term investments	15,061	
Accounts receivable	358,329	245,570
Inventory		650
Prepaid expenses and other	4,030	3,711
Total current assets	636,261	300,235
PROPERTY		
Laboratory equipment	107,724	74,975
Furniture and fixtures	92,533	38,794
Leasehold improvements	11,003	3,050
Equipment held under capitalized lease	30,862	30,862
Vehicles		26,004
Total	242,122	173,685
Less accumulated depreciation	65,687	42,354
Property—net	176,435	131,331
OTHER ASSETS		
Notes receivable—net of discount	8,669	
Other	9,668	
Total other assets	18,337	
TOTAL	$831,033	$431,566
Liabilities	431,814	191,653
(Capital) net worth	399,219	239,913
	239,913	
Increase	$159,306	

$$\text{Growth} = \frac{159{,}306}{399{,}219} = 39.9\%$$

FIGURE 3.2 Assets, Scientific Software Services, Inc.

LIABILITIES AND STOCKHOLDERS' EQUITY	December 31 1993	1992
CURRENT LIABILITIES		
Accounts payable	$ 35,087	$ 33,944
Accrued interest	5,598	2,566
Accrued payroll and related taxes	5,467	1,323
Accrued employee stock grants		26,400
Accrued profit-sharing contribution	74,124	20,620
Income taxes payable	110,294	12,284
Notes payable (current portion)	42,827	34,483
Accrued vacation benefits	26,536	
Other	3,475	
Total current liabilities	303,408	131,620
LONG-TERM LIABILITIES		
Notes payable (amounts due after one year, net of discount)	128,406	57,618
Accrued interest		2,415
Total long-term liabilities	128,406	60,033
COMMITMENTS		
STOCKHOLDERS' EQUITY		
Common stock—$.20 par value; authorized, 500,000 shares, issued, 409,380 shares in 1993 and 249,940 shares in 1992, outstanding, 409,180 shares in 1993 and 198,400 shares in 1992	81,876	49,988
Paid-in capital		110,273
Retained earnings (profit)	317,743	82,229
Total	399,619	242,490
Treasury stock—at cost; 200 shares in 1993 and 51,540 shares in 1992	(400)	(2,577)
Stockholders' equity—net	399,219	239,913
TOTAL	$831,033	$431,566
Shareholder equity	−399,219	−239,913
Liabilities (not including shareholder's equity)	$431,814	$191,653

FIGURE 3.3 Liabilities and stockholders' equity, Scientific Software Services, Inc.

FOR THE YEARS ENDED DECEMBER 31	1993	1992
REVENUES		
Sales	$2,632,586	$1,297,319
Other	4,648	1,350
Total	2,637,234	1,298,669
EXPENSES		
Cost of sales	1,004,194	569,098
Selling, general and administrative [Over 50% of revenues: should be about 10%]	1,237,086	651,388
Interest	11,986	6,746
Other	1,042	—
Total	2,254,308	1,227,232
INCOME BEFORE PROVISION FOR TAXES ON INCOME	382,926	71,437
PROVISION FOR TAXES ON INCOME		
Current	110,405	400
Deferred	1,300	(1,300)
Total	111,705	(900)
NET INCOME	$ 271,221	$ 72,337
Overall profitability factor = $\frac{\text{net income}}{\text{net worth}}$	.68	.30
Company profit rate, net/expenses =	12.03%	5.89%
EARNINGS PER COMMON SHARE	$.66	$.18

FIGURE 3.4 Statement of income, Scientific Software Services, Inc.

company's profit percentage. In this example, the liabilities (exclusive of stockholders' equity) are subtracted from the company's assets at the end of each of the two previous years to find the company's net worth at the end of 1992 and 1993. The net worth increase for 1993 was found by subtracting the net worth at the end of 1992 from the net worth at the end of 1993. This net worth increase was divided by the 1993 net worth to provide an indication of growth during the past year. Phenomenal growth of 40% occurred, as shown on the bottom of Figure 3.2. Obviously, SSS is a highly successful, growing, profitable company.

The profit percentage rate for Scientific Software Services can be computed for the two years by analysis of the statements of income shown in Figure 3.4. The profit percentage rate is found by dividing the net income by the total expenses and multiplying by one hundred to convert to a percentage figure. Note that SSS not only experienced a 40% growth, but its profit percentage rate increased from 5.89% in 1992 to 12.03% in 1993.

If it is the goal of Scientific Software Services to continue with a high profitability index as a result of continued growth, the company must assure itself that the new work it acquires has a profitability potential equal to or greater than its past performance. Since Scientific Software Services is in the enviable position of being in a fast-growing field that has a continued high demand for services, its bid committee may reject many high-profit ventures in favor of those with a high profitability factor. Note that the profitability factor increased from .30 in 1992 to .68 in 1993.

The important message of the profitability analysis is this: a company cannot expect to have continued high or growing profitability if new projects are accepted that have a lower profitability than those projects carried out in the past. Further, the growth rate of a company cannot be expected to continue if its new work activities do not contribute to growth rate as much as those projects completed in the past contributed.

In addition to a numerical financial analysis such as that described above, it is also necessary for the marketing function to make other information available to company management and to the bid committee before an intelligent bid/no-bid decision can be made. This information includes answers to questions like: "Is it likely that other new work will grow out of this new work acquisition?" Often a company acquires a significant competitive advantage for the next job if it performs the current job in an outstanding manner. The proposed work may offer managers at the company the opportunity to make new friends and new contacts that will result in expanding their product line and customer base. Further, performance of the proposed work may just give the company the edge in experience needed to rout potential competitors.

THE DECISION TO BID

In addressing the decision to bid, one must first determine the goals of bidding and performing the work. There are "no-win" bids and "win" bids. Why would a company want to bid on a job knowing or believing that it was not going to win the contract? There are several reasons. One is that the company may have just entered the field and may need experience in putting together and submitting a proposal. The proposal activity itself usually provides a good experience base for going after new work. Another reason is that the company may want to gain recognition or publicity as a potential supplier. It may merely want to let the customer know that it is in the running (competition) for a particular service or product line. Pricing policies for proposals that are submitted on a no-win basis are flexible, because the price can be made high enough to cover personnel, facilities, or equipment acquisition to do the job as well as to theoretically earn a sizeable profit.

For this discussion, however, we are assuming that the bid committee will be basing its bid decision on the criterion that the company should have a high potential if not a certainty of winning the proposed contract.

It is necessary to determine the reasons for wanting to win. Because the proposal style and content will depend upon them, it is necessary to convey these reasons to the proposal team. Reasons for wanting to win the contract fall into three categories: (1) immediate cash-flow improvement, (2) keeping the team going, and (3) long-term growth and prosperity. In Category 1 proposals, the purpose is to make an immediate gain, profit, or increase in cash flow to save the company from impending financial difficulties or to raise cash for an important new project or for capital acquisition. Little thought is given to follow-on effort, continued business with the same customer, or long-term profitability. Category 2 proposals are submitted for work that will keep the team employed, active, and available until more attractive or applicable work can be found. Even less consideration to long-term growth potential, profitability, and expansion is given in these cases. Category 3 proposals—needless to say, the most desirable—are to assure that the company is acquiring work that will bring long-term and continually increasing growth and prosperity. These proposals can include those that will capture a portion of the future market as well as of the existing market.

The proposer's bid committee must consider the company's competitive advantage, technological advantage, geographical advantage, political advantage, and price advantage.

Determining Competitive Advantage

It is very difficult to precisely assess the potential of capturing any specific new work activity or work output, but one can conceive of and describe variables that have an effect on capture potential, and one can organize these into a systematic method of comparison. This numerical or quantifiable method of comparison can then be used in combination with the expert judgment of the bid committee to arrive at a bid decision. Two quantifiable factors that can usually be assessed or estimated fairly accurately before a competition are (1) the number of probable bidders and (2) the experience of the company and key personnel in the proposed work. A *competitive ratio*, *R*, can be envisioned, which is an indicator of capture potential derived from numbers of bidders and the company's experience base. The equation for the competitive ratio is shown in Figure 3.5. The company should strive toward

---

1. Competitive ratio is for comparative analysis only.
2. It is a function of experience level and number of bidders.
3. It does not take into account geographical, technological, and political advantages.
4. Inputs and computation:

   N = number of bidders

   $E1$ = percent of project team that has the exact amount of *generalized* experience required

   $E2$ = percent of project team that has the exact amount of *specialized* experience required

   $E3$ = percent of the job that represents *company* experience

$$R = \frac{E1 \cdot E2 \cdot E3}{1{,}000{,}000N}$$

5. If R = 100%: "We're going to get the job."
   If R = 0%: "We're *not* going to get the job."
6. Competitive ratio is an indicator of capture potential based on number of bidders and experience.

FIGURE 3.5 Determining the proposal's competitive ratio, *R*.

high competitive ratios by entering markets where there are not too many competitors and by entering markets where their experience is directly applicable to the proposed work.

The potential for capturing one of a number of jobs on which proposals have been submitted can be computed by adding the competitive ratios of the several proposals to get the composite competitive ratio, as shown on Figure 3.6. If a desired *capture ratio* (the percent of proposals that must be successful) has been established for a company's proposals, the composite competitive ratio can be used to compare the projected capture potential with desired capture potential.

Determining Technological Advantage

Of the five factors that the proposer's bid committee must evaluate in making a decision (competitive, technological, geographical, political, and price advantages), the technological factor is the one that can, in itself, cause a unanimous "no" vote. Therefore, it is considered first in the bid committee's deliberations. The technological advantage will depend principally on whether the firm has a superior product when measured from a standpoint of performance, capacity, quality, speed, accuracy, or any of many other specified characteristics. A unique design or configuration of a product, or uniquely useful features of a service, can give a company an immediate technological advantage over its competitors. The bid committee's assessment in this area should also evaluate the technology that is being used to produce the product or deliver the service. Use of modern, advanced equipment and facilities designed for speed of delivery and responsiveness to customer requirements will rate high when assessing technological advantage. Questions such as: "Is the company employing the latest computer-based technologies for design, manufacturing, testing, materials handling, logistics, and management reporting?" should be answered and assessed. Like the other areas of assessment described below, these answers will necessarily be subjective in nature. But some recognition of the proposers state of the art in performing the work relative to that of its competitors' will be needed to fully assess technological advantage.

1. Sum the competitive ratios (R) for proposals 1, 2, 3 . . . N to determine the *potential** of receiving work in any given calendar period covered by these proposals.

   W = *Potential** of receiving work

   $$W = \frac{R1}{100} + \frac{R2}{100} + \frac{R3}{100} + \frac{R_N}{100} \times 100$$

2. Compare W/R with desired (or required) capture ratio**

* *Potential* is not to be confused with *probability*, a statistically derived factor.

** Capture ratio is the percent of proposals that must be successful.

FIGURE 3.6 Calculating the composite competitive ratio.

Determining Geographic Advantage

The location of the proposer's physical plant or activity may be of significant value in any given new acquisition proposal activity. Location in an area where skilled personnel are plentiful is an important factor in being able to attract the quality of worker required to do the job. Proximity to the customer may be an advantage, but not necessarily an overriding one. Geographic advantage may also give a price advantage because of lower labor rates, lower transportation and travel costs, lower energy costs (due to climate), or lower communications costs. A geographic advantage may also turn into a political advantage, particularly in underdeveloped, distressed areas with high unemployment that are in desperate need of jobs, new industry, and new work.

In determining the geographic advantage, certain quantifiable factors can be developed that will give an objective view of the proposer's possible advantage over competitors. Communications costs to one location versus another location can be computed in dollars and cents, as can travel costs and transportation costs. Labor costs in various geographic areas, obtained from the Bureau of Labor Statistics, can provide quantitative information, verifiable by a government source, regarding labor rates in various locations. The company located in a mild climate or one that has inexpensive energy sources has an advantage because of these factors. Proximity to major suppliers may offer cost and schedule advantages, as well as geographical advantages.

Determining Political Advantages

The proposer's bid committee should also assess the political overtones, if any, of the proposed procurement. Does the new procurement fall into a depressed industry, or is it one in which skills of the now unemployed can be effectively used? Does the work offer an opportunity for minorities, veterans, or the handicapped to participate? Are there political factors such as congressional support, public support, media support, or special interest group support, which could incline the customer toward having your company performing the work, as opposed to the competition? Likewise, through market research information, it can be determined if the company is in any way favored by the customer. Perhaps the proposing company already has provided outstanding services or products to a customer of the desired proposal acceptor. All of these factors are politically oriented inputs to the bid/no-bid decision.

Determining Price Advantage

Even though a detailed cost estimate has not been developed at the time the bid/no-bid decision has to be made, there are factors other than those already mentioned above that enter into the potential price advantage in the competition. Of these factors, the overhead rate is significant. If overhead is traditionally low, this one factor (all other factors being equal) could be a price advantage. Other factors, mentioned earlier, are low labor rates, availability of inexpensive energy sources, and proximity to raw material sources.

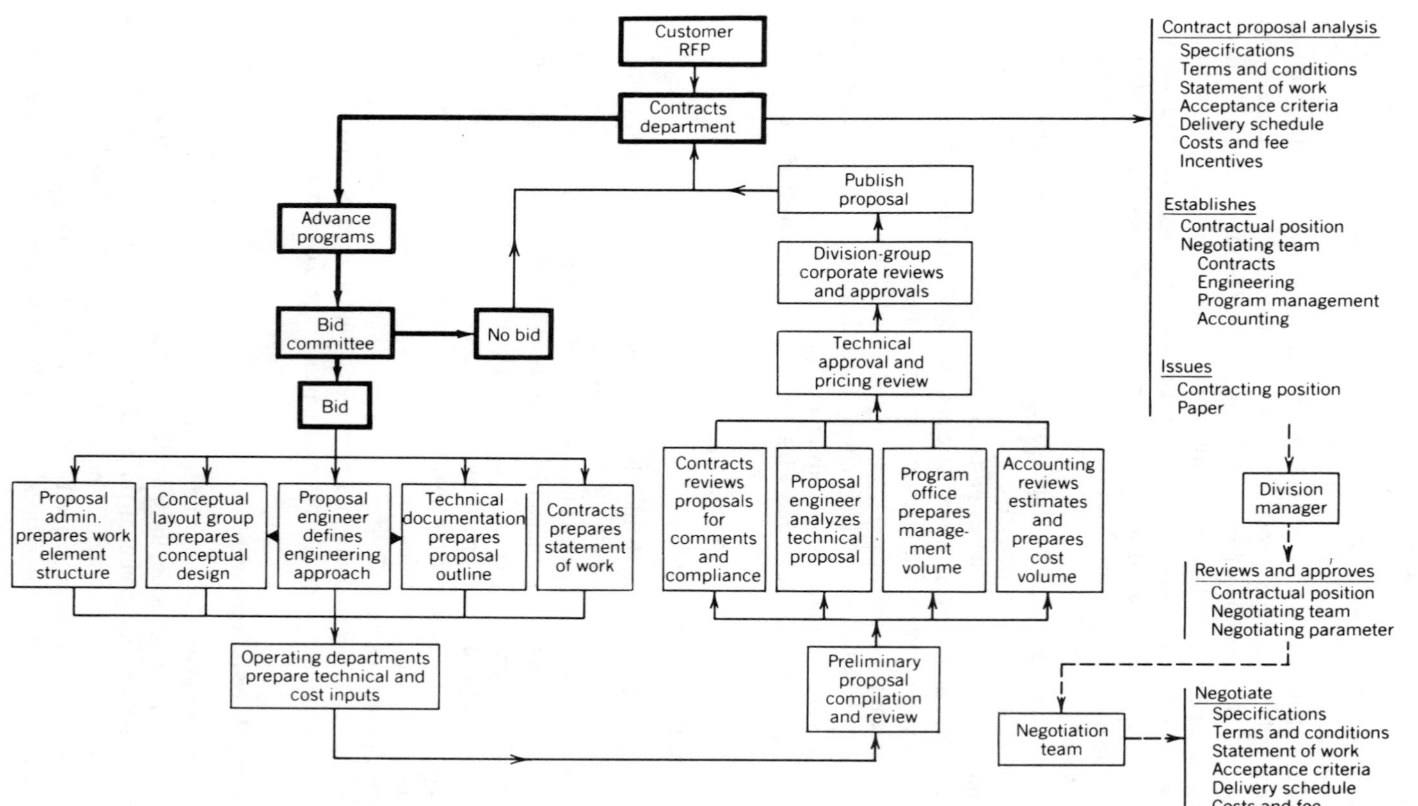

FIGURE 3.7 Proposal flow chart showing bid/no-bid decision. RFP, request for proposal.

Estimation of a rough order-of-magnitude price prior to the bid/no-bid decision will allow evaluation of:

1. Cost in relation to price
2. Price in relation to known budgets
3. Price in relation to supply and demand
4. Price in relation to the competitor's most probable prices.

A cursory evaluation of these four areas will provide the bid committee with quantitative financial data on which to determine overall the most probable pricing advantage.

WHEN THE "WINNER" IS REALLY THE LOSER

Armed with excellent marketing information, the bid committee can make a knowledgeable decision whether to bid or to forego the opportunity to bid in favor of the pursuit of other business opportunities. It has often been said that there is nothing more courageous than a knowledgeable no-bid decision. We might add also that there is sometimes nothing more profitable than an informed, courageous no-bid decision. (See Figure 3.7 for bid proposal flow chart.) Many companies have bid and won coveted contracts only to find out that their preproposal marketing techniques were so lacking that major pitfalls in the proposed contract were obscured. Obscure contract fine print, a propensity to include unanticipated changes, contractually covered actions by the customer that delay the work, or untimely cancellation or termination of parts or all of the work can cause a company to regret having won the competition. A company should rely on the courage of its convictions when it comes to bidding on marginally profitable or ill-defined work, and long-term profitability will improve rather than decrease. If the required marketing homework is done, however, certainty of winning will justify the very best in the proposal preparation effort.

4

PREPARING FOR PROPOSAL ACTIVITIES

Be strong and courageous, and get to work, do not be frightened by the size of the task.

—I Chronicles 28:20

Some who prepare proposals have the misconception that proposal preparation starts with receipt of a request for proposal and is completed when the proposal is submitted to the customer. Nothing could be further from the truth! As emphasized in chapters 1, 2, and 3, the proposal activities of any company are inextricably enmeshed in its overall business acquisition activities and business acquisition plans. A company's overall marketing strategy, its research and development program for new products and services, and its capital acquisition and expansion plans all form a part of the business acquisition process, of which proposal preparation is only a part. Because of this integral relationship of the proposal with other business-related activities, it is often difficult to identify exactly where proposal preparation begins. When an entrepreneur conceives an idea, he or she must then start taking actions that will ultimately result in the manifestation of a final work activity or work output. The steps taken to bring the idea from the conceptual stage to a place where it will fulfill a real need and produce real prosperity can be thought of as part of the proposal preparation process. Although the entrepreneur, scientist, engineer, or business person may have no thought (at the time he or she conceives and starts developing the idea) of submitting a formal written proposal document, he or she invariably will have to do so in some form sooner or later if the idea is to be put into practice. Hence, any description or treatment of the overall proposal preparation process must include a comprehensive analysis of the overall business acquisition process from the viewpoint of how the activities preceding and following actual proposal submission affect the proposal itself, and how these activities influence the potential of receiving an authorization, funds, or contract to proceed with the work. This chapter will describe some of the often neglected and always important steps in the process and will point out ways that a company can enhance, even assure, a winning record. When it is decided which of these methods or techniques will be used, they must be scheduled, staffed, and funded to complement the proposal effort.

PREDEVELOPMENT, DESIGN, BREADBOARD, AND PROTOTYPE TESTING

Straight-line control in preparing proposals for any new work activity or work output is enhanced by forethought put into the design of the product or service through in-depth design, analyses, operations research studies, literature reviews, market surveys, and customer-attitude polls. All of these activities represent vitally important predefinition activities. These preproposal activities provide hard evidence that the product will work properly in its intended environment or that the service can be performed in the time period and with the quality desired by the customer.

For convincing evidence of the viability of a new work activity or work output, there is no substitute for preliminary activation of the work process to produce a simulation, prototype, breadboard, or dry run of the proposed work output or work activity. This convincing evidence, once planned and completed, must then be skillfully documented in anticipation that this evidence may be the deciding factor in the selection of the proposing organization to do the work. If the proposal is based strictly on theory, or what is theoretically possible, it will have little opportunity for success. If it is based on demonstrated results, there is a far better chance of success. All other factors being equal, the more thoroughly simulated, tested and demonstrated work output or work activity will be the one that has the best opportunity of being chosen to fulfill the stated need, provided that the simulations and demonstrations are adequately and skillfully documented in the proposal.

If the proposed work output is a product or project, it is desirable to demonstrate in the proposal that a prototype has been designed, electronically simulated in three-dimensional computer-aided design programs, built, tested, or operated under conditions similar to or more stringent than those that would be expected to be encountered in actual use. If it has not been feasible or practical to build a complete prototype, it is often beneficial to show that working models or operating models of the components or subsystems of the final product have been successfully tested or that testing has been simulated in the environment in which they are expected to operate when combined with the total system. This type of testing is often accomplished in systems modeling on the computer or by what is commonly known as a *breadboard* or *brassboard*.

A breadboard or brassboard is a laboratory or shop working model of the total system that may or may not look like the final product or system, but that will operate in the same way as the final system. An example of a breadboard would be an electronic circuit in which all components and interconnecting wires are placed in their proper electronic position relative to each other, but are not packaged in the same way that they will be packaged in the final product. A breadboard frequently does not conform to size or weight requirements or convenience to the customer's specifications in arrangement of components, but is constructed for convenience, visibility, component removal and replacement, and operational testing. A brassboard is a similar arrangement for hydraulic, pneumatic, or mechanically interconnected components. A breadboard or brassboard design may be arranged horizontally as individual components on a test bench, while the final product that consists of the same components may be packaged in a small, integrated three-dimensional "black box." The breadboard or brassboard can be used for wide ranges of operational testing to evaluate the effect of performance changes and component interactions. The value of the breadboard or brassboard to the proposal activity, then, is to provide tangible evidence (test data) that proves that the product or system works.

The design, construction, and documentation of a "soft" or "hard" mockup is a good investment to supplement the breadboard or brassboard tests and to show that planning and forethought have gone into the final configuration. A soft mockup is one built of wood, soft metals, plastic, foam, or even cardboard or paper. A skillful maker of soft mockups can develop a three-dimensional model that, when painted and photographed, looks like the actual completed product. Soft mockups can be built to full-scale size or to reduced-scale size, depending on the preproposal time and the funds available. Reduced-scale soft mockups or models are economical and, when photographed with suitable scale-model size references, such as a scale model man, car, or house, will look like the real thing. Hard mockups are usually constructed of the same or similar materials to those to be used in the final product, but the model may or may not be a working or operational model. Both types of mockups have the advantage of allowing the customer to visualize through a photograph in the proposal what the completed product will actually look like, which lends credibility, realism, and visual reinforcement to the confidence of the proposal evaluator, because the work output can be realistically depicted in physical form.

Processes and services (work activities) are more difficult to verify during the preproposal process through physical evidence of workability; however, there are ways that tangible evidence can be generated and documented for the edification of the potential buyer. One way to demonstrate a work activity, of course, is to have actually carried out the activity on a small scale prior to proposal submittal. This is similar to reducing to practice a patented process prior to putting it into full-scale production. Laboratory results, time-and-motion studies of a dry run of the process, videotapes, or computer demonstration diskettes of the work activity being accomplished, as well as live demonstrations of the work activity, are all ways of supplementing the proposal material with physical evidence of speed or quality of results.

PRELIMINARY UNSOLICITED PROPOSALS

One way to establish relationships with potential customers is through preliminary unsolicited proposals. Since these proposals can go directly to the people who will make the decisions rather than through the normal contracting or procurement organization, they can be very useful in providing a direct input to the customer's technical decision-makers during the time of formulation of requirements. Through the use of preliminary unsolicited proposals, information can be informally provided to the customer that will be useful in preparing the request for proposal. When a company has provided input that has been used in developing the request for proposal, it is aware of the approach and perhaps even of the content before the request for proposal is formally issued on the open market. Preliminary unsolicited proposals also have the advantage of making the decision-makers in the customer's organization aware at an early date in the procurement process of the capabilities, interest, and the quality of work the proposing firm can provide. Such a revelation to the customer could result in a sole-source procurement, provided that the customer is convinced that the company is the only one that is truly qualified to do the task on schedule and within the costs allocated.

It is necessary to preplan the submittal of any preliminary unsolicited proposals. These proposals should be integrated into the overall new business acquisition strategy, their value considered, and their use adopted if it is believed that they will enhance a competitive position. (One note of caution in submitting preliminary unsolicited proposals: an unethical potential customer might find it advantageous to leak information from the proposal to competitor proposers. The proposer should beware of this possibility and should determine the inherent risks versus benefits before submitting the preliminary unsolicited proposal.)

Since it is often difficult to discern where preproposal marketing leaves off and proposal preparation begins, it is necessary to schedule and plan preproposal conferences with the customer as a continuation of the marketing process and to provide continuity of communication with the customer in this manner. Once the customer has been led through the four steps of the marketing process (attention gained, interest created, desire stimulated, and action taken), continued contact and communication with the customer will be required until the sale has been consummated through the signing of the contract. Since the marketing or sales department's job is essentially completed when the customer has been stimulated to take the action of sending a request for proposal or a request for quotation, the proposal team must take over the job of customer contact and communication and continue this intercourse into the more detailed technical, organizational, and cost aspects of the project.

This is when the marketing department phases in the operational team, which will be defining the more subtle and detailed aspects of the work activity or work output. The proposal manager and the sales engineer (see Chapter 5) will be key individuals in this process and should coordinate the continuity of customer contacts early in the overall proposal preparation process. Detailed communication with the customer should proceed as far as possible into the proposal submission process. If the proposal is competitive with other proposals, sometimes the customer will institute a *blackout period*, during which further communications with the potential suppliers are prohibited. If communications with the customer are continuous and effective, the maximum amount of benefit can be gleaned from customer contact prior to the blackout period.

Once the blackout period is instituted, it cannot be violated or circumvented. The customer will not appreciate those who try to extract information concerning the status of the procurement while the evaluation activity is in process. By the time the blackout period is started, all of the information needed to prepare an effective and winning proposal should have been collected from the customer. Once the blackout period has begun, information is passed on equally to all bidders through formal communications procedures instituted by the customer. These formal procedures usually serve to provide the customer with supplemental information about the capabilities for the work to be performed; they are carried out during the evaluation process after the proposal has already been submitted.

Prior to proposal submission, and subsequent to issuance of the request for proposal, the customer will often arrange a formal preproposal conference to which all bidders are invited and where they are allowed to ask questions to clarify the intent and content of the request for proposal. Attendance at these conferences by marketing personnel and key proposal team members is usually desirable because these conferences offer the opportunity to hear the types of questions that are being asked by competitors. These meetings or conferences frequently do not yield substantive information because competitors will not want to reveal their key concerns or lack of knowledge. But some questions, usually procedural, will be asked, and each potential bidder will be allowed to submit written as well as oral questions about the procurement at this time. The written questions, as well as some of the oral questions, may not be answered by the customer at the time of the bidders' conference, but will be answered in writing later; answers to all questions go to all bidders. The cardinal rule of a bidders' conference at which competitors are present is to ask only those questions that are necessary and to ask intelligent questions rather that ones that may unknowingly reveal potential weaknesses to the competition. The one-on-one meetings already held with the customers prior to the formal phase of the competition will be far more valuable in gaining significant information, which will be useful in preparing the proposal.

The key point in all communications with the customer is to listen carefully! The customer knows what he or she wants, and it is the purpose of a proposed work activity or work output to fulfill that need. Key information will be missed if a company is more concerned with what it would rather supply than with what the customer actually wants; this will result in failure to get the job. Having established in the marketing phase that the job has a high enough potential profitability to cause one to want it, it is essential to listen carefully to the customer's requirements and react responsively and sensitively to them.

WRITTEN QUESTIONS TO THE CUSTOMER

In planning the overall steps of the proposal-preparation process, it is necessary to devote specific time and resources to the function of providing written questions to the customer about the procurement if provision has been made in the request for proposal for a formal question submission. Thoughtful and careful preparation of these questions is necessary because these questions actually serve two functions.

The more obvious of these functions is to get answers or clarifications about the request for proposal content. The other function of carefully prepared written questions is to indicate intelligence, competence, insight, and interest in prospective work. Although competitors may also formally receive the same answers, the customer usually does not indicate which bidder has asked the question. Therefore, much less is revealed to a competitor than in the oral preproposal question-and-answer period. Oddly enough, an absence of questions from the bidder may erroneously be interpreted as a lack of interest, competence, or understanding of the project requirements, rather than a complete understanding. The strategy of formal written communication with the customer, then, should be a subject of preproposal discussion and policy within a company.

PROPOSAL–CLARIFICATION CONFERENCES

For large procurements, particularly those that are of the negotiated-procurement category, the customer may call in one bidder at a time to obtain additional information about his or her proposal after it has been submitted. In this type of conference, the best people should be there and they should be well-briefed on the answers to provide and the supplemental information to present. The customer has usually formulated written questions about the proposal and submitted them to the proposer several weeks prior to the proposal conference, to allow time for preparation of suitable oral answers as well as written replies or proposal addenda. Since this meeting will be one-on-one with the customer without competitors present, there is an opportunity to reveal much more valuable information to the customer and probably even glean some overall information from the customer as to major concerns, doubts, leanings, and trends in the selection. There may or may not be an opportunity to submit supplemental written information to the customer after the conference. If an erroneous statement has been made or an incomplete answer to a question has been given in the conference, a follow-up letter to clarify the answer, whether solicited or not, will be appreciated by the customer.

PREPROPOSAL PREPARATION ACTIVITIES

Preproposal preparation activities involve laying out and dedicating the time required for proposal preparation, allocating and identifying the facilities required, appointing or designating the proposal preparation and review team members, and estimating and authorizing the resources required to do the job. A well-planned proposal preparation activity is essential for the development of a high-quality proposal. The proposal preparation activity itself, particularly for larger proposals, must be meticulously planned and estimated, just as the activity being proposed must be meticulously planned and estimated. In estimating the time, facilities, team, and resources required to complete the proposal preparation process, it is handy to use a checklist like that shown in Figure 4.1.

I Review RFP Requirements

1. Identify each item called for by the RFP through use of a checklist.
2. Identify company approach and develop marketing strategy.
3. Provide inputs or feedback to the customer if appropriate.

II Define Proposal Effort

1. Identify tasks required for proposal preparation.
2. Estimate number of personnel and time required for each task.
3. Establish a proposal schedule.
4. Determine security classification and information protection and handling techniques.

III Perform Preproposal Functions

1. Establish specific approach.
2. Determine extent of proposal (austere, standard, or elaborate).
3. Define work activity or work output to be proposed.
4. Develop a proposal outline and an estimate of the number of pages.
5. Establish proposal preparation labor-hours by task.
6. Identify team members and work area location.
7. Initiate proposal funding request and obtain management approval.
8. Select critique committee (red team).
9. Establish a target price for the work to be proposed.

IV Hold Plan-of-Action or Kickoff Meeting

1. Present program to be proposed to proposal team.
2. Distribute work package.
   a. Proposal preparation schedule.
   b. Proposal outline.
   c. Program outline (preliminary work statement if developed to this extent at this time.)
   d. Assumptions and ground rules.
   e. Preliminary designs and specifications.
   f. Work breakdown structure and dictionary.
3. Assign individual responsibilities and allocate labor-hours.
4. Marketing manager or proposal manager briefs team on customer, program history, and competitive situations.

V Prepare Rough Draft

1. Expand proposal outline.
2. Develop and refine preliminary work statement.
3. Prepare materials and drawing lists.
4. Prepare labor-hours estimates for the proposed work.
5. Develop new facilities requirements.
6. Monitor facility requirements for RFP compliance.
7. Review program schedule for compliance with work statement and RFP.
8. Develop introduction, summary, and a draft letter of transmittal.
9. Maintain proposal funds budget surveillance to control labor-hour expenditures.
10. Obtain line management ideas early.

FIGURE 4.1 The proposal-preparation checklist.

11. Review labor-hour estimates and materials costs to be sure they are consistent with the RFP and expected customer funding.

VI Detailed Review of Technical and Cost Volumes

1. Study preliminary cost breakdown.
2. Eliminate any unncessary or excess tasks and/or labor-hours.
3. Assure a competitive price and technology that is consistent with RFP and customer funding.

VII Critique Committee Review All Volumes

1. Obtain critique committee or red team comments and extent of rewrite required.
2. Rework proposal in accordance with critique committee or red team comments.
3. Submit for final type by sections.
4. Review final corrected copy.
5. Check figure sequence and page sequence.

VIII Produce the Proposal

1. Finalize letter of transmittal and obtain signature.
2. Complete final artwork.
3. Complete final copy editing.
4. Print the proposal.
5. Collate, assemble, and bind volumes.
6. Check over all bound volumes and all copies for completeness.

IX Package the Proposal

1. Assemble volumes into complete sets.
2. Provide the required number of copies.
3. Package into convenient size envelopes or boxes.

X Mail or Hand Carry the Proposal to the Customer

FIGURE 4.1 *(Continued)*

ABOUT REVIEW TEAMS

The conduct of one or more in-depth critical and searching reviews is an important step in producing a winning proposal. In very rare instances can a single individual, or even a proposal team, construct a superior proposal at the first writing. Even small proposals benefit tremendously from objective and constructive peer reviews. In large proposals, where there is more opportunity for inconsistencies and weaknesses to appear in initial drafts, several reviews have proven to be necessary to provide a smooth-flowing, convincing, and invulnerable argument for selection of the team's proposed technical, cost, and management approach. To accomplish this objective, a literal rainbow of review teams has evolved and is generally considered to be mandatory in guiding a company's proposal into the evaluators' "winners" column. The number of members of review teams is not fixed, but should be more than 2 and not more than 12 or 13. Pink, blue, red, green, and gold team reviews, and sometimes other colors, proliferate in the proposal preparation process. Each team has its purpose and time-oriented position in the proposal preparation schedule. Although all proposers do not necessarily agree on the same order, sequence, importance, and team composition of this rainbow of colorful review teams, the proposers invariably employ internal, external, and peer reviews to improve proposal flow, to weed out inconsistencies, to build in strengths, and to minimize potential proposal weaknesses. One of many possible review team arrangements, and a few alternative definitions, follow.

The Pink Team Review

The pink review team consists of individuals who are closely associated with the potential work output or work activity, but who are not usually members of the proposal team. The pink review team conducts the first review of initial proposal themes, graphics, approaches, and outlines. The pink team will often be the reviewer of storyboards if the graphic representation of work (GROW) approach is employed in proposal preparation. (Storyboards and their use in the GROW approach are discussed in detail later in this chapter.)

In many cases, member of the pink team are the same as those of later review teams. Continuity of review-team membership is beneficial to some degree and will help provide consistency of guidance and direction to the proposal team during the proposal preparation process. On the other hand, new viewpoints injected by bringing in more independent and outside reviewers can be beneficial in providing fresh new insights into the improvement of proposal quality. The exact approach taken to pink team membership in relation to the membership of teams that come into play later in the process will depend on the personalities of those on the team, their technical and management expertise, and their organizational location. Potential members of the pink team could be the vice-president of marketing, key personnel of potential associate contractors and teaming partners, independent proposal preparation and writing consultants, experienced proposal creators from other projects, principal investigators, surrogate customer advocates, and representatives of other independent corporate technical and nontechnical organizations, such as quality control, engineering, manufacturing, industrial engineering, and systems engineering.

The Blue Team Review

The blue team in the proposal-preparation process can take on one of several functions. One potential use of the blue team has been advocated in classical approaches to proposal preparation: the blue team exhaustively explores the competition and serves as a surrogate competitor in developing counteractive and adversarial positions and approaches that must be addressed by the proposal team in overcoming potential competition. Although this function is theoretically an ideal and potentially a very helpful adjunct to proposal preparation, there is often not enough time to develop exhaustive alternative competitors' strategies. The number of competitors and the intensity of competition is often too great to permit the construction and conduct of a cost-effective mini-evaluation process of surrogate proposals. A more common use of the blue team is to conduct an informal review of the first draft of the proposal. The blue team, as used for this purpose, is a "friendly" review team assigned to provide constructive comments and suggestions as the first draft of the proposal takes shape. Blue team membership can consist of persons within and outside of the proposal team.

The Red Team Review

The red team review is usually the most critical, searching, and stringently controlled of all of the reviews. If there is only one review conducted, it is usually the red team review. Red team membership always includes division or corporate management and can include independent consultants. The persons who are providing the proposal preparation resources, those who will perform the work, and those who will gain most from the win are key members of the red team. As mentioned earlier, the red team may include many of those who were members of the pink team. Usually, higher levels of the organization are also represented on the red team.

The Green Team Review

There is some disagreement on the color, but not on the function of this team, which conducts the final review of the proposal before it is released. Some companies call this team "the gold team." Whatever its color, this final team usually consists of fewer persons; it is principally composed of corporate top managers, who pass on the final content, format, and cost or price quoted. Because of its late entry into the review process, the green (or gold) team must limit its inputs to substantive issues rather than minor proposal revisions. Little time is left to change proposal details, but minor repricing of the proposal or rewriting of the executive summary or letter of transmittal may be necessary to better conform to corporate objectives or last-minute marketing intelligence. The green team gives the final signal to publish and distribute the proposal.

The degree of use of the various review teams will have a significant impact on the time required for proposal preparation, our next subject.

TIME, FACILITIES, AND OTHER NECESSITIES FOR PROPOSAL PREPARATION

Time. The competitive nature of the procurement process makes the factor of time important in planning the proposal preparation process. A fast-response unsolicited or solicited proposal may result in one firm's being the only bidder for a job and may result in a sole-source procurement. In most solicited competitive procurements, the time allocated for proposal preparation is only a small fraction of that required for actual performance of the job being proposed, yet some of the aspects of job performance must be at least addressed if not initiated during the proposal preparation process. The construction of models, mockups, breadboards, and prototypes and their testing and documentation usually take more time than that usually allowed for proposal preparation, therefore, this work must be completed prior to receipt of the request for proposal. The short time allocated for proposal preparation precludes last-minute decisions and haphazard planning. Many companies, through inadequate proposal planning, have entered the competition too late and have had to play a game of catch-up throughout the proposal competition. Because marketing feeds information to the proposal team continuously, even during the proposal preparation process, there is usually a scarcity of time during the proposal preparation process. Hence, it is vital that detailed time-planning, including all team reviews, be done for the proposal preparation process. Deadlines must be established for each phase of the proposal preparation process, and these deadlines must be met. There can be no room for slippage or delays, because the submission date is usually fixed. Time delays must be made up within the schedule (rather than by extending the schedules) by the use of overtime, additional personnel, or multishift operation. Since the proposal team is an integral unit (see Chapter 5), the last two of these alternatives are less attractive and oftentimes become the only way to make up for proposal preparation schedule slips. A courageous no-bid decision is much better than accepting a proposal preparation job with insufficient proposal preparation time.

Facilities Required for Proposal Preparation. The principal facilities required for the actual proposal preparation process are (1) an isolated and secure office space or area for the proposal preparation team, furnished with chairs, desks, conference tables, display boards, and computer-aided proposal preparation equipment and (2) a publication facility. These facilities must be supplemented in the preproposal phase by any shops, test laboratories, and test facilities required to develop tangible evidence of work quality and credibility through physical demonstrations of prototypes and dry runs. Action must be taken to assure that these facilities are made available to the team on a dedicated basis during and throughout the proposal preparation period.

The Proposal Team. Preplanning of proposed activities is an essential element in identifying the personnel required for the proposal team and determining the skill categories and skill levels required for these personnel. A look forward to Chapter 5 will show that some very specialized and competent people are required for proposal preparation. Planning by company management to identify, allocate, and assign these personnel to the team during the proposal preparation process is necessary to assure that other ongoing work being performed by the company is not adversely affected by the reassignment of these highly qualified team members to the urgent job of proposal preparation during the new-work acquisition process.

Resources Required to Propose. Proposal preparation requires money and manpower. In the preproposal marketing and planning process, a detailed cost estimate should be made of the costs of preparing the proposal, and this estimate should be used to develop a budget, which is then placed under the control of the proposal manager. Proposal preparation costs can vary in relation to the the costs of the proposed work from 1% to 10% of the costs of the proposed work. Therefore, a rough estimate of proposal preparation costs is usually not acceptable to company management. The same principles spelled out in Chapter 9, "The Cost Proposal," can be used in estimating the cost of preparing the proposal itself. This is a vital part of the proposal planning activity and will help to identify all of the activities and actions that must be accomplished to do the total proposal preparation job.

FOLLOW-UP ACTIVITIES TO PROPOSAL PREPARATION

Because members of the proposal preparation team may become negotiators and/or performers of the work itself, the job is not finished when the proposal is submitted to the customer. The proposal activities include other follow-up tasks that must be planned and carried out before a contract is signed. These tasks, although not a part of the physical preparation of the proposal document, must be included in planning the proposal activities because they require the continued allocation of time, facilities, the proposal preparation team members, and monetary and manpower resources. The two principal postproposal activities are preparation of the best and final offer (if one is allowed for in the procurement process) and negotiation of contract provisions, scope of work, contract price, and fee.

The Best and Final Offer

In many negotiated procurements a provision is made for the offer or to submit a "best and final offer" (BAFO) prior to completion of the evaluation process and the selection of the winning firm. A deadline date is usually provided in the request for proposal, beyond which best and final offers will not be accepted. Although best and final offers vary in content and depth, the principal constituent of a best and final offer is always a final quoted price for accomplishment of the work. Often the procuring company or agency requires little backup material for this revised bid price, and it is usually not necessary to resubmit the entire cost proposal. Even though backup material is often not required, it is wise and prudent for the proposing firm to do a detailed revision of its cost estimate and of its proposed pricing adjustments to this revised cost estimate. The reason for this is that the work activity or work output must be adjusted to match the newly proposed price by modifying the work content, skill categories or levels, timing of schedule elements, or specifications. Time and effort must be allocated in the overall proposal activity schedule to accomplish modification of company work plans to meet the new quoted price.

Contract Negotiations

It is the practice of some procuring organizations to conduct parallel negotiations with several bidders before making a final source selection decision. This reverse-auctioning process is often time- and manpower-consuming, because company funds must be expended on a job before the assurance is available that the job is won. Proposal preplanning should recognize this fact in these instances, and careful planning should be done to assure the availability of team members and resources to carry out the various phases of the negotiation process.

Planning for Negotiation of the Statement of Work. Whether source selection is completed before or after negotiations, both technical and management proposal team members may be involved in negotiation of the contract work statement and specifications. The key to negotiating the contract work statement and specifications will be to go over all words and numbers in the contract very carefully to ensure that the work contracted for can be accomplished with the resources available to do the work. Many losses have been incurred and companies have been ruined because they failed to evaluate fully the subtle wording or numerical requirements during contract negotiations. It is essential for the proposer to understand what is being sold and for the customer to know what he or she is buying. Sufficient manpower and resources should be allocated to the new-work acquisition effort to assure a thorough and systematic approach to work scope and specification negotiations.

Planning for the Negotiation of Contract Price. Negotiation of contract price (or cost if a cost-reimbursable contract is contemplated) should involve individuals from the technical, cost, and management proposal teams who can support company negotiators in dealing with the customer. Principally, it is the cost proposal team members who will be the most useful in this negotiation, because they have access to and can explain detailed labor-hour and material estimates from a detailed cost proposal. The cost proposal team members are also most likely to know if and how much contingency or allowance for cost growth has been included in each cost element, within each work element, and within each calendar time period. Although these contingencies and allowances for cost growth may or may not be revealed to the customer, it is vital that the negotiator have this information at his or her fingertips. The most common ploy of customers who are negotiating on a limited budget is to try to eliminate or at least to reduce the contingency costs and allowances for cost growth. Unless a work output or work activity is thoroughly defined in painstaking detail, elimination of cost growth allowances could be a precursor to an eventual overrun in cost, resulting in a loss in profit (fee) or even in lack of reimbursement for materials and direct labor. Very few jobs are so well-defined that it is possible to eliminate all contingencies and allowances for cost growth. Sufficient manpower and resources must be planned for as early as the preproposal phase in order to assure that the proper individuals will be available for this aspect of the negotiation process.

Planning for Negotiation of the Fee. Early in the proposal preparation process it is necessary to establish a policy on what profit or fee will be required to make the whole venture worthwhile. As pointed out in Chapter 3, the quality of profitability has to be considered and evaluated early in the process and used as a basis for a bid decision. If the venture has an extremely high profitability (long-term or life-cycle profit and growth advantage), then a lower immediate profit or fee can be accepted. On the other hand, if profitability as related to growth and long-term benefits are limited or nil, a high front-end fee or profit should be demanded. Because the fee or profit negotiation is usually performed by corporate or company negotiators or company management, little support of the proposal preparation team except perhaps the proposal manager and cost manager will be required. But time must be allocated in the overall business-acquisition cycle to negotiate profit or fee. If the fee structure is to be based on an award fee or incentive fee arrangement, fee negotiation time must be increased in proportion to the fee structure complexity.

ESTABLISHING A PROPOSAL PREPARATION SCHEDULE

When the proposal cycle is expected to take several weeks or months, and when inputs will be required from various organizations and/or disciplines, an essential tool is a detailed proposal schedule. The minimum key milestones in a proposal preparation schedule are (1) a kickoff meeting; (2) a review-of-ground-rules meeting; (3) a technical and resources input and review meeting; and (4) summary meetings and presentations. Descriptions of these meetings and their approximate places in the proposal preparation cycle follow.

The Kickoff Meeting. The very first formal milestone in a proposal preparation schedule is the kickoff meeting. This is a meeting of all the individuals who are expected to have an input to the proposal. It usually includes individuals who are proficient in technical disciplines involved in the work to be performed; business-oriented individuals who are aware of the financial factors to be considered in developing the proposal; project-oriented individuals who are familiar with the project ground rules and constraints; and, finally, the proposal manager and the proposal preparation team.

Sufficient time should be allowed in the kickoff meeting to describe all project ground rules, constraints, and assumptions; to hand out copies of selected portions of the request for proposal, technical specifications, drawings, schedules, and work element descriptions and resource estimating forms; and to discuss these items and answer any questions that might arise. It is also an appropriate time to clarify proposal preparation assignments among the various disciplines represented, in the event that organizational charters are not clear as to who should support which part of the proposal.

The Review-of-Ground-Rules Meeting. Several days after the kickoff meeting, when the participants have had the opportunity to study the material, a review-of-ground-rules meeting should be conducted. In this meeting the proposal manager answers questions regarding the proposal preparation, assumptions, ground rules, and assignments. If the members of the proposal preparation team are experienced in developing technical descriptions and resource estimates for their respective disciplines, very little discussion may be needed. However, if this is the first proposal preparation cycle for one or more of the team members, it may be necessary to provide these team members with additional information, guidance, and instruction.

The Technical and Resources Input and Review Meeting. Sometime after the kickoff and review-of-ground-rules meetings, each team member that has a technical and resources (labor-hour and/or materials) input is asked to present his input before the entire proposal team. Hence starts one of the most important parts of the proposal preparation process: the interaction of team members to reduce duplication, overlap, and omission in technical descriptions and resource data.

The proposal manager should, in this meeting, make maximum use of the synergistic effect of team interaction. In any multidisciplinary activity, it is the synthesis of information and actions that produces wise decisions rather than the mere volume of data. In this review meeting, the presenter responsible for each discipline area has the opportunity to justify and explain the rationale for technical descriptions and estimates before his or her peers—an activity that tends to iron out inconsistencies, overstatements, and incompatibilities. Occasionally, inconsistencies, overlaps, duplications, and omissions will be so significant that a second input and review meeting will be required in order to collect and synthesize all proposal inputs.

Summary Meetings and Presentations. Once the proposal inputs have been collected, adjusted, and priced or costed, the proposal is presented in a package on a dry run to the proposal team. This dry run can reveal further inconsistencies or errors that have crept into the proposal during the process of consolidation and reconciliation. A final review with the company management could also bring about some changes in the proposal because of last-minute changes in ground rules.

As shown on Figure 4.1, there are other steps leading up to and surrounding these key meetings. These other steps are developed and explained more thoroughly throughout this book, but are summarized briefly here in a form that will allow the proposal activity planner and estimator to plan and estimate the resources required to carry out a total proposal preparation activity.

ESTABLISHING DETAILED PROPOSAL PREPARATION ACTIVITIES

Review of Request for Proposal (RFP) Requirements. The review of RFP requirements (or of internal company requirements if the proposal is to be unsolicited) initiates the detailed process leading to the publication and delivery of a completed proposal document. The first step in this RFP review is to develop a summary or checklist of the RFP. Since many RFPs are long and complex, this step eliminates much confusion and provides in-depth understanding as well as an overview of RFP requirements. Concurrent with this step, assuming that a decision to bid has already been made, is the development or formulation of an overall company approach and marketing strategy that will be used in developing the proposal. Then the results of the RFP review should be evaluated for places where supplemental information is needed either from marketing or from the customer. If the customer has made provision for written or oral questions from potential bidders, a formal list of questions is prepared and submitted to the customer (the results or answers to these questions will be returned in the form of a letter or an addendum to the RFP). Keep in mind that competitors will also be afforded an opportunity to receive the answers to any questions.

Define the Proposal Effort. This and the following step include the activities that are spelled out in this chapter, including identification of all the tasks required for proposal preparation; estimation of the number and types of personnel, equipment, facilities, and time required to prepare the proposal; establishment of a proposal preparation schedule; and determination of security classifications, logistics, and handling techniques.

Preproposal Functions. Preproposal functions commence with the establishment of a specific proposal approach that fits within the request for proposal requirements and the overall company marketing approach. A determination is made of the extent of the proposal; that is, will it be an austere, standard, or elaborate proposal? This decision will be guided by the assessment of marketing and company management concerning the worth to the company of winning the contract as well as the resources available to submit the proposal. The work activity or work output to be proposed must then be defined in detail. This is done by preparing a work breakdown structure and a work element structure dictionary. A detailed outline of the proposal is developed along with an estimate of the number of pages, figures, tables, and appendices that will be included. Proposal preparation labor hours are estimated and budgeted by task, and work team members, and the work area location is identified. A funding request or proposal task authorization is initiated to obtain management approval to proceed with the proposal preparation activity. A critique committee, red team, or "murder board" is selected to review the final proposal draft before it is sent to production, and a target price is established for the work to be performed.

Plan of Action or Kickoff Meeting. As mentioned earlier, all key proposal team members are invited to a plan-of-action or kickoff meeting. In this meeting the proposal manager, company marketing or sales manager, or other knowledgeable company officer describes the work activity or work output to be proposed. A work package is distributed, which includes a proposal preparation schedule; a detailed proposal outline; a preliminary work statement (if available); and a list of proposal team assignments, responsibilities, and labor-hour allocations. Also distributed are proposal assumptions and ground rules, preliminary designs and specifications, the work element structure and dictionary, and blank forms for submission of technical and resource data. The marketing manager or sales manager also briefs the proposal preparation team on the customer's characteristics and desires, program history, and competitive situations.

Preparation of Rough Drafts of Proposal Volumes. When the proposal team has been thoroughly briefed, work commences on expansion and deepening of the proposal outline and development and refinement of the preliminary work statement.

Materials and drawing lists are prepared, and labor-hour estimates are prepared for the proposed work. Concurrently, facilities and equipment lists and descriptions are developed. Both the facility requirements and the program schedule are monitored for request for proposal compliance. The introduction, summary, and draft letter of transmittal are prepared.

Throughout the time period of rough draft preparation, surveillance of the proposal funds budget is maintained to control labor-hour expenditures, line management ideas are interjected into the draft, and estimated labor-hour estimates and material cost estimates are reviewed to be sure they are consistent with the RFP and the expected customer funding.

Detailed Review of Technical and Cost Volumes. Once the preliminary or rough draft of the proposal is completed, a detailed, in-depth review is made of the technical and cost volumes by studying the preliminary cost breakdown and eliminating any unnecessary or excess tasks and/or labor hours, thereby assuring a competitive price and technology that is consistent with the request for proposal and the available funding.

Proposal Writing and Publication. Finalizing proposal writing and production starts with a polishing of the letter of transmittal and approval and signature of the letter of transmittal by company management. Final artwork and copy editing are completed, and the proposal is sent to press. When printing is completed, collation, assembly, binding, and checking of all copies are performed by the production staff.

Packaging and Delivering the Proposal. Proposal volumes are assembled into complete sets, copy counts are verified, and the proposals are packaged into convenient-sized envelopes or boxes. Often one or more copies of the proposal are handcarried to the customer to personally assure that the deadline date is met. The remaining copies are mailed or shipped by a reliable and fast mail or parcel carrier.

PROPOSAL SUPPLEMENTS

For very large or complex proposals it is sometimes both necessary and desirable to supply additional material besides the actual written proposal volumes. Scale models, mockups, briefings, brochures, videotapes, computer software demonstration disks or tapes, and other tangible material may have to be developed, produced, and delivered to the customer at appropriate points in the new-work acquisition cycle. Planning for the production and delivery of these items must be done at the outset of the proposal cycle to permit effort to proceed parallel to the written material.

Scale Models and Mockups. Earlier it was mentioned that photographs of models and mockups may be desirable to display realism gain credibility for the proposal document. The bidder should also consider the possibility of providing an actual scale model, working model, or mockup to the customer as part of the proposal. A three-dimensional representation of a completed product or project helps the customer visualize what the final configuration will look like better than detailed engineering drawings and specifications do. Also, the customer may need to use such a scale model or mockup to illustrate the principles, advantages, or operation of the item to his or her customer; to management; or even to the public, through the media.

Scale models are particularly appropriate in large architect–engineer tasks, construction tasks, public works projects, and high-technology hardware, such as aerospace and defense systems. Also to be considered is the value of a scale or working model, a cutaway model showing operation, or an operational breadboard or brassboard to supplement a proposal of any size. The production of the model itself must be well planned, organized, and carried out with forethought and prudence in order to make it a truly useful item in the marketing of a proposal.

Motion Pictures, Videotapes, and Software Simulations. When a process or a work activity is the subject of the proposal, a motion picture, videotape, slides, compact disks, software simulations, or audio tapes may also supplement the proposal. The expense of these media should be weighed against potential benefits. Technology improvements in these areas of presentation are reducing costs significantly, and a well-produced videotape, graphic presentation, software demonstration, or slide show might be the additional icing on the cake that will assure palatability of the entire proposal. Audiovisual or digital imagery production should be planned and carried out parallel to the preparation of written material, to assure consistency and compatibility of the written and audiovisual presentations.

Computer Demonstration Programs. When computer technology is all or part of the work activity or work output being marketed, an essential element to provide with the proposal is a hands-on computer demonstration. Because computer technologies are continuing to expand rapidly, it is often necessary to educate the customer on data systems, information technologies, and software capabilities. When the potential customer is unfamiliar with emerging technologies, there is no substitute for a real-time, live demonstration of a preliminary version, or even of a similar version of the technology that is being marketed. Communications links enable onsite demonstrations of information systems technologies on data terminals. Cost and time required to develop demonstration packages to illustrate operations to the customer should be considered. This expenditure of resources and development effort may well be the final element that convinces the customer to select your firm for the job.

A TYPICAL PROPOSAL FLOW

On Figure 4.2, each organization involved in the proposal preparation process is represented, and the activities of each of these organizations is briefly depicted.

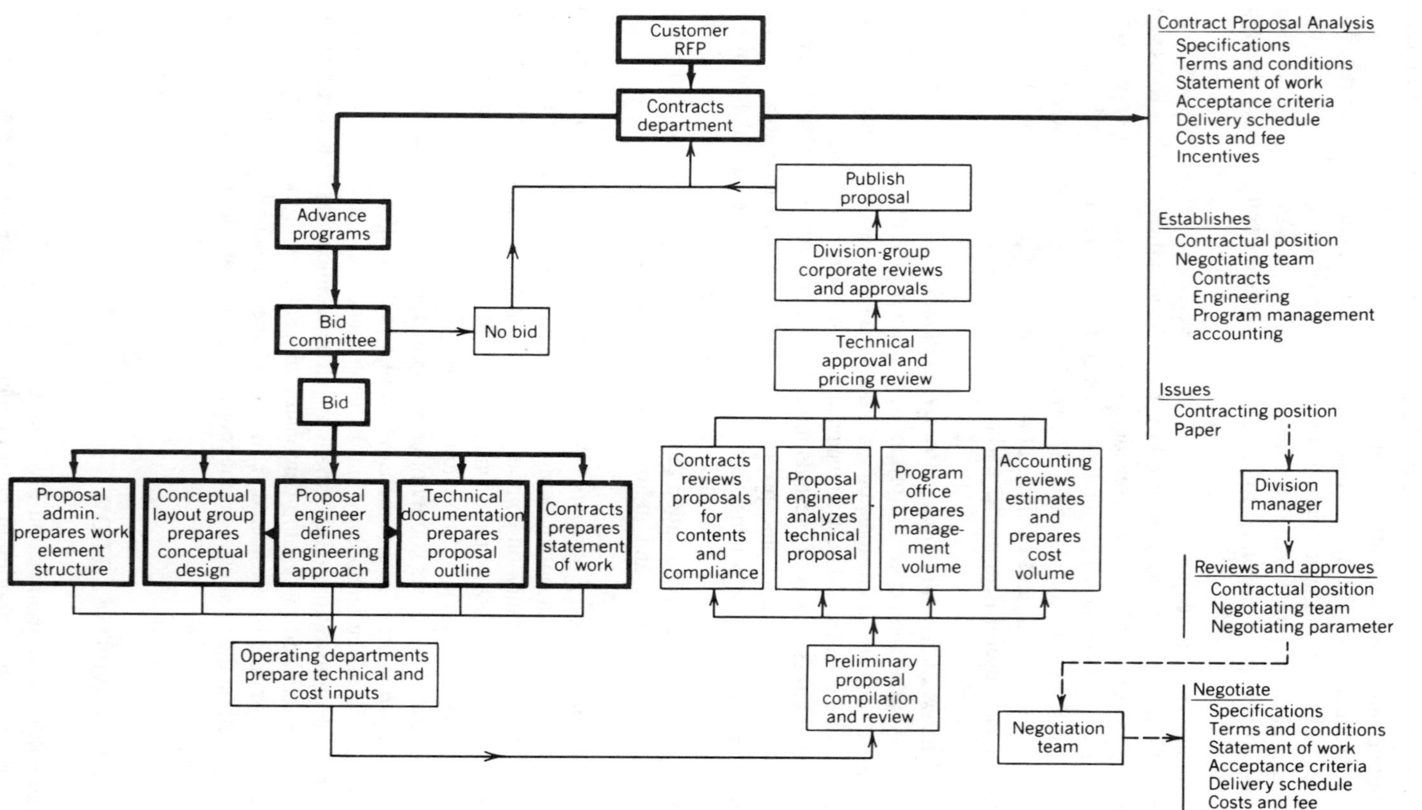

FIGURE 4.2 Proposal flowchart showing typical organization and responsibility.

The flow chart lets each organization know where it fits into the overall proposal preparation picture and indicates to these organizations from whom they will be receiving their information and to whom they will be providing information. This chart will vary with organizational structure and with the type of proposal being submitted. The proposal flow chart will also identify all major elements of the proposal preparation process and will be an effective aid to the identification and planning of resources required to effectively carry out the total proposal effort.

To be sure all involved segments of the organization are aware of the actions required, a proposal checklist similar to that shown on Figure 4.3 is prepared and submitted to each organization expected to contribute to the proposal effort.

A GENERIC 45-DAY PROPOSAL SCHEDULE

Typical preparation times from RFP to submission of very large proposals are 45 days, 60 days, or (in rare instances) 90 days. The 45-day proposal preparation schedule presents the biggest challenge to the proposal preparation team, as it compresses the time required to prepare and publish one or more major documents into about 8% to 16% of the time it would normally take to prepare and publish a commercial peer-reviewed document of the same proportions. It is for this reason that a precise and closely controlled milestone schedule needs to be produced and implemented for the preparation of large proposals. Figure 4.4 is a generic 45-day proposal schedule that shows the major milestones, reviews, and activities that must occur to complete and submit a proposal on time. The chart is given in terms of "Days After RFP Release." Although minor changes can be made in this schedule to meet the needs of your company, little flexibility exists if rigorous reviews are conducted and if resulting revisions and updates are to be input on a timely basis. The first 10 to 12 days of this process are critical. If the proposal team is to have a chance for success, scheduled activities during these first 10 to 12 days must be rigorously followed. Figure 4.5 illustrates this first critical period and shows the time frame required to prepare "storyboards" upon which actual writing of the text of the proposal can be based.

GRAPHIC REPRESENTATION OF WORK (GROW) APPROACH AND STORYBOARDS

Through many years of preparing rapid-response proposals for large projects, we have found that the most effective technique in responding to RFP requirements is the graphic representation of work (GROW) approach for writing proposals. This technique is equally applicable to small, 10- to 20-page proposals and to huge, multivolume responses. In fact, even if the proposal requires only a 1- or 2-page response or input, the GROW approach is helpful. All members of the proposal team should be pre-indoctrinated and pretrained in the use of this procedure before any small or large proposal effort is contemplated or undertaken. This procedure

PROPOSAL CHECKLIST — Date

Type of Proposal	Contract Effort			Proposal Volumes
___ Cost Type ___ ______	___ Engineering Study	___ Production	___ Field Support	___ Letter ___ ______
___ Fixed Price ___ ______	___ Breadboard	___ Spares	___ __________	___ Technical ___ ______
___ ROM ___ ______	___ Prototype	___ Data	___ __________	___ Management ___ ______
				___ Cost ___ ______

Proposal Title: ______________________ Notes: ______________________

Proposal Manager: ______________________ ______________________

Proposal Administrator: ______________________ ______________________

Attachments:	Yes	No		Yes	No
W.B.S.	___	___	Technical Summary	___	___
Program Schedule	___	___	Estimating Instructions	___	___
Proposal Schedule	___	___	Task Descriptions	___	___
__________	___	___	__________	___	___

Effort Required	Administration	Program Staff	Data and Information	Security Documents	Adv. Tech. Laboratory	Reliability Engineering	Mechanics	Electrical Engineering	Systems Engineering	Design & Draft.	Manufacturing Engineering	Test	Product Assurance	Data Management	Contracts	Accounting
Attend kick off meeting																
Prepare technical write-up																
Prepare cost estimate																
Prepare bill of materials																
Prepare spares list																
Itemize capital equipment																
etc.																
etc.																

FIGURE 4.3 Checklist for responsibility assignments, WBS, work breakdown structure.

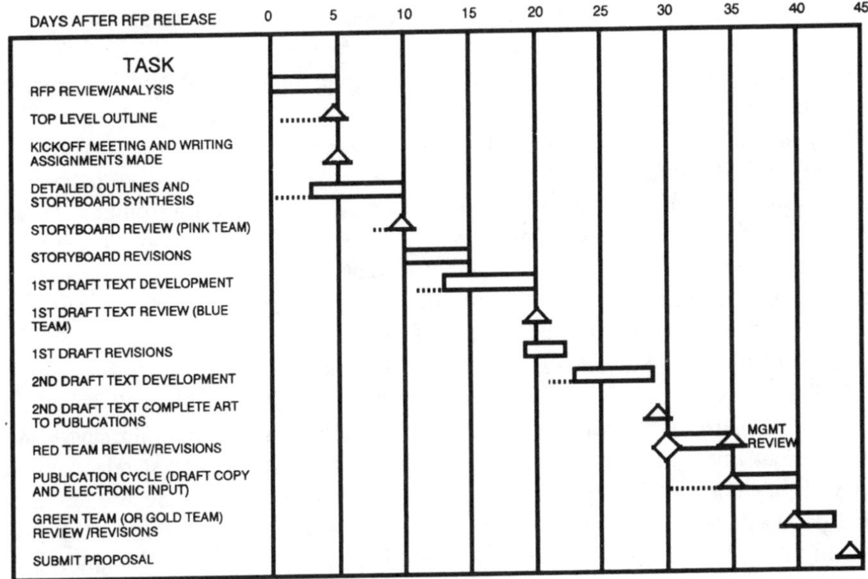

FIGURE 4.4 Generic 45-day proposal schedule. Hatched line: preliminary work; rectangle: activity. Triangle: event. Diamond: red team review.

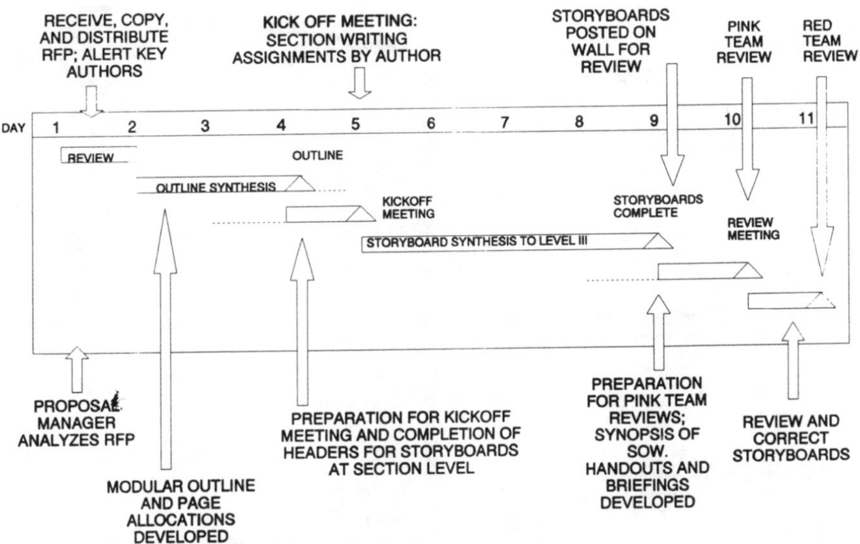

FIGURE 4.5 Storyboard schedule based on a 45-day proposal preparation schedule.

ensures straight line control and focused attention to the achievement of highest scores on one of the most important proposal evaluation criteria: *Understanding of the Requirement*. Figure 4.7 shows how your team should "GROW" a proposal from the work statement into a viable, credible, supportable winning proposal. The storyboard and the final proposal should include the graphics created by this method.

The GROW approach includes the development of sketches, graphs, block diagrams, or illustrations, and a thorough review and iteration of these illustrations *before* writing is started. There should usually be at least one graph developed (*prior to writing*) for each 1½ to 2½ pages. This illustration may or may not end up being actually used in the proposal document, but its very existence helps the author focus rapidly on the essentials of what needs to be said.

The medium used in implementing the GROW approach is the storyboard. Figure 4.6 is a storyboard form used to develop initial outline thoughts and the initial illustration for 1½ to 2½ pages of proposal material. The left portion of the storyboard is filled out first with brief outline information and supporting key statements that may or may not appear in the final proposal text. These statements are key thoughts and points that must be brought forth strongly in the proposal text, and they are usually devised from a *close scrutiny* of the request for proposal. A graphic representation of work is prepared for each discrete task in the work statement. One

MOBILE DATA SERVICES

PROPOSAL NAME: ______

SECTION ______
SUBSECTION *(For each work task/subtask)*

RFP REQUIREMENTS ______ WRITER (SB) ______ /TEXT ______
SOW # ______ PHONE EXTENSION ______
EVAL CRI. ______ DUE DATE ______
APPROVED: ______

SECTION OBJECTIVE AND CONTENTS

OBJECTIVE *Respond directly to each paragraph of the work statement to demonstrate understanding.*

THESIS SENTENCE

This sentence describes the approach that will be taken by the proposer, as a potential performer, to meet the objective stated above.

SUPPORTING STATEMENTS

1. *Understanding of the problem and key technical issues: (why does the customer need this task performed within context of the overall effort.)*
2. *Approach and methodology in respondng to the statement of work requirements and technical issues.*
3. *Examples of other similar efforts, and their implementation versus the implementation of this effort.*
4. *People, tools, and experience available to do the job; and synopsis plus pointers to other sections.*

SKETCHES OF ART

FIGURE NO. ______
FIGURE TITLE: ______
(Initial Caps)

FIGURE 4.6 Storyboard form with guiding instructions. SB: storyboard, RFP: request for proposal, SOW: statement of work, EVAL CRI: evaluation criteria.

phrase or a sentence-long objective is placed at the top left corner of the storyboard to guide the construction of text ideas and graphic ideas. A thesis sentence is then composed that sums up the entire idea that is being described in response to the objective. Supporting statements are then composed and listed under the thesis sentence. The objective, thesis sentence, and supporting statements represent a framework of ideas and thoughts that must come through strongly in the supporting text that will be composed *after* the storyboard is reviewed and approved for commencement of the proposal-writing process.

Next, a simple, straightforward sketch is placed on the right, ruled side of the storyboard. This sketch can take the form of a block diagram, flow diagram, Gantt chart or schedule, illustration of a part or component; or a sketch of an assembly, test, inspection, or software concept. It is a visual representation of an idea, concept, or device that fulfills the need of the thesis and supporting statements.

To provide an illustration of how such storyboard sketches can evolve from the RFP statement of work, we present a series of plates (Figures 4.8a–m) in which sketches represent the words from an actual request for proposal, one that was issued in the early 1990s for support to NASA's Earth-Observing System program. Simple diagrams are developed that explain, in graphic format, the meaning of each statement of work section.

Keep in mind that a paragraph's ideas can be diagramed in myriads of forms. In Figure 4.8, one of a virtually unlimited number of illustrations is provided as an example for each paragraph. With some study of these examples and some practice on your own, it will become second nature to picture and develop a diagram of what is happening in each paragraph. The diagraming process is started with the RFP to gain an understanding of the requirement. A diagram derived directly from the RFP's text is an excellent start for a responding proposal's graphics. From this basic "RFP in graphic form" sketch, the proposer can then develop and embellish the artwork with the innovative ideas and approaches needed for a winning proposal. This, again, is the straight-line control approach. You are taking text words *directly* from the request for proposal, converting them into clear and understandable sketch (which represents your understanding of the requirements) and then adding information in the illustration that shows how your firm is planning to fulfill this requirement in a way that will provide the best and most cost-effective solution (see Figure 4.8a–m).

The plates of Figure 4.8 were provided to illustrate how easily one can diagram a thought or idea. Try this yourself. Pick a paragraph at random from any text and practice making sketches of the ideas, concepts, people, devices, or thoughts depicted in the paragraph. Once you have practiced it, you will find that it is both easy and enjoyable.

Once the requirements have been thoroughly understood, storyboard preparation can start. Figure 4.9 is a sample storyboard on how to prepare storyboards. This graphic serves two purposes: (1) it shows sample storyboard entries for objective, thesis, supporting statements, and graphics; and (2) it provides a diagram of the steps required to prepare a storyboard. This figure is truly synergy in action, as it represents straight-line control toward helping you prepare storyboards and helping

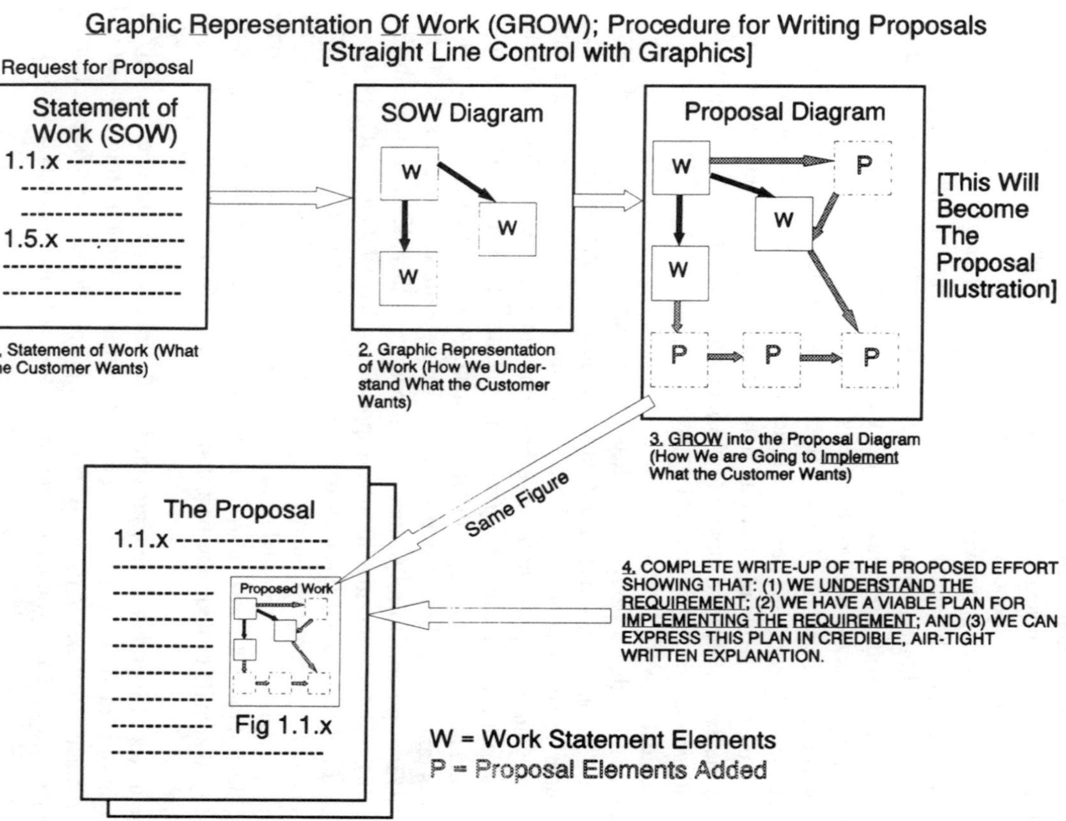

FIGURE 4.7 The "GROW" process for storyboard preparation.

PLATE I

STATEMENT OF WORK PARAGRAPH:

1.1 Overview

The Earth Observing System (EOS) Program Office is located in the Earth Science and Application Division (ESAD), National Aeronautics and Space Administration (NASA) Headquarters. This office is responsible for the management of the EOS Program in the planning, development, procurement, integration, test, launch, in orbit check-out, and the operations of the Earth Observing System for remotely sensed data. It is also responsible for the supporting information system necessary to develop a comprehensive understanding of the way the Earth functions as a natural system.

Sketch:

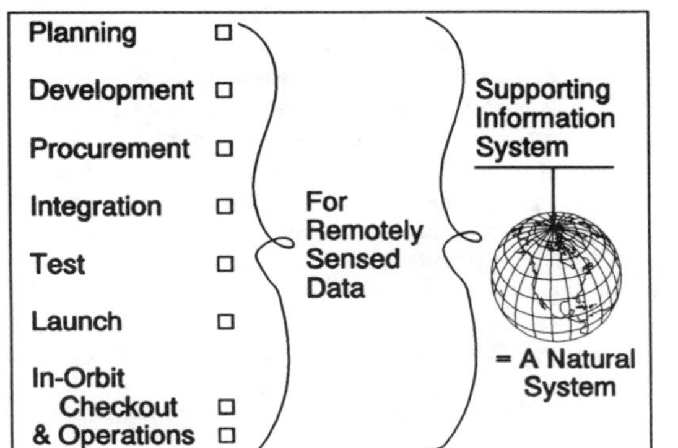

PLATE II

STATEMENT OF WORK PARAGRAPH:

1.2 Objectives

EOS will provide a) an observing system to acquire essential global Earth science data on a long-term, sustained basis and in a manner which maximizes the scientific utility of the data and simplifies data analysis and b) a comprehensive data and information system to provide the Earth science research community with easy, affordable, and reliable access to the full suite of Earth science data from U.S. and International Partner platforms.

Sketch:

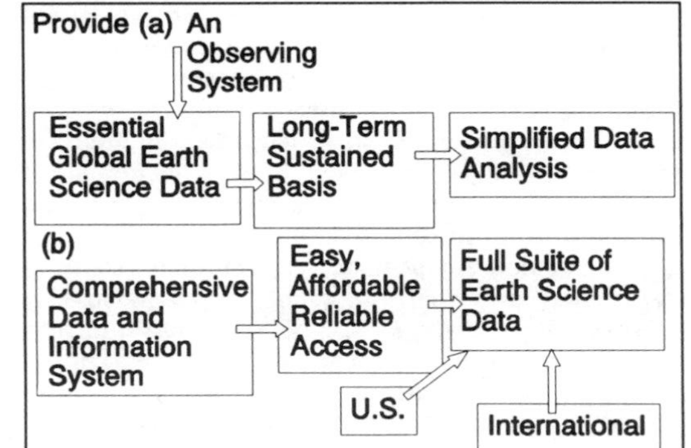

FIGURE 4.8b. Objectives.

FIGURE 4.8a–m. Series of storyboards from a request for proposal for NASA's Earth-Observing System (EOS) program. a: Overview.

PLATE III

STATEMENT OF WORK PARAGRAPH:

1.3 Schedule

The EOS instruments are planned to be flown on a minimum of six NASA polar platforms, European Space Agency (ESA) polar platforms and the Japanese National Space Development Agency (NASDA) polar platform. The basic six NASA EOS platforms will consist of two series (EOS-A and EOS-B) with complementary payloads. The first of three platforms comprising the EOS-A series is planned to be launched no earlier than the fourth quarter of 1997. The first of three platforms in the EOS-B series will be launched no earlier than 2 1/2 years after EOS-A. Each EOS platform will be replaced every five years to achieve the 15-year mission lifetime goal.

Sketch:

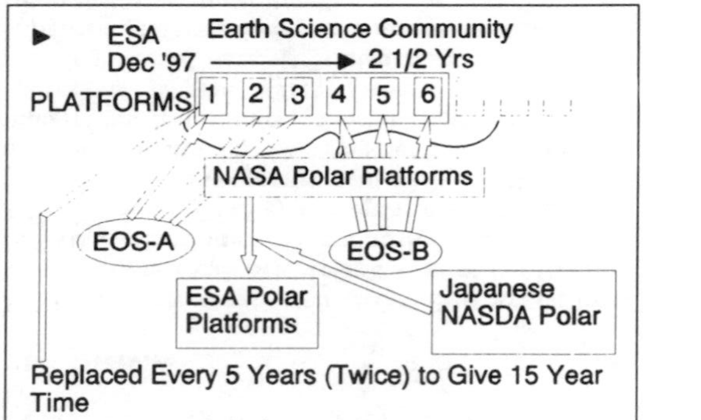

FIGURE 4.8c. Schedule.

PLATE IV

STATEMENT OF WORK PARAGRAPH:

1.4 Payload Groupings and Interfaces

The proposed payload groupings are the result of extensive analysis of accommodations, science requirements, international partner plans, cost, schedule, and other factors. The first ESA platform is scheduled for launch in December 1997. The Japanese platform is scheduled for launch in mid-1998. NASA's attached payloads are also scheduled for launch in mid-1998. The overall program goal is to obtain a 15-year data set, with 10 years of overlapping coverage with platforms from other nations. The use of common instrument interfaces will be maximized to facilitate interchange of selected instruments among the EOS, NASDA, and ESA platforms as well as the National Oceanic & Atmospheric Administration (NOAA) free-flyer spacecraft.

Sketch:

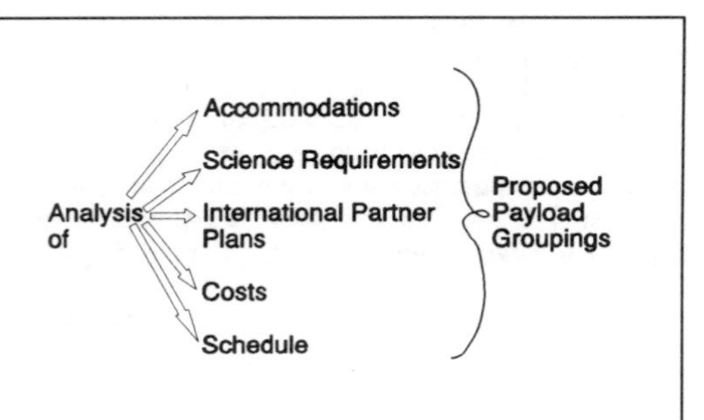

FIGURE 4.8d. Payload groupings and interfaces.

PLATE V

STATEMENT OF WORK PARAGRAPH:

1.5 Information Systems

The Earth Observing System Data and Information System (EOSDIS) will be NASA's most complex, advanced data system for support of science. It will provide facilities for mission operations, including remote scheduling and generating command sequences, real-time communication with the platform and anomaly resolution, instrument health monitoring, and distributing quick-look and engineering data to instrument investigator teams.

Sketch:

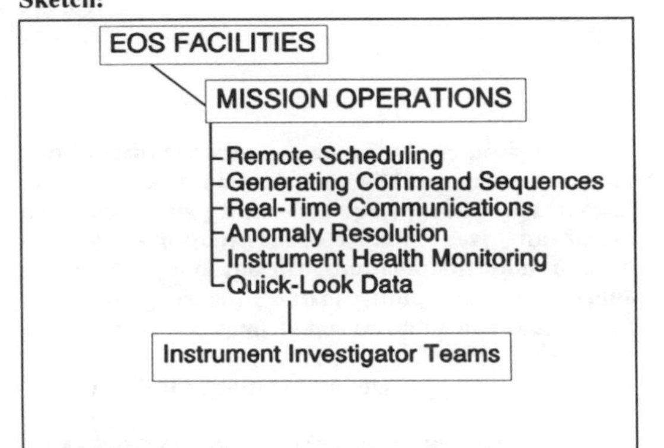

FIGURE 4.8e. Information systems.

PLATE VI

STATEMENT OF WORK PARAGRAPH:

1.6 Mission Support Components

The EOS Mission Support role is split into Program Integration and Assessment (PI&A) and Program Control. The PI&A role is being performed by the MITRE Corporation. The Program Control role is the scope of this contract.

Sketch:

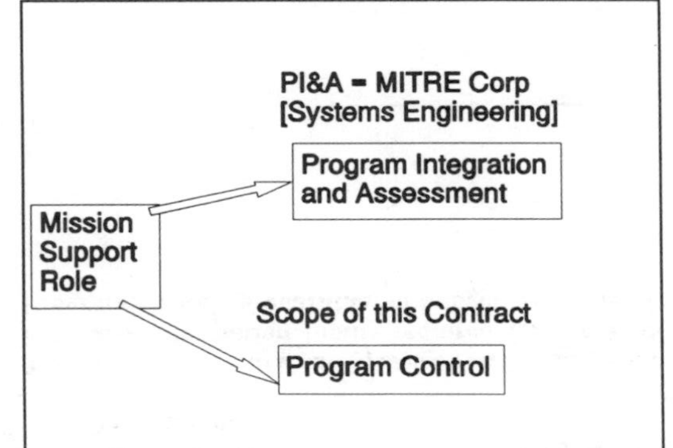

FIGURE 4.8f. Mission support components.

PLATE VII

STATEMENT OF WORK PARAGRAPH:

1.7 Program Control Support

The Contractor shall make recommendations to the Manager, Program Control Branch, for establishing and maintaining the Program configuration, budget, control milestone/schedule and cost integrated baselines. The Contractor shall identify parameters for measuring performance against these baselines and procedures used for changing the baselines.

Sketch:

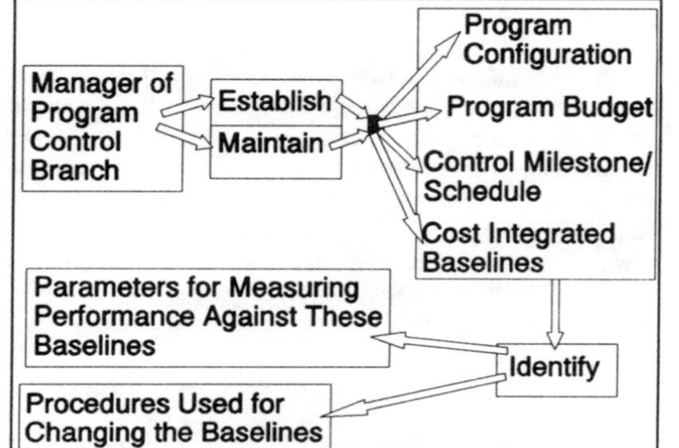

FIGURE 4.8g. Program control support.

PLATE VIII

STATEMENT OF WORK PARAGRAPH:

2.1 Budget Guidelines, Coordination, Analysis, and Verification

The Contractor shall identify, collect, and organize the data and information inputs required for the EOS Program Office's preparation of budget guidelines.

Sketch:

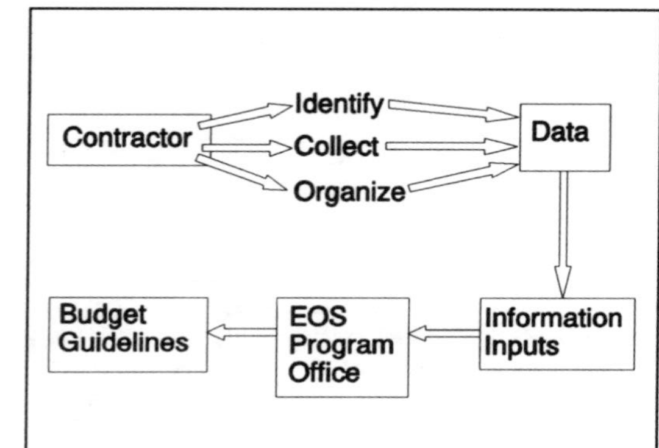

FIGURE 4.8h. Budget guidelines, coordination, analysis, and verification.

PLATE IX

STATEMENT OF WORK PARAGRAPH:

2.2 Resource/Cost Containment Plan and Reports

The Contractor shall provide, coordinate, and maintain a Resource/Cost Containment Plan (DR No. 01) for tracking, controlling, and reporting EOS Program costs. The Contractor shall provide the EOS Program Director with input in the preparation of the Financial/Cost Containment Report (DR No. 02) by summarizing the current EOS status by Work Breakdown Structure (WBS) elements/Unique Project Numbers (UPNs).

Sketch:

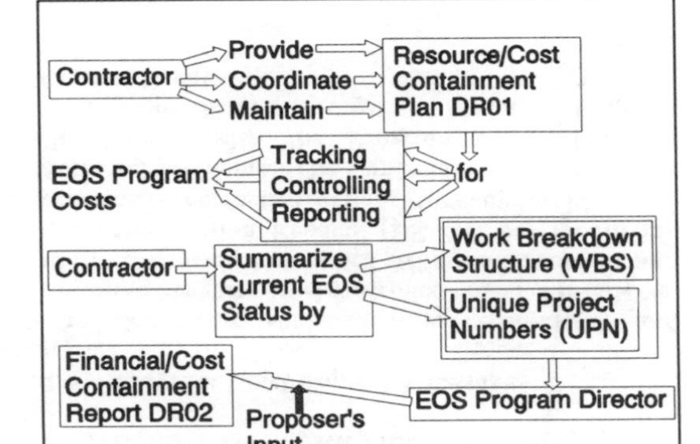

FIGURE 4.8i. Resource/cost containment plan and reports.

PLATE X

STATEMENT OF WORK PARAGRAPH:

2.3 Cost Analysis

The Contractor shall identify and/or develop methodologies and Models for Cost Analysis [eg. Cost Estimating Relationships (CER), PRICE model, etc.] (DR No. 03), and shall provide Independent Cost Assessment Reports (DR No. 04) delineating the EOS program development and life cycle costs. The Contractor shall provide input to and coordinate with the PI&A Branch in the development of cost criteria and assumptions.

Sketch:

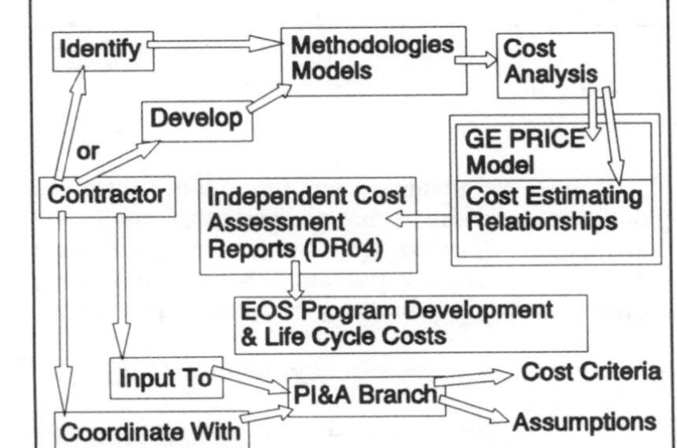

FIGURE 4.8j. Cost analysis.

PLATE XI

STATEMENT OF WORK PARAGRAPH:

2.4 Program-Level Control Milestones

The Contractor shall identify, recommend, and maintain Program-level controlled milestones (DR No. 05) and provide Controlled Milestone Problem and Corrective Action Reports (DR No. 06) identifying alternative corrective actions for any problems or potential problems. The Contractor shall coordinate the reports with the Program Integration and Assessment Branch through the Manager, Program Control Branch.

Sketch:

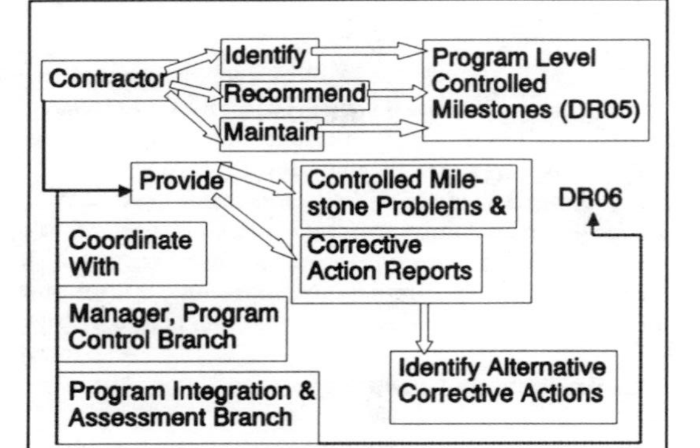

FIGURE 4.8k. Program-level control milestones.

PLATE XII

STATEMENT OF WORK PARAGRAPH:

2.5 Configuration Management Plan

NASA will provide a Program-level Configuration Management (CM) Plan which covers configuration identification, configuration control, configuration status, and configuration audit. The Contractor shall coordinate, maintain, and document the Plan.

Sketch:

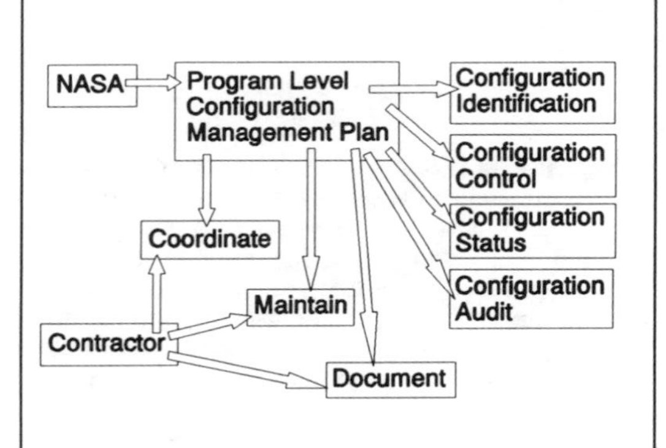

FIGURE 4.8l. Configuration management plan.

PLATE XIII

STATEMENT OF WORK PARAGRAPH:

2.6 Review Reports

The Contractor shall provide recommendations for planning and implementing Program-Level management and technical reviews. Management reviews are held at NASA Headquarters and are monthly, quarterly or annually depending upon the level of the manager holding the review. In coordination with the PI&A Branch, the Contractor shall provide administrative support for technical reviews and technical audits and shall prepare related Program Management and Technical Review Reports (DR No. 07).

Sketch:

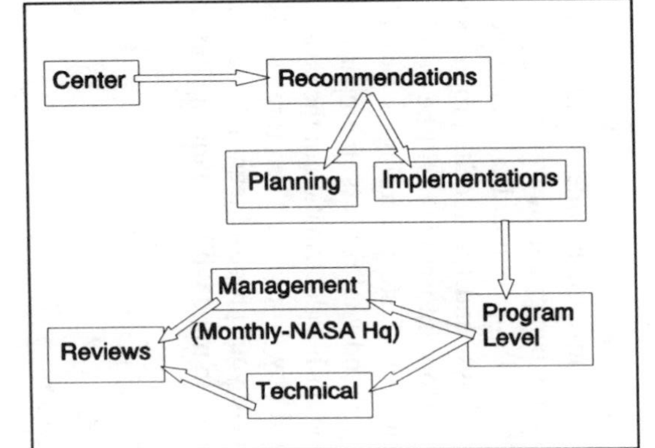

EXERCISE

As an exercise, review each of the graphic representations to determine if corrections or changes are needed or desirable. Does each graphic truly represent what is described above in words? Draw your own version if you can improve upon the graphic. Remember, there are many potential solutions. Use your imagination and creativity to develop your own graphic.

Sketch:

FIGURE 4.8m. Review reports.

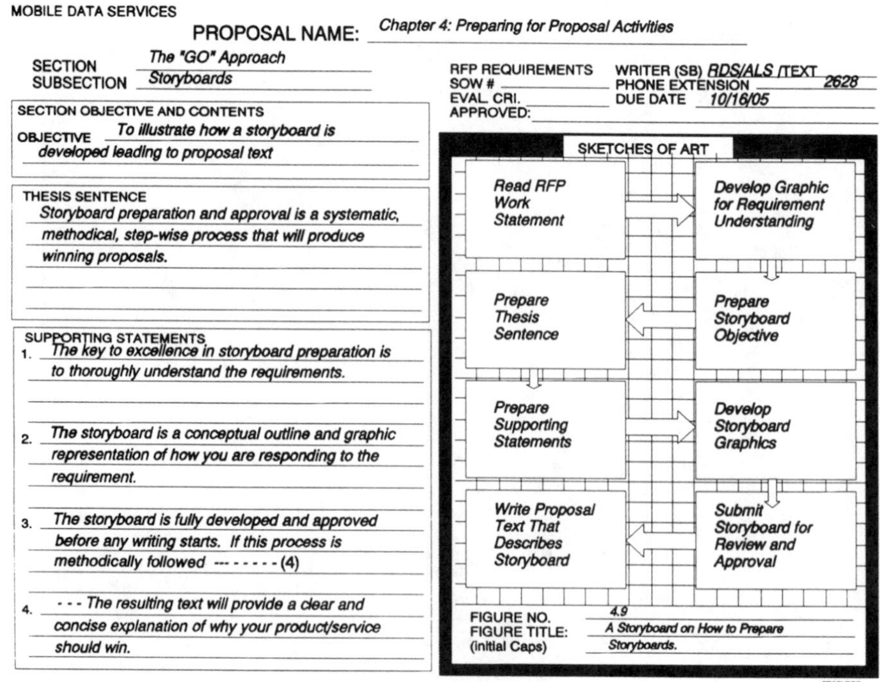

MOBILE DATA SERVICES

PROPOSAL NAME: *Chapter 4: Preparing for Proposal Activities*

SECTION *The "GO" Approach*
SUBSECTION *Storyboards*

RFP REQUIREMENTS
SOW #
EVAL. CRI.
APPROVED:

WRITER (SB) *RDS/ALS /TEXT*
PHONE EXTENSION *2628*
DUE DATE *10/16/05*

SECTION OBJECTIVE AND CONTENTS
OBJECTIVE *To illustrate how a storyboard is developed leading to proposal text*

THESIS SENTENCE
Storyboard preparation and approval is a systematic, methodical, step-wise process that will produce winning proposals.

SUPPORTING STATEMENTS
1. *The key to excellence in storyboard preparation is to thoroughly understand the requirements.*
2. *The storyboard is a conceptual outline and graphic representation of how you are responding to the requirement.*
3. *The storyboard is fully developed and approved before any writing starts. If this process is methodically followed -------- (4)*
4. *--- The resulting text will provide a clear and concise explanation of why your product/service should win.*

SKETCHES OF ART

Read RFP Work Statement
Develop Graphic for Requirement Understanding
Prepare Storyboard Objective
Prepare Thesis Sentence
Prepare Supporting Statements
Develop Storyboard Graphics
Submit Storyboard for Review and Approval
Write Proposal Text That Describes Storyboard

FIGURE NO. *4.9*
FIGURE TITLE: (Initial Caps) *A Storyboard on How to Prepare Storyboards.*

STORYBRD

FIGURE 4.9 A storyboard on the topic "How to Prepare Storyboards."

you recognize their value. In fact, the storyboard shown in Figure 4.9 was used to write this paragraph.

When the storyboard is completed, reviewed, modified as required as a result of reviews, and fine-tuned, proposal text writing can begin. The writing will be easy. More information will be provided about proposal writing in Chapter 10. A discussion of storyboarding is provided here because storyboarding is such an integral and vital process in preparing proposals. Part of the *training* for preparing proposals is indoctrination of the proposal preparation team as to what storyboarding is and how to use it. Therefore, we have provided, in this chapter, the means to prepare your proposal team for the potential use of the GROW process. Keep this process in mind while organizing the proposal team, the subject of the next chapter.

5
ORGANIZING THE PROPOSAL TEAM

Do you see a man skilled in his work? He will stand before kings.
—Proverbs 22:29

Since proposals are prepared by people, a large amount of attention must be given to the types, numbers, and skill levels of people assembled to do the proposal preparation and proposal review jobs. Proposal preparation teams can range in size from one person for a small grant proposal to 50 or 100 people for a mammoth multifaceted system or service. Each member of the team must be selected based on his or her qualifications to provide the specific information needed in the proposal, and he or she must be provided with an accurate and complete description of the work to be performed. We will initiate our description of the proposal preparation process in this chapter by referring to a medium-sized proposal, one which requires multiple proposal volumes.

Once the bid committee has made the affirmative bid decision, a proposal task authorization is signed, the proposal manager is selected, and a proposal schedule is prepared. A typical proposal task authorization form is shown in Figure 5.1. If the proposal manager is selected beforehand, which is the recommended practice, he or she attends the bid committee meeting and takes part in the above actions; otherwise, the data package, including the request for proposal and all materials used to support the bid decision, are forwarded to the selected proposal manager for immediate action.

It is the responsibility of the proposal manager to put together a cohesive team that possesses all the skills required to produce the proposal. The team must be functionally oriented and designed to create the specific proposal. (The review teams, however, are usually selected by upper-level managers who are independent of the proposal manager.)

Proposal Task Authorization

Work order	Task	Rev. No.

Title		Prime division
Log No.	Customer	Business area
Rep. No. or cust. msg. No.	Security classification	Department having prime responsibility

Proposal work statement

Business potential (estimated contract price)—is requirement budgeted? Yes ☐ No ☐ Unknown ☐

Marketing Strategy
- Contract possibility
- Past relationship with customer
- Prime competitors
- Conversion date
- Contract type
  ☐ CPFF ☐ CPIF ☐ FFP ☐ T&M ☐FPI ☐Other

Proposal task budget
- Previous auth. ______
- This auth. ______
- Total auth. ______ (Incl. this PTA)

End Item User	DOD	NASA	Other gov't	Comm'l	Comm'l purchaser supplying gov't

Proposal budget														Total	
Department		Jan	Feb	Mar	Apr	May	Jun	Jul	Aug	Sep	Oct	Nov	Dec	MM	$
	Lab $														
	N/L $													X	
	Lab $														
	N/L $													X	
	Lab $														
	N/L $													X	
	Lab $														
	N/L $													X	
	Lab $														
	N/L $													X	
Total Lab $															
Total N/L $														X	
*Totals $														This auth.	

Does this contract require significant additional capital equipment or facilities (describe)		$ Amount cap. eq.
Identify key contributors required for this proposal	Proposal manager	Submission date

Marketing (1)	Date	Dir. adv. plan (2)	Date	Dir. adv. prog. (3)	Date
Director (1)	Date	Gen. mgr. (2)	Date	Controller (4)	Date
Div. controller (1)	Date	Asst. controller (3)	Date	Sr. V. Pres. (4)	Date

*If in excess of $1,000 requires bid comm. approvals—
total authorization determines approval level:
(1) $1,000 or less ☐ (2) over $1,000 ☐ (3) over $10,000 ☐ (4) over $25,000 ☐

FIGURE 5.1 A proposal task authorization (PTA) form. CPFF: cost plus fixed fee. CPIF: cost plus incentive fee; FFP: firm fixed price; T&M: time and materials; FPI: fixed-price incentive; N/L: non-labor; MM: man-months.

SKILLS REQUIRED FOR PROPOSAL PREPARATION

A number of different skills are needed in the preparation of the proposal. Whether these skills are possessed by one person or by an organization, it is necessary to arrange for their availability and application to the proposal preparation process during the appropriate time phase of the proposal activity. The quality of the mix of skill types and the skill levels used to develop proposals has a great bearing on the overall credibility, accuracy, and completeness of the resulting proposal. The key generic skills used in proposal preparation follow, along with a description of the functions they perform and an indication of the skill levels required for credibility and quality in proposal preparation.

Business and Finance Skills. Business and finance skills are an essential part of the proposal preparation process, particularly in the preparation of the cost volume. A knowledge of accounting procedures and techniques and an awareness of changing economics and business policies are needed. For example, a person with this knowledge will have a full appreciation of many of the "hidden costs" that must be covered by the cost proposal such as: (1) direct charges that are added to basic direct costs by "factoring," (2) overhead costs, (3) general and administrative costs, and (4) profit or fee. Since the purposes of bidding on a proposal and winning the contract are to make a profit, the cost volume must be constructed in a way that will do more than merely recover the costs of labor and materials. Business and finance skills are mandatory to understand these facets, so that all costs of the work output will be included in the cost volume, with sufficient allowance for profit.

Engineering and Technical Skills, and Skills of Functional Managers. Engineering and technical skills, acquired by actual on-the-job experience, are the basis for a sound, competitive, credible, and realistic proposal. A completed proposal must be based on a practical knowledge of the work activity as well as on the theory of design of the work activity or work output. Although educational background and knowledge of theory are important, this theoretical knowledge must be supplemented by actual hands-on prior experience in producing a similar or identical work output; therefore, experts in the technical field required by the request for proposal must be available to the proposal team. In addition to the technical experts, functional line managers should be made available to the proposal team on at least a part-time basis. These are the people who will be supervising the technical aspects of the work and who will be able to contribute realism and credibility to the technical approach as well as to the estimates of resources required to do the job.

Manufacturing and Assembly Skills. For work activities or work outputs that involve manufacturing and assembly operations, detailed knowledge of each manufacturing, assembly, test, and/or inspection function is essential. This detailed knowledge requires people who have had experience in manufacturing and assembly operations. The most valuable attribute of these individuals is their ability to originate and organize the manufacturing and assembly plan for the proposed work output and to plan the effort to eliminate gaps, overlaps, and duplications. Should the proposal involve production line operations, these skills are even more important. The most common fault in manufacturing plans is the omission of essential steps in the process. Simple steps such as receiving and unpacking raw materials or parts, inspection of incoming parts, in-process inspection, attaching labels and markings, and packaging and shipping of the final product are often inadvertently omitted. Team members skilled in manufacturing and assembly will assure proposal accuracy and credibility in these areas.

Management Skills. Part of any proposal team's expertise must consist of abilities in the area of project management. A skilled and experienced project manager will be able to best correlate the need for workers, material, equipment, and systems with the proposed work output or work activity. The manager will be able to envision and plan the management tools, resources, and expertise required to effectively carry out the proposed job and will be able to effectively communicate the management control, schedule control, and cost control aspects of the job to the customer.

Mathematical, Statistical, and Data-Processing Skills. Higher mathematics, the application of statistics, and data-processing skills are not always required in the development of credible and supportable proposals, but in high-technology and multidisciplinary work activities and work outputs, these skills have become essential. Often, a design will not be fully developed and various mathematical or statistical techniques will be necessary to develop data for the technical and cost volumes. When new products are designed and new services are envisioned, it is always best to verify the performance and cost projections by use of mathematical and statistical techniques or computer simulations. Data-processing skills are also required for the creation of the proposal itself (see Chapter 11, "Computer-Aided Proposal Preparation").

Production-Planning and Industrial Engineering Skills. Production-planning and industrial-engineering skills are closely related to the manufacturing and assembly skills mentioned earlier, but these skills are usually learned and applied at a higher organizational level. Where the manufacturing and assembly skills used in proposal preparation are derived from hands-on experience by workers or their immediate supervisors, production-planning and industrial engineering skills are acquired from an overall knowledge of the workload and work flow in an office, factory, or processing plant. Production-planning and industrial engineering skills are particularly important for work activities or work outputs that involve high rates or large quantities of production. Knowledge of automation and labor-saving techniques in the shop, factory, or office become important in these applications.

Writing and Publishing Skills. Since the proposal is primarily a sales document, it must present the best possible picture of the proposing company. Writing style, contents, quality of graphic reproduction, even the choice of cover or binding may have an effect on the evaluating team. Individuals capable of writing and editing material while working under pressure are essential. It is necessary for the proposal team to have available a knowledge of the mechanics of the writing and publishing process, including expertise in storyboarding, proposal layout and design, desktop publishing, reproduction, and binding.

In soliciting the skilled personnel required to work on proposals, the recruiter should remind the participants that the proposal preparation process is often regarded as an essential step in developing the careers of future project managers, business managers, and corporate management. Because an in-depth knowledge of the company and one or more of its products or services is developed during preparation of a proposal, proposal team participation has historically been a vital asset in the career paths of future managers. Management usually puts its best people on proposals and therefore expects these best people to grow into positions of higher responsibility and authority.

PROPOSAL TEAM COMPOSITION

The proposal team usually consists of (1) a proposal manager, (2) a proposal administrator, (3) a sales engineer, (4) a technical volume manager, (5) an organization-and-management volume manager, (6) a cost-volume manager, and (7) a publications manager. Positions 4, 5, and 6 are often called "book bosses." In addition, the proposal team often includes representatives of performing organizations, such as engineering, manufacturing, purchasing, property, and contracts. See Figure 5.2 for a typical organization chart for the proposal team of a large proposal. The activities of the team will be performed far more effectively if all members are located in the same area. Since the proposal administrator is charged with day-to-day control of the team, coordination will be more efficient if the team location is adjacent to his or her regular work area.

An effective proposal team is the result of bringing together individuals from a wide variety of unrelated organizations. Their action can be performed most effectively in accordance with a well-thought-out and well-documented plan or schedule. The overall proposal schedule is usually developed by the proposal manager prior to the formation of the team. Within the overall schedule, each team member should be informed in writing of the task or tasks assigned and the expected schedule of performance. This internal control within the team not only provides the team members with a clear definition of what is expected, but also forces the proposal manager and proposal administrator to define clearly the results expected.

Proposal Manager

The proposal manager is responsible for the planning, organization, direction, and control of the entire proposal process. Good, competent, aggressive leadership is crucial to the preparation of winning proposals. Proposal managers should be selected before the request for proposal is received so that vital preliminary work can be done: (1) members of the team can be selected; (2) routine proposal sections can

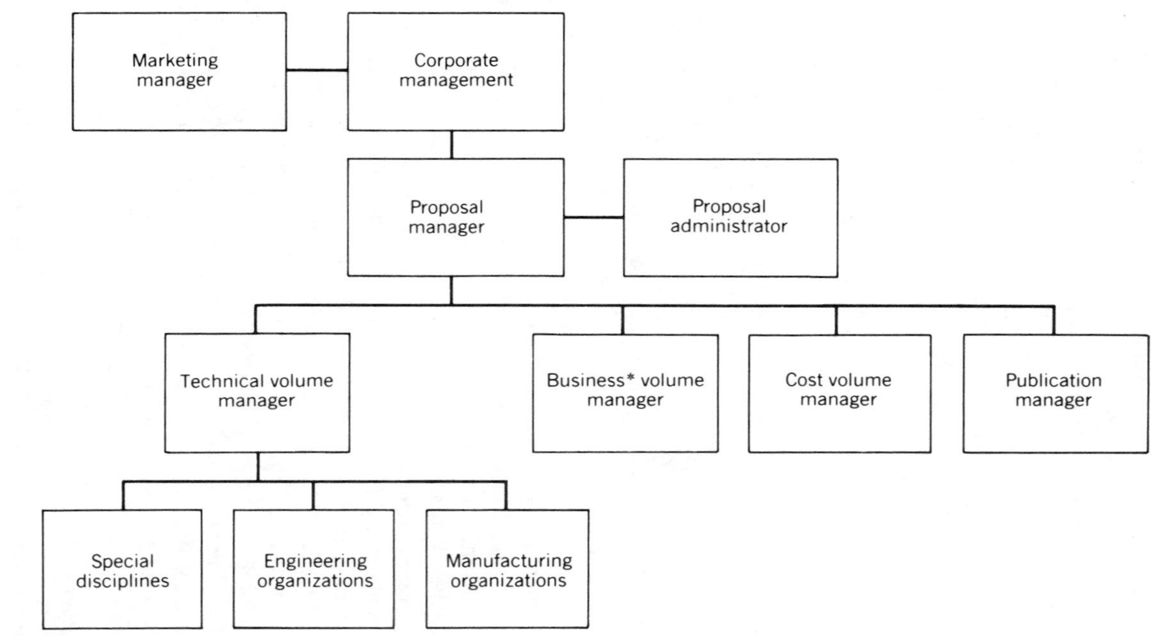

FIGURE 5.2 Proposal team organization.

be prepared in advance, thus reducing the hectic and sometimes frantic demands of the preparation period; (3) résumés of potential key personnel can be collected; (4) a synopsis of contracts and jobs already performed that are similar to the requirements in the request for proposal can be prepared; and (5) a concise history of the proposer company can be specifically tailored to the needs of the request for proposal.

The first task of the proposal manager is to define the theme or the proposal approach by preparing a master storyboard and a preliminary proposal outline. This storyboard and outline should be responsive to the request for proposal. A typical master storyboard, which was used in the preparation for the case study shown in Chapter 14, is shown on Figure 5.3. The storyboard and the preliminary proposal outline will aid in collecting and measuring the output and performance of the members of the team. The proposal outline should include a table of contents, as well as short descriptions of each section of the proposal, along with number of pages, due date, person responsible, and authorized hours. See Figure 5.4 for a sample outline of a single-volume proposal. The proposal outline will indicate how much emphasis is placed on each particular topic, thereby further defining the theme for this particular proposal. Above all, it should reflect responsiveness to the request for proposal.

The proposal manager then arranges for a proposal preparation room where the team members may work together in a secure and undisturbed environment. For large proposals, this room (or set of rooms) should have plenty of blank wall space for mounting of storyboards, outlines, sketches, and other pertinent information. Many of the independent reviews will be held in this room, using the storyboards as visuals.

Once the master storyboard and outline are established (time is of the essence), the proposal manager convenes the kickoff meeting, sometimes called *the plan of action meeting*. The first meeting of the proposal team is held in the proposal room and quite often occurs within the 72-hour period following the authorization to propose or following receipt of the request for proposal. Attendees include the proposal administrator, the sales engineer, technical volume manager, organization-and-management volume manager, cost-volume manager, publications manager, and representatives from all contributing, operating, and support departments, as determined by the proposal manager. At this meeting all pertinent information is distributed and discussed. The following are examples of the type of materials supplied to each attendee:

1. Proposal control form (see Figure 5.5)
2. Copy of request for proposal or request for quotation
3. Overall proposal preparation schedule (see Figure 5.6) and budgets
4. Marketing intelligence and proposal strategy
5. Basic proposal theme and technical and costing guidelines
6. Master storyboard and proposal outline
7. Detailed proposal preparation schedules for all proposal contributors

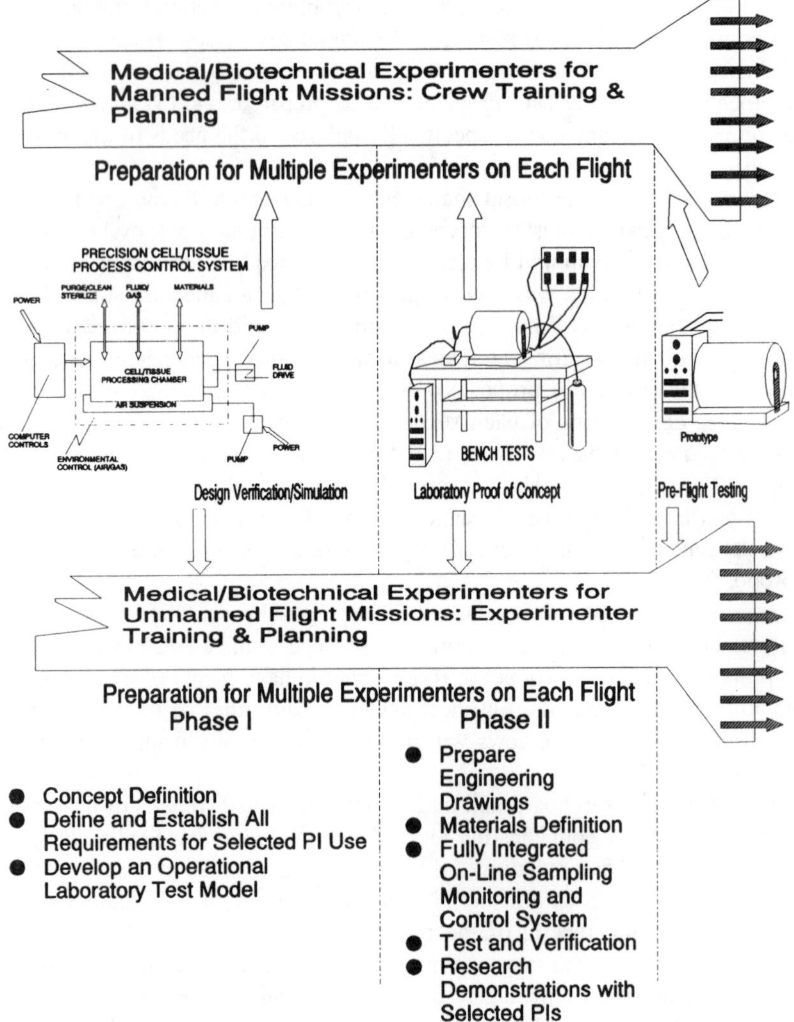

FIGURE 5.3 Sample master storyboard for a long-duration precision cell-tissue experimental processing system, submitted by Rantek, Inc. (See Chapter 14 for more details.) PI: principal investigator.

8. Assignment of tasks, as appropriate
9. Coverage of any other proposal effort items, as applicable.

At this time, the proposal manager appoints a technical team to study the technical aspects of the proposal and do a conceptual analysis and design, including risk areas and program scope.

It is also the responsibility of the proposal manager to conduct other general team meetings. These meetings occur periodically during proposal preparation to

Manned Space Flight Facilities (Space Station)

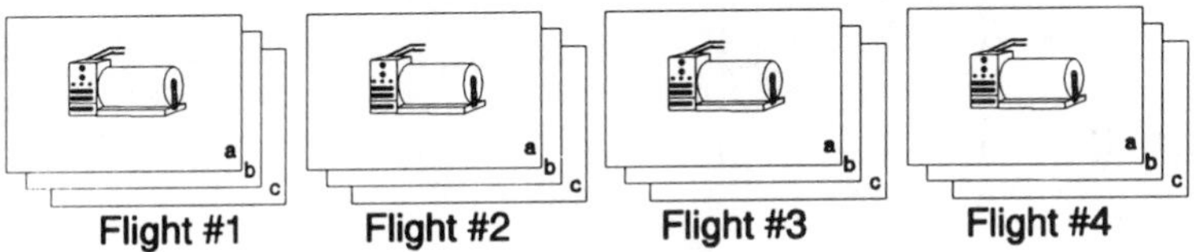

Unmanned Space Flight Facilities (Long Duration Satellites)

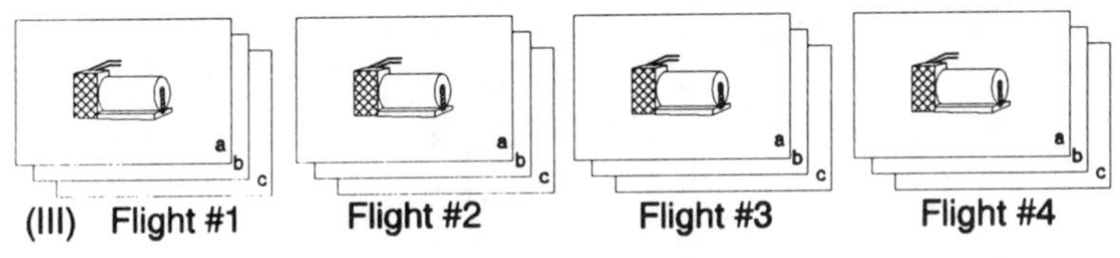

Phase III

- Define Requirements for Flight Units
- Design and Development of Flight Units
- Verification in Flight
- Several Models of the Commercial Unit in the NASA Inventory

FIGURE 5.3 (*Continued*)

assure adherence to the proposal outline and proposal schedule and to review material in progress. Contradictions, overlaps, and items of questionable value thus come to light *before* presentations are made to the formal review teams. These other meetings include an evaluation meeting (technical and cost evaluation), review meetings (critics' review), and a management-approval meeting. These meetings are general in nature and involve the whole team; other meetings are also conducted that bring together the smallest possible group of team members needed to solve specific proposal problems.

The primary concern of the proposal manager is quality. It is the proposal manager's responsibility to integrate all outputs of the team into a high-quality, persuasive, flowing sales document. Technical quality is measured by the responsiveness of the proposal to the request for proposal. The proposal manager must be prepared to answer a number of questions: Does the proposal describe the desirable results the customer will receive by purchasing the proposed solution to the problem? Do these results reflect performance equal to or better than the requirements described in the request for proposal, and are they likely to be favorable when compared to competitors' proposals? Are the key technical results outlined

PROPOSAL OUTLINE

1. Expand or simplify outline to meet specific proposal requirements.
2. Cross-check to assure that all RFP requirements are covered.

Headings—Outline	Contents	Emphasis
Title page	Title, classification, and preparing organization.	
Foreword	Authority for submittal of this proposal.	
Table of contents	List of sections, headings, figures, tables, and appendices.	
Introduction and summary	Summarize the proposal in 3 to 5 percent of total pages in the proposal. This should contain the justification for award of contract to the company.	
	Section I	Strong
Work statement Scope Objective List by task of work to be performed Delivery commitments	The Work Statement is the contractual commitment. (Do not explain here how work will be performed.)	
	Section II	Strongest
Technical approach	Prepare detailed outline required to be responsive to RFP. Introductory statement presenting requirements to be met and problems associated therewith. (Drawings and sketches are helpful.)	
Schedule	Time-phased milestones are related to work statement and logic network.	
Engineering plan	Analysis and design. (Many subparagraphs may be added in this area.)	

Material section	Types of material and reasons for selecting.	
Tooling plan	Design and fabrication.	
Manufacturing plan	Fabrication. (Production flow diagrams may be useful.) [Analysis/coding plan if software]	
	Assembly.	
	Shipping.	
Quality Assurance plan	Relate to customer's objectives.	
Logistics plan	Show how this benefits the customer.	
Spares		
Support		
	Section III	Strong
Management	Define: scope of management required; the program management group; where group fits; program control, cost control, and personnel qualifications.	
	Section IV	Stronger
Related experience	Pertinent, applicable, and true.	
	Section V	Stronger
Capabilities and facilities	Those needed and available, and source for those not presently available.	
	Section VI	Strong
Costs	Detailed cost estimate with rationale.	
	Section VII	
Appendices	Derivations, statistics, and technical detail.	

FIGURE 5.4 Format for a Single-Volume Proposal.

Proposal Title: ____________________

RFP title and number: ____________________

Customer name and address: ____________________

Proposal charge number: ____________________

Proposal manager: ____________________

Proposal administrator: ____________________

Contract administrator: ____________________

Financial representative: ____________________

Technical writer: ____________________

Courier expediter: ____________________

Critical dates: ____________________

Technical section to tech writing: ____________________

Cost section to accounting: ____________________

Sign-off meeting: ____________________

Delivery to publication activity: ____________________

Mailing date: ____________________

FIGURE 5.5 Proposal control form.

in the proposal those perceived as most wanted by the customer? Does the management volume clearly describe the personnel responsible for each particular task and how and when they will know when they have accomplished each task? Does the cost volume depict a cost that is fair and reasonable? Are supporting cost data sufficiently detailed and presented in a readable and traceable format? Is the cost volume consistent with the technical and management solutions proposed? These questions and many more of the same type have been developed in tabular form and are shown in Appendix 2. It is up to the proposal manager to evaluate and shape the document to meet all these criteria.

Duties of the proposal manager or the selected project manager include responsibility for formulating the proposed project organization, describing how this project organization will work within the overall company framework, and selecting key personnel who are qualified to fill the proposed positions.

In addition, the proposal manager's expertise includes a knowledge of the customer's organization and procurement procedures. Perhaps the most important qualification, however, is the ability to direct and control the activities of others. Since proposal teams often work long hours for extended periods of time, morale and team spirit play an active part in the success or failure of the organization. Therefore, it is essential that the proposal manager have their full confidence in directing their day-to-day activities.

Perhaps one of the most important tasks of the proposal manager is the preparation of a convincing, well-thought-out letter of transmittal that will motivate the recipient

	Days after RFP
Market research and preliminary bid decision	0–45 days prior
Organize proposal team	0–45 days prior
Preliminary review of RFP or requirement	1
Kickoff meeting and issue proposal task authorization	1
Preliminary feedback and discussion of ground rules	2
Initial input response from team and first cut pricing	45
Team input review meeting	49
Instructions developed for revisions	51
Final input from team members	53
Analysis, pricing, rework, and management reviews	53–56
Publication	56–59
Submittal	59
Follow-up	60 until contract award

FIGURE 5.6 Sample proposal preparation schedule, based on a 60-day response time.

to adequately and thoroughly review and study the proposal. Supervision of the activities required to produce, package, and transmit the proposal is the final task of the proposal manager, assisted by the proposal administrator.

The proposal team is created to respond to a particular proposal request; it is shaped by the characteristics of the request for proposal or request for quotation and by the characteristics of the company itself. Since it must draw on a wide variety of diverse skills and resources throughout the company, it is all the more important that a team spirit prevail. The creation, motivation, and control of the team demands dynamic leadership. Winning proposals are the result of team effort, and it is up to the proposal manager to inspire the team to do the best job possible in an incredibly short time. There is no room for petty arguments or disagreements. First and foremost is the objective to win the contract award, to translate the data accumulated into a sale for the company.

Proposal Administrator

For large proposals, the proposal administrator is often selected from a permanent proposal preparation group trained in proposal preparation procedures. Business-oriented and familiar with company objectives from a broad standpoint, this person has a thorough knowledge of the company plans, the accounting system and possible constraints, and the availability of company resources. He or she is fully informed on the status of current and proposed programs. Functions involving day-to-day administration of the proposal team are delegated to the proposal administrator by the proposal manager. Such items as cost control, schedule preparation and monitoring, and technical details of graphic and reproduction support are usually delegated to the proposal administrator. He or she is thoroughly knowledgeable in communication skills, particularly in oral, written, and graphic media. Not only is the proposal administrator able to instruct the team, but he or she is also familiar with the mechanics involved in the proposal preparation process. As stated earlier, locating the proposal team close to the office of the proposal administrator (or locating the proposal administrator in or near to the proposal work area) will improve communications among the team members and make the proposal administrator's job of controlling the output of the team much easier.

Among the specific duties that may be included in the proposal administrator's job are the management and tracking of labor hours and dollars spent during preparation of the proposal, the monitoring of team performance in adherence to schedules, and the control and integration of all text, tables, and illustrations used in the proposal. The proposal administrator also ensures the security of proprietary and classified material. Should the proposal require supplements such as brochures, briefings, or models, they are prepared under the direction of the proposal administrator in conjunction with the sales engineer.

Sales Engineer

Describing what the customer wants is the primary task of the sales engineer. A close personal working relationship with the customer during the preproposal period gives this person an opportunity to find out exactly what the customer is looking for in relation to the contents of the technical, organization and management, and cost volumes. He or she may have an opportunity to influence the contents of the request for proposal, thus giving the company an advantage over the competition. Once the request for proposal is issued, the sales engineer is able to expedite delivery and arrange for attendance and favorable consideration of proposal team members at the bidder's conference. Because of a unique position as a contact with the customer, the sales engineer may be useful in oral briefings or presentations, should they be needed. Such activities give the sales engineer an opportunity to further the relationship with the customer, thus making this person even more valuable to his or her company later on in the negotiation and contract-award phases of procurement.

Early in the proposal process, the sales engineer prepares an information package that provides significant information and data about the customer. It contains a detailed analysis of the customer's organization as well as specific details about the background, education, and personal biases of those who will evaluate the proposal.

Another contribution of the sales engineer is the preparation of a program history package. He or she provides the proposal team with information on the history and background of the program, both within the customer's organization and within their own company. It includes the business plan and estimates on future development plans, thus enabling the proposal manager to determine the growth potential of the work activity or work output. The proposal manager is then able to make educated decisions based on these facts and give the team an overview of the entire project at the kickoff meeting so that each member will be well-informed and better able to perform the tasks assigned.

One area in which the sales engineer is particularly useful is market intelligence. Who is the competition? What are the strengths and weaknesses of the competitors, and just how hard are they competing for the job? Have any of the competitors worked for the customer before and, if so, what kind of relationship did they have? What is the relationship of the competitor with the customer now? By closely observing the marketplace, the sales engineer can usually find the answers to these and other questions that the proposal manager may have. It behooves the proposal manager to make the best use possible of his or her sales engineer in gathering any information that will make a proposal a winning one.

Technical Volume Manager

The prime responsibility of the technical volume manager is the preparation of the technical volume in accordance with the outline prepared by the proposal manager. Duties are to carry out all of the functions and meet all of the ground rules (which are spelled out later in Chapter 7). Technical specialists are available in engineering, design, manufacturing, or any other functions of the company that have the expertise required by the request for proposal. The technical volume manager integrates and coordinates the inputs of these specialists to ensure that they meet the specifications of the outline.

Labor-hour estimates, materials lists, and skill mixes are developed, which are sent to the cost-volume manager to develop the cost estimate for the proposal. The technical volume manager is uniquely qualified to develop the test plan and schedule in which the type, scope, and rationale of all tests recommended in the work activity or work output are defined. Spare parts and logistics support recommendations and program schedules, flow diagrams, technical change procedures, and facilities requirements are prepared by the technical volume manager.

The technical volume manager uses all of the specialists within the organization to accomplish the job. Each individual is required to describe the contribution of a unique specialty to the total project in writing and show the relationship to the solution of the customer's problem. With so many diverse specialties, the technical volume manager must integrate the material carefully; this can best be done by following the proposal manager's outline consistently. Any deviations or adjustments should only be made with the concurrence of the proposal manager and, frequently, with key members of the team.

Organization-and-Management Volume Manager

The organization-and-management volume manager has the responsibility for carrying out all of the activities and adhering to all of the guidelines spelled out in the request for proposal and described in detail in Chapter 8. Intimately familiar with the company organization, this individual must know how work activities or work outputs are initiated, proposed, negotiated, and carried out by the company. He or she (1) must be generally familiar with plant facilities and equipment and the method that the company uses to capitalize these assets; (2) must be familiar with past company experience and capabilities and able to bring together a convincing argument that this company can do a better job than any of the competitors of successfuly completing the proposed work; (3) must be familiar with methods of achieving efficiency, economy, and effectiveness in an organization; and (4) must be able to credibly apply these methods to perform the proposed work on a timely basis.

Cost-Volume Manager

The cost-volume manager should be skilled in estimating techniques, methods, and procedures in general as well as in the specific pricing and estimating principles, policies, and procedures of the company. He or she is responsible for establishing and acquiring the physical tools required to do the complete cost-estimating and pricing job. This includes the development or acquisition of cost-estimating computer programs and the rental, purchase, or use of computers and software to perform the computation and desktop publishing of cost information. The cost-volume manager must assure that all rationale is credible, up-to-date, consistent with the resource estimates, and included in the cost-proposal volume or in either of the other two proposal volumes, if appropriate. Although the cost-proposal volume manager's job comes next to last in the proposal cycle, this job is vital and is crucial to the success of the overall bid. The cost-volume manager, a person who has a background in business matters, publication techniques, and technical operations of the company, must be ready and willing to spend the dedicated hours required to complete a cost volume that is consistent within itself as well as with the other proposal volumes.

Publications Manager

The real hero in proposal preparation is the publications manager. For large proposals this person must accomplish in 6 to 9 weeks (or sometimes even less), all of the activities that a publisher of textbooks or handbooks performs in 6 to 9 months. The publications manager is responsible for the editing, artwork, tables and figures, typesetting, review, reproduction, assembly, binding, and shipping of a multivolume document with multiple copies. This is done under enormous pressure, because a missed deadline will mean a missed contract. With the financial health of tens, hundreds, or even thousands of workers at stake, to say nothing of the financial health of the company itself, the publications manager is looked to as the person who will produce an attractive, professional, error-free document. This must be done with the added burdens of deadlines missed by individuals or organizations that precede this final step in the proposal cycle, the constant overview and critique by management, and the steady harassment by several bosses: the three proposal volume managers or "book bosses." This hectic life is abundantly supplied with the multiple gremlins of missed meals, lost sleep, and the pesky last-minute change. Needless to say, the proposal publications manager must have a thick skin, an even-tempered disposition, and a determination exceeded only by his or her knowledge of the multiple facets of the publication production process.

Some large companies divide the responsibilities described here for the proposal manager, proposal administrator, and publications manager (three people) differently and distribute them to two people, the proposal manager and the proposal operations manager. Other companies with smaller staffs and budgets fold all of the duties into one person. The important point here is that all of the described functions must be done by a designated capable and responsible team member.

CLOSING COMMENTS

It cannot be overemphasized that the most important feature of the proposal team is team spirit. If the proposal manager has the leadership qualities needed, he or she will be able to motivate the members of the team and keep them on schedule. Preparing a winning proposal should be the objective of the entire team. Sufficient time should be taken at the beginning of the process to ensure that each member knows exactly what is expected of him or her and when it is expected to be done. Once a schedule is established and each task is defined and assigned, it is up to the proposal manager and proposal administrator to follow the process through to its conclusion.

The underlying theme of the proposal team should be "go to the experts." Whatever section is to be prepared, the best person possible should do it: the one who knows how to do the job and knows when it should be accomplished in the course of the work activity or work output, where the most suitable location is to do it, and why it should be done a certain way. It is essential to the team that the proposal manager institute quality-control milestones providing the members with targets and a clear rationale upon which to pace their efforts.

Once the proposal has been published and sent to the customer, a good proposal manager will keep the team members informed of progress as it is being evaluated. All pertinent facts, whether the proposal wins or loses, will be provided to the team. This personal touch not only indicates that the proposal manager is aware of their contribution to the project, but also shows the team members what they did right or wrong in that proposal so that they can correct the situation the next time. This follow-up is also necessary in the event that responses to detailed customer questions or best and final offer requests are required. In summary, the following rules of thumb can be applied to the proposal team:

1. Use proposal specialists for those proposal tasks that are similar from one request for proposal to the next.
2. Use those who will actually perform the proposed work to write the technical volume and provide cost-estimating inputs.
3. On major proposals, the proposal team should be located together and should be dedicated to the proposal effort for 100% of the working hours in each workday. Overtime must also be provided, if required.
4. Anyone that will be involved in subsequent negotiations should be involved in the proposal effort from the start.
5. Corporate management should ensure team availability as required after initial proposal submission.

6

THE REQUEST FOR PROPOSAL: WHAT TO EXPECT

Give to him who asks of you, and do not turn away from him.

—Matthew 5:42

Unless the proposal is to be an unsolicited proposal, it most likely will be prepared in response to a request for proposal or request for quotation. A request for proposal (RFP) or request for quotation (RFQ) can consist of anything from a telephone call or a letter to a large, multipage, multivolume document. The requestor will always ask for a price and may also ask for some backup material for pricing, as well as the terms and conditions of any resulting contracts and detailed specifications of the work activity or work output that is proposed. The cardinal rule to follow in the treatment of a request for proposal or request for quotation is to give the customer what is asked, but skillful treatment of unreasonable requests may be the key that spells success. The customer must fully understand the cost, schedule, and performance impacts of each request-for-proposal requirement.

Before discussing RFP requirements in detail, it is important to note that there are key points in the procurement process where one can influence these requirements.

OPPORTUNITIES TO INFLUENCE THE REQUEST FOR PROPOSAL

For proposals resulting from a request for proposal or request for quotation, the proposer may interact with the procuring agency or company in a more complex manner than merely receiving a request for proposal and submitting a proposal. Figure 6.1 shows, in flow-diagram format, twenty steps that may occur in the customer–supplier interaction process from the time the customer formulates the project requirements until the time the contract is awarded. If all of these steps are employed in a procurement cycle, there will be no fewer than seven opportunities to provide additional information to the customer, which may enhance competitive position.

Opportunity 1: Influencing Initial Requirement. Remember that step zero in the overall process is to plant the seed for the idea in the customer's mind. Presuming this has been done, and that the "customer requirement" is the result of an initial sales effort, there has already been an opportunity to provide a significant input to the customer before the process even starts.

Opportunity 2: The Draft Request for Proposal. Some customers informally solicit comments on their requests for proposals from prospective bidders prior to formal issuance of the request for proposal (steps 3 and 4 of Figure 6.1). This is an ideal time to provide significant inputs relative to work scope, work timing, and work performance standards. An active participation in providing inputs at this stage in RFP preparation will enhance one's knowledge of what to expect and may even result in the interjection of specifications that will favor one's product or service, or one's firm.

Opportunity 3: The Prebid Meeting. When a prebid meeting is held, the questions that are posed by potential bidders can influence the issuance as well as the content of any amendments or revisions to the request for proposal. It is recognized that any questions from potential bidders provide clues to competitors regarding technical or management approaches, since all bidders are usually represented in a bidder's conference held for clarification of RFP requirements. But the value of the information transmitted to the customer and the value of the customer's responses may far outweigh the disadvantages of revealing one's strategy relative to the proposal approach. Intelligent questions about the RFP requirements convey information to the customer about the proposer's (1) interest in providing a responsive proposal (indicated by well-thought-out questions); (2) knowledgeability of the project or subject; and (3) desire to fulfill all project requirements. The answers to these three points will help formulate a more responsive proposal.

Opportunity 4: The Proposal. Step 9 on Figure 6.1, submission of the initial proposal, is the principal opportunity to provide comprehensive information on why a particular firm should be chosen to perform the work. This step is the culmination of the preparation activity for the initial formal proposal.

Opportunity 5: Proposal Clarifications. Proposal clarifications requested by the customer and special meetings established to allow a proposer to clarify the content of the proposal (step 11 on Figure 6.1) are opportunities to provide still more information to the customer about competence and responsiveness. By analyzing the phrasing and content of the customer's questions, one can also gain valuable information on evolving customer requirements that, when combined with the original request for proposal and amendments, will give a company an edge on the competition.

Opportunity 6: The Best and Final Offer. The best and final offer, step 16 on Figure 6.1, is the final place where an input can be made to the customer prior to source selection. This best and final offer can include not only a final price but also

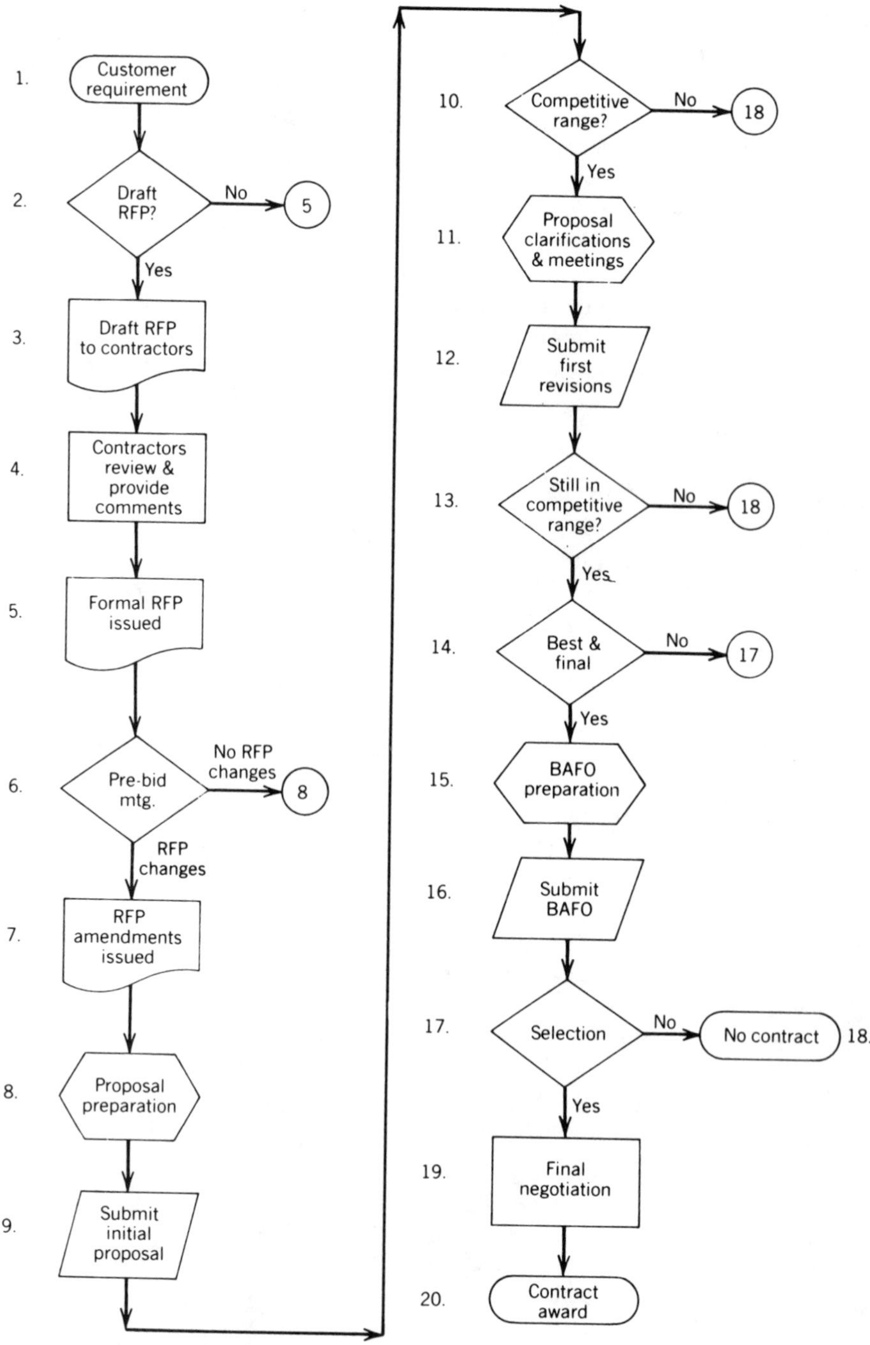

FIGURE 6.1 The request for proposal (RFP) procurement flow chart. BAFO: best and final offer.

supplemental technical, schedule, management, or organizational data based on information gained from the previous contacts with the customer.

Opportunity 7: The Final Negotiations. After a company is selected to perform the work, of course, there will once again be an opportunity to provide information to the customer as part of the final contract-negotiation process (step 19 on Figure 6.1).

The above listing of opportunities illustrates that there are various channels of communication during the request for proposal and selection process that will enable the customer to spell out requirements and the supplier to present capabilities. To know in advance of RFP submission that these channels exist will help in understanding the degree and method of responsiveness required to respond to the request for proposal itself.

THE CONTENT OF A REQUEST FOR PROPOSAL

It is necessary for proposers to gain, in advance of proposal preparation, an insight into the many facets and elements of a request for proposal. Whether the request is from a government agency, a private concern, a university, or a nonprofit corporation, many of the same elements will be included in proposal or quotation requests. The following information can be used as a checklist for the types of information that may be required in a proposal for either government or industry. It will also demonstrate some methods and techniques of dealing with proposal requests, even if they are of the more detailed type normally encountered when state, local, and federal government agencies are requesting products, projects, or services. A detailed request for proposal will normally be sent to bidders as an enclosure to a letter of transmittal that invites a bid and describes the procurement in general terms. The letter of transmittal may also contain instructions as to how many copies of the proposal should be submitted, the deadline date for proposal submittal, whether the proposal material can be sent in electronic format (see Chapter 11), the address to which the proposal is to be sent, and any special proposal packaging and handling provisions. The letter of transmittal will also ask for the proposer's acknowledgment of receipt of the request for proposal within a given time period if there is an intention to bid.

A large request for proposal normally includes five basic parts: (1) general instructions to bidders; (2) a proposed contract draft or schedule; (3) general contract provisions; (4) the statement of work; and (5) appendices. Smaller requests for proposals or requests for quotations may contain only part of this information. We have in our files five recent United States government requests for proposals, three for services and two for high-technology development projects. They range from small to medium, large, and very large documents: 75 pages (small), 114 pages and 166 pages (medium), 506 pages (large), and 760 pages (very large). (Any request for proposal consisting of 1,000 pages or more can only be termed "enormous," "mammoth," "voluminous," or "huge"!) For the purpose of this discussion, we will describe the content and approach to the very large request for proposal. This discussion will provide information on the major aspects of a request for proposal or request for quotation that one is likely to encounter in doing business with industry or government in either large or small procurements.

General Instructions to Bidders

The first part of a very large request for proposal is the section that includes general instructions to the bidder. This part of the request for proposal contains instructions, conditions, and notices to the bidder; certifications, representations, and other statements that must be made and/or agreed to by the bidder when the proposal is submitted; and a description of proposal evaluation factors and methods that will be used in grading the proposal. In the typical very large request for proposal that was evaluated, the first part of the request for proposal occupied 105 pages, or about 14% of the total number of pages in the request for proposal.

Instructions, Conditions, and Notices. The "Instructions, Conditions, and Notices" section of the request for proposal (perhaps called by another suitable or appropriate title) contains general information about the project and the proposed contract, site visits by the proposer to the place of work performance, preaward onsite reviews by the potential customer, desired proposal validity times, and other general clauses and conditions required by law or by the customer's procurement policy. Some typical parts of this section are described below.

Contract Planning. The contract-planning section describes the program, the program phases, the type(s) of contract, funds available (if appropriate), and specific procurement provisions. The following is an example from a potential request for proposal for a cargo transfer vehicle (CTV).

> The Cargo Transfer Vehicle is intended to be carried up in the space shuttle orbiter along with cargo destined to be delivered to Space Station *Freedom*. The CTV is to be released from the launch vehicle and to rendezvous and berth with the space station. It is planned that the Cargo Transfer Vehicle will be implemented in two phases: an Advanced Development Phase to be followed by a Design and Development (Phase C/D). This RFP solicits proposals for both phases. A firm contract is planned for the Advanced Development Phase, with a firm option for the Design and Development Phase.
>
> The Advanced Development Phase will be a predevelopment type contract with a period of performance of sixteen (16) months.
>
> The Design and Development Phase will be a mission type contract with a period of performance of approximately forty (40) months. If the option for this phase is exercised, the contract will be initiated immediately following the Advanced Development Phase. The contractor is advised that initiation of the Design and Development Phase is contingent upon congressional approval for a new start.
>
> The Advanced Development procurement will be awarded utilizing a cost-plus-fixed-fee (CPFF) contract. The Design and Development (C/D) procurement will be negotiated as

> a firm option, utilizing a cost-plus-award-fee (CPAF) contract. Any resulting contracts will contain all provisions required by law and regulations and such other provisions as may be required to adequately protect the interests of the Government. The clauses and the provisions applicable are set forth in the sample contract contained herein.
>
> Competition has been limited to the two (2) Phase B study contractors, one of which will be selected as a result of this RFP to perform both the Advanced Development Phase and, if contracted, the Design and Development (C/D) Phase of the CTV Project. However, if NASA determines [that] the proposed concepts inadequately fulfill mission need objectives, all design concepts together with the result of agency in-house study effort may be made available to industry to propose on any concept design or development solicitation.

The contract-planning section for a typical project includes a bar chart schedule similar to that shown in Figure 6.2 to establish the time relationships between major program elements.

Estimating and Pricing. General instructions to the bidder relative to submission of estimating and pricing information are included in this section. The forms to be used are referenced and a requirement is included that pricing and estimating techniques will be described in the proposal. One recent request for proposal states that "All pricing or estimating techniques should be clearly explained in detail (projections, rates, ratios) and should support the proposed cost in such a manner that audit, computation, and verification can be accomplished. Price visibility and traceability to the lowest practical cost level is essential." The meaning of *lowest practical cost level* is understood better when the cost proposal instructions, contained in another section of the request for proposal, are taken into consideration.

Proposal Instructions. Specific proposal instructions provided in the request for proposal include a number of areas of policy that the customer wishes to convey to the bidder. One such clause cautions against the use of unnecessarily elaborate proposals. An example is given below:

> Unnecessarily elaborate brochures or other presentations beyond those sufficient to present a complete and effective proposal are not desired and may be construed as an indication of the offeror's lack of cost consciousness. Elaborate art work, expensive paper and bindings, and expensive visual or other presentation aids are neither necessary nor desired.

The policy on whether or not proposal preparation costs can be covered in the proposed contract estimate is also included in this part of the request for proposal. Government requests are usually strong in pointing out that sending a request for proposal to a prospective bidder does not commit the government to pay any costs incurred in submission of the proposal or in "making necessary studies or designs for the preparation thereof." They further state that no costs may be incurred in anticipation of the contract with the exception that "any such costs incurred at the proposer's risk may later be charged to any resulting contract to the extent that they would have been allowable if incurred after the date of the contract and to the extent

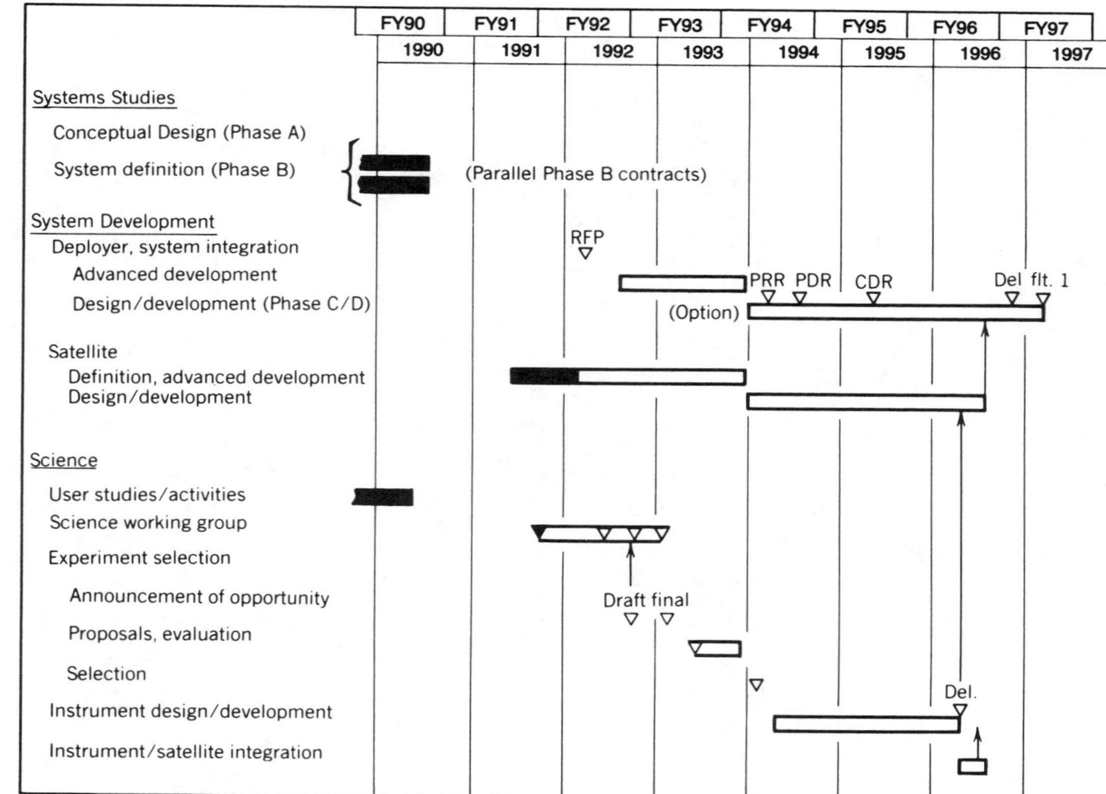

FIGURE 6.2 Typical long-term project schedule. Black bars indicate completed work; white bars indicate work to be done; triangles indicate events; PRR: preliminary requirements review; PDR: preliminary design review; CDR: critical design review; Del: delivery; Flt.1: flight number one.

authorized by the contracting officer." This appears to be a loophole that will allow some proposal preparation costs to be charged to the resulting contract under certain specific conditions.

Treatment of proposal data by the customer is covered in most requests for proposals to assure the bidder that the information presented in a proposal, particularly if it is proprietary in nature, will be protected from transmission to the proposer's competitors or to any company or organization outside that of the customer's. Generally, the customer will and should use a bidder's technical, organizational, and cost data for evaluation purposes only. Stating this in the request for proposal assures the customer of maximum disclosure (to the customer) of a company's plans, ideas, and capabilities.

A *proposal validity period*, usually 90 days after proposal submission, is specified in most requests for proposals. The bidder can easily change this period, but would be wise not to shorten it unduly because evaluation procedures frequently extend beyond their prescheduled dates, forcing a delay in selection and contract negotiations. Requests for proposals also state that the issuance of the request for proposal does not obligate the customer to sign a contract with any of the bidders. In the event that the customer is not satisfied that any of the bids will meet the stated requirements, the procurement can be canceled completely. Each request for proposal also contains a policy on late proposals and withdrawal of proposals. (Customers usually reserve the right to accept late proposals.)

Miscellaneous Clauses. Miscellaneous clauses are also included in this section even though some of them are amplified, clarified, or repeated for emphasis in other sections of the request for proposal. Typically, these clauses pertain to the following subjects:

1. Security classification of the work
2. Technical direction of the work
3. Contracting officer's responsibility
4. New technology policy
5. Inventions and patents policy
6. Equal opportunity provisions
7. Small and disadvantaged business use
8. Visits by bidders to the work performance site
9. Onsite visits by the proposal evaluation team
10. Customer-furnished equipment assumptions
11. Information required on proposed consultants:
    a. Résumés—including education, experience and technical publications
    b. Unique capability possessed by the individual(s) that makes his/ her/their use on this program necessary and advantageous (basis for selection)
    c. Estimated hours and rate.

   d. Estimated cost and supporting data
   e. Estimated contract duration
12. Safety and health policies

Representations, Certifications, and Other Statements of the Offeror. In most government requests for proposals, and in some industrial and commercial requests for proposals, the bidder is required to make certain certifications concerning adherence to specific laws, policies, and directives that have been enacted and are in force in state, federal, or local governments. Most of these laws, policies, and directives have resulted in the provision of socioeconomic objectives. The representations and certifications are merely forms to be filled out and returned to the government prior to or along with the proposal to help assure that all state, federal, and local laws will be complied with in the performance of the work. Typical representations and certifications are listed below, along with a brief description of what each is designed to accomplish.

Buy-American Trade Agreements Certificate. The proposer is asked to list, in this certification, all end products that are not American-made. (The certificate implies that preference will be given to a supplier who is providing the most American-made products.) In a related certification, the proposer must estimate what percentage of the proposed contract price represents foreign effort.

Certificate of Contingent Fee. The proposer states in the certificate of contingent fee whether or not any company or person other than a full-time bona fide employee working for the proposer has been employed to solicit or to secure the contract. The proposer also states whether or not he/she has paid or agreed to pay any company or person (other than an employee) any fee, commission, percentage, or brokerage fee contingent upon or resulting from the award of the contract. A positive response to either of these questions will more than likely bring further questions from the procurement officer, addressed to the company or person accepting the fee, to the proposing company, and/or to the customer. The purpose of this certification is to help ensure that there is no conflict of interest in the possible award of a contract.

Equal Opportunity. In this certification, the proposer states whether or not all equal-opportunity compliance reports required by the government's equal-opportunity programs have been filed.

Affirmative Action Program. The proposer states in this certification whether or not the proposing company has developed and maintained an equal opportunity affirmative action plan at each geographical plant location.

Small Business, Small Disadvantaged Business, and Nonsegregated Facilities. In these certifications, the proposer certifies whether or not the proposing firm is a small business concern, a regular dealer–manufacturer, an individual, partnership, or corporation; and whether or not the proposing company is a small business concern owned and controlled by socially and economically disadvantaged individuals. Definitions of a small business and a disadvantaged small business are included in Section 8(a) of the U.S. Small Business Act. Some government and industrial requests for proposal are directed specifically to small businesses. In others, small businesses are favored. In still others there is no preference one way or the other but the customer must merely know whether or not the bidder is a small business to complete its reports to the government on the percentage of small businesses in the overall contract or subcontract structure.

In a related certification, the proposer certifies that the proposing company does not have or condone facilities such as waiting rooms, work areas, restrooms and washrooms, restaurants and other eating areas, transportation facilities, or housing facilities that are segregated on the basis of race, color, religion, or national origin. This certification draws attention to and helps the enforcement of equal opportunity laws or clauses in the proposed contract.

Clean Air and Water Certification. This certification is used to assure the customer that the proposer is using facilities that conform to the Clean Air Act and the Federal Water Pollution Control Act.

Certification of Current Cost and Pricing Data. This certification is used by the proposer to authenticate the fact that cost and pricing data used in the proposal are accurate, complete, and current as of the date of contract negotiation. Therefore, this certification is not submitted until the contract award and negotiation are complete.

Cost Accounting Standards and Practices. Bidders who are proposing to the government as prime contractors or major subcontractors are required to execute a series of forms that certify that their cost-accounting systems meet certain standards; that they will disclose the principles, practices, and procedures used in cost accounting; and that they will inform the customer if any changes are made in their cost-accounting practices. Whether these certifications must be filed or not depends on the total amount of business a firm does with the government and the size of the procurement.

Government Facilities. Contractors or subcontractors to the federal government must certify if they are using government facilities, if any of those facilities are to be used in the proposed contract, or if additional government facilities are required. If the proposer is a government contractor, the proposer must certify as to whether property control and accounting procedures have been approved by the government.

Keep in mind one thing about representations and certifications: a large part of them will not be necessary for most commercial and industrial procurements. In the private sector, most of these practices are maintained merely as a matter of preserving company dignity and prudence and are an integral part of good business sense. Public sector contracts have fallen under legislative controls and guidelines that require special documentation of these certifications because of the failure of a few to hold to continued good business practices or because of pressures to correct past political and social inadequacies.

Evaluation Methods and Procedures. Requests for proposals will normally give the proposer a clue as to how the proposal will be evaluated. In some instances a complete section of the request for proposal will be devoted to evaluation methods, procedures, factors, criteria, and to cost evaluation. If the customer plans to use a source evaluation board, which is usually the case for large, multimillion dollar government procurements, the customer will preestablish mission suitability factors, which are weighted based on a total score of, say, 1,000 points. Within each factor are criteria to which a further breakout and apportionment of the weight of each mission suitability factor are applied. Mission suitability factors are normally scored, while cost factors and certain other factors such as experience and past performance are generally not scored by the source evaluation board. They are, however, taken into account in the overall evaluation process. (See Chapter 12, "How the Proposal Is Evaluated.")

Although weightings assigned to mission suitability factors and criteria are not listed in the request for proposal, a very interesting paragraph is usually contained in the request for proposal which will give a series of clues that can be used to derive an approximate breakout of the weighting. The reader must patiently study this paragraph, analyzing the interrelationships of the various factors and criteria, to derive what specific numerical weighting has been applied to each factor and criterion. Through a process of iteration, one can come reasonably close to the most probable breakout or weighting of the total allocated points.

For example, a typical request for proposal contains a listing and description of the following mission suitability evaluation factors and criteria:

A. Excellence of Proposed Design Criteria
    1. System Engineering and Analysis
    2. Design and Development
    3. Manufacturing Test and Operations
B. Organization and Management Criteria
    1. Team Authority and Control
    2. Team Organization and Structure
C. Key Personnel Criteria
    1. Project Manager Experience and Commitment
    2. Key Personnel Experience, Capabilities, and Commitment.

A typical paragraph in the request for proposal reads as follows:

> For evaluation purposes, *Excellence of Proposed Design* is the MOST IMPORTANT factor and is of substantially higher value than the sum of the other factors. *Organization and Management* is equal to *Key Personnel*. In the evaluation criteria, *Systems Engineering* is MOST IMPORTANT and significantly higher in value than *Design and Development*, which is VERY IMPORTANT and moderately higher in value than *Manufacturing, Test,*

and Operations. Manufacturing, Test, and Operations is IMPORTANT and moderately higher in value than either *Team Authority and Control* or *Project Manager Experience and Commitment*, which are also important and equal in value. *Key Personnel Experience, Capabilities, and Commitment* is LESS IMPORTANT than either *Team Authority and Control* or *Project Manager Experience and Commitment*, by a moderate value, and is equal in value to *Team Organization and Structure*.

Through an iterative process, by arbitrarily assigning various weights to each factor and criterion within a framework of 1,000 points, the authors developed Table 6.1 as a plausible and probable weighting.

Although the point distribution in Table 6.1 but may not be exactly what was assigned by the source evaluation board, the interrelationships stated in the request for proposal hold; therefore, the actual assigned weights cannot be too far off from the above. Assuming that a proposal will be scored against factors and criteria, a proposer can develop a weighting distribution. The proposer will then gain a valuable insight into the emphasis to be placed on these criteria and will be able to structure a proposal with emphasis in the corresponding subject areas.

Technical, Management, and Cost Proposal Format and Contents. The RFP section on "General Instructions to Bidders" also includes a description of the expected contents of the technical proposal, the organization and management proposal, and the cost proposal. This section may go so far as to specify the number of volumes in a proposal, the maximum number of pages in each volume, the number of copies of each volume required, the type of line spacing required (double, single, or 1½ spaces), font size, page-numbering system, policy on the use of foldouts and double-sided printing, and even how many proposal copies should be included in each shipping carton. The main information derived from this part of the request for proposal, however, is the desired content of the technical, management, and cost volumes.

TABLE 6.1. Probable Weighting Derived from Text

A. Excellence of Proposed Design Criteria		650
1. System Engineering and Analysis: Most Important	375	
2. Design and Development: Very Important	150	
3. Manufacturing, Test, and Operations: Important	125	
B. Organization and Management Criteria		175
1. Team Authority and Control: Important	100	
2. Team Organization and Structure: Less Important	75	
C. Key Personnel Criteria		175
1. Project Manager Experience and Commitment: Important	100	
2. Key Personnel Experience and Capabilities and Commitment: Less Important	75	
Total		1,000

Technical Proposal Volume(s). Instructions as to the content of technical proposal volumes will include a requirement for (1) a description of the work activity or work output being proposed and (2) implementation plans and schedules that describe how and when the work is to be performed. A large part of the technical work to be performed manifests itself in a data procurement document, data requirements list, or technical reports listing, which usually appears as an appendix to the request for proposal. In this data procurement document, the customer lists and describes the various drawings, specifications, progress reports, manuals, training documents, operating plans, and maintenance procedures that will be required under the proposed contract. Content, format, frequency of distribution, and number of copies required are included in this document for each report that is to be required under the contract. Since documentation is a large cost in itself, and since the documentation of work must reflect work content, a thorough analysis of the documentation requirements in the request for proposal is a must. A data procurement document for a recent very large government request for proposal consisted, in itself, of three hundred pages.

Organization and Management Proposal Volume(s). The request for proposal specifies what type of information should be provided to explain and substantiate a company's experience, past performance, and corporate interest and investment; the company and project organization that will be assigned to do the work; the qualifications, experience, and training of key project personnel; and how the work will be managed.

Other information that is often required (or desirable) to include in the organization and management volume or volumes consists of: (1) personnel practices and labor relations experience; (2) procurement policies, practices, and procedures; (3) make-or-buy policies; (4) configuration management; (5) quality, reliability, and safety management; (6) cost and schedule reporting; (7) logistics; (8) design reviews; and (9) documentation to be submitted during performance of the resulting contract. The request for proposal will usually require some discussion of: (1) how the work of the organizations and personnel who will be participating in the project will be integrated; (2) the method of general planning and master scheduling to be used; (3) how task assignments will be made and followed up; (4) the method to be used for internal review, job tracking, and setting priorities; and (5) how you will detect problems, determine the appropriate corrective action, and assign it to a responsible individual or organization.

Cost Proposal Volume. A cost volume must contain resources estimates, costing methodology, and rationale sufficient to allow the customer to establish the fact that the work will be cost-effective and that cost estimates are realistic.

The request for proposal will usually specify the exact format required in the cost proposal. To date, there has been very little uniformity and consistency in the format and content of cost proposal requests among various government and industrial agencies and organizations. There is no reason for this to continue, since the basic resource information required to do a complex task or series of tasks can be presented in a relatively uniform or consistent format. Some government agencies have become aware of the need for greater standardization, consistency, and condensation of cost data in proposals to the government, but, at this writing, there is still much to be done in both government and industry, to simplify and standardize cost data requirements in requests for proposals. Increased simplification and uniformity in cost-estimating and cost-reporting systems would reduce the bidder's labor hours required to prepare proposals, the customer's labor hours required to evaluate proposals, and would enhance the collection of actual cost data, comparison of actual costs between projects, and comparison of actual costs with estimated costs.

Greater standardization in cost proposals has been difficult in the past because each company has its own unique accounting system that has more than likely grown with the company for a number of years. With the advent of computerized financial management and cost estimation and cost-reporting systems, it has become feasible to convert cost data from a company's traditional accounting system into almost any format required in a very short period of time. The easy-to-use and transportable computer applications software now available has paved the way toward greater possibility of standardization and comparison of cost data. Until a relatively standard approach is developed for cost accounting, estimating, and reporting, however, the potential bidder for both governmental and industrial contracts may expect to see a wide variation in cost proposal requirements in requests for proposal. A flexible and adaptive computer system that will format and output cost information in a large variety of ways is a must for any organization that expects to bid to different agencies or companies.

The cost proposal requirements in requests for proposals include ground rules for cost estimating, suggested or required cost-estimating methodology, formats for proposal of resources, and descriptions of the bidder's estimating methods and rationale. The request for proposal provides definitions of various categories and elements of cost to permit a close correlation of the cost proposal presentation with customer understanding. The request for proposal will also specify different levels of depth of cost proposal data for different phases of the project.

The Contract Document or Schedule

The contract document or contract schedule is the legal document that is negotiated and mutually agreed to by buyer and seller prior to commencement of the work. The contract document is included in the request for proposal to assure that the proposer understands the legal terms and conditions of the contract and to provide a basis for specific comments, exceptions, deviations, or additions that may surface during the proposal preparation process. The contract schedule contains the following articles, which describe the legal agreements made between the seller and the buyer.

Statement of Work. The statement of work (SOW), which is usually referred to or included in Article I of the contract schedule, includes an introduction and scope, mission objectives, roles and responsibilities, and specific contractor tasks. Included in the statement of work or as references or appendices are work activity or work output specifications, the data procurement document, the work breakdown structure or work element structure dictionary, product assurance and safety requirements, and other technical and/or programmatic information.

The statement of work, if prepared in a detailed manner by the customer and included in the request for proposal, can and should form the basis for the technical proposal volume, as it covers generally as well as specifically all of the work to be performed as the principal deliverable under the proposed contract. If the statement of work is not prepared and included in the request for proposal, the bidder will have to prepare a detailed statement of work as part of the proposal. The statement of work defines all work contained in the work element structure. Ideally, there is a close if not identical correlation between the statement of work and the work element structure dictionary, with an identical numbering system for both if feasible. The statement of work not only lists all deliverable hardware, software, and services, but also spells out all the work that will be required to deliver these items and to prepare and deliver all required documentation. Hence, the statement of work is a document that will eventually serve as a legally binding contract document that integrates and interrelates the work elements with deliverable and nondeliverable hardware, software, documentation, and services. Prepared with great care, it is the basis for the entire cost and resources estimates and is integrally tied to the time schedules for the conduct of work. It is recognized that the development plan portion of the technical proposal volume as well as the organization and management proposal will bear heavily on the content of the statement of work, which is the detailed written description of what is going to be accomplished under the proposed contract. Thus, it is an essential part of a proposal.

Statement of Work Introduction and Scope. The contract schedule or document, as amplified by the statement of work introduction, states that the selected contractor "shall" furnish the management, labor, facilities, and materials required to design, develop, procure, manufacture, assemble, test, deliver, and/or operate (as applicable) the item described in the design and performance specification. The design and performance specification is usually included as a reference or appendix at this point in the statement of work. The statement of work also identifies specific deliverable items and states when they are to be delivered.

Mission Objectives. If the work output or work activity being proposed is a part of a larger work activity or work output, the overall mission objectives of the larger work activity or work output are described in the statement of work. This information is provided not only to give a clear understanding to both parties as to the relationship of the work to an overall plan of action, but to assure that the resulting contract starts and stays in tune with an overall mission objective, which may consist of parts other than those being proposed.

Roles and Responsibilities. A clear and concise description is included in the statement of work to define specific roles and responsibilities of the proposed supplier and the customer. If there are to be any customer-supplied direction, approvals, parts, labor, or facilities, they are physically described or referenced along with the dates or times the customer commits to provide these items, services, or approvals.

Detailed Contractor Tasks. Detailed tasks of the proposed contractor are spelled out and keyed to the work breakdown structure and work breakdown structure dictionary. The request for proposal usually provides a sample work breakdown structure and work breakdown structure dictionary. Figure 6.3 shows a typical work breakdown structure contained in a request for proposal.

Options. More often than not, the contract document or contract schedule will contain procurement options that are contingent upon acceptance and/or successful performance of the proposed work. These options will be described and priced in the responding proposals but may not be placed under contract at the same time as the initial work. The request for proposal will usually specify a time period within which the option is expected to be exercised, and the customer reserves the right to exercise the option in this time period.

Period of Performance. The contract document will normally contain a section or article on the exact period of performance to be covered under the proposed contract. This period of performance will be used in time-phasing the proposed contract activities and in determining the allocability and acceptability of costs incurred.

Allowable Cost, Fixed Fee, and Payment Provisions. The overall total contract agreed-to cost and fee will be specified along with the method and timing of payment for the proposed work and for any contract options. The method and timing of payment should be carefully reviewed, understood, and negotiated to assure that adequate cash flow will be available to do the work. For government contracts, it should be recognized that there are delays in payment because of unforeseen legal and procedural obstacles, and that yearly incremental funding for a multiyear contract is often delayed because of failure to enact budget legislation of continuing funding resolutions on a timely basis.

Special Fee Structure and Fee-Payment Provisions. Increasingly, government and industrial contracts are incorporating more complicated fee structures and fee payment provisions. These more complex fee structures are brought about by the desire to motivate the contractor to stay within cost, to stay on or exceed schedule requirements, and to adhere to or surpass performance requirements. Various contract forms such as fixed-price incentive (FPI) fee, cost plus incentive fee (CPIF), and cost plus award fee (CPAF) are used in these fee payment structures. Incentive fees are usually based on a mathematical or schedule relationship between contracted and actual work performance and can often be complex in structure, with varying increments of fee allocated for various increases or decreases in cost, schedule, or performance.

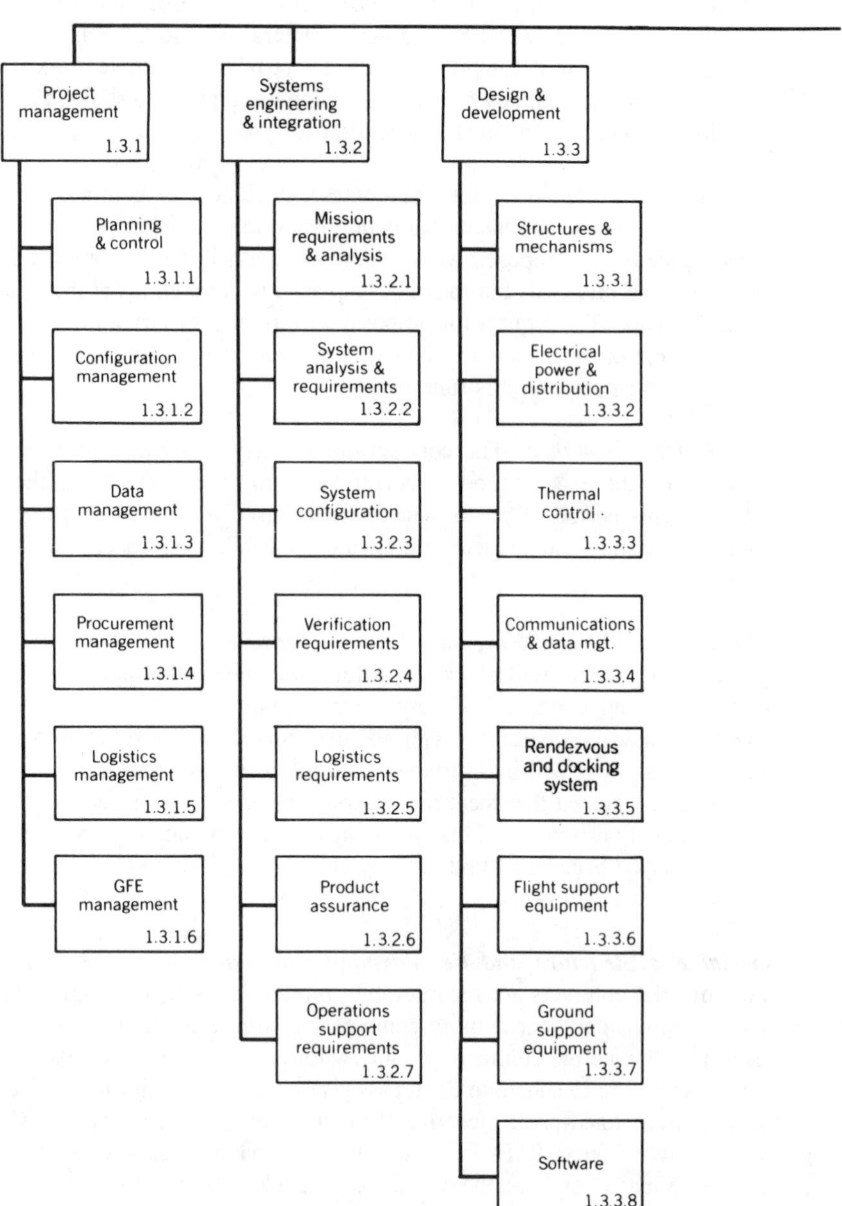

FIGURE 6.3 RFP work breakdown structure for cargo transfer vehicle (CTV). C/O: checkout.

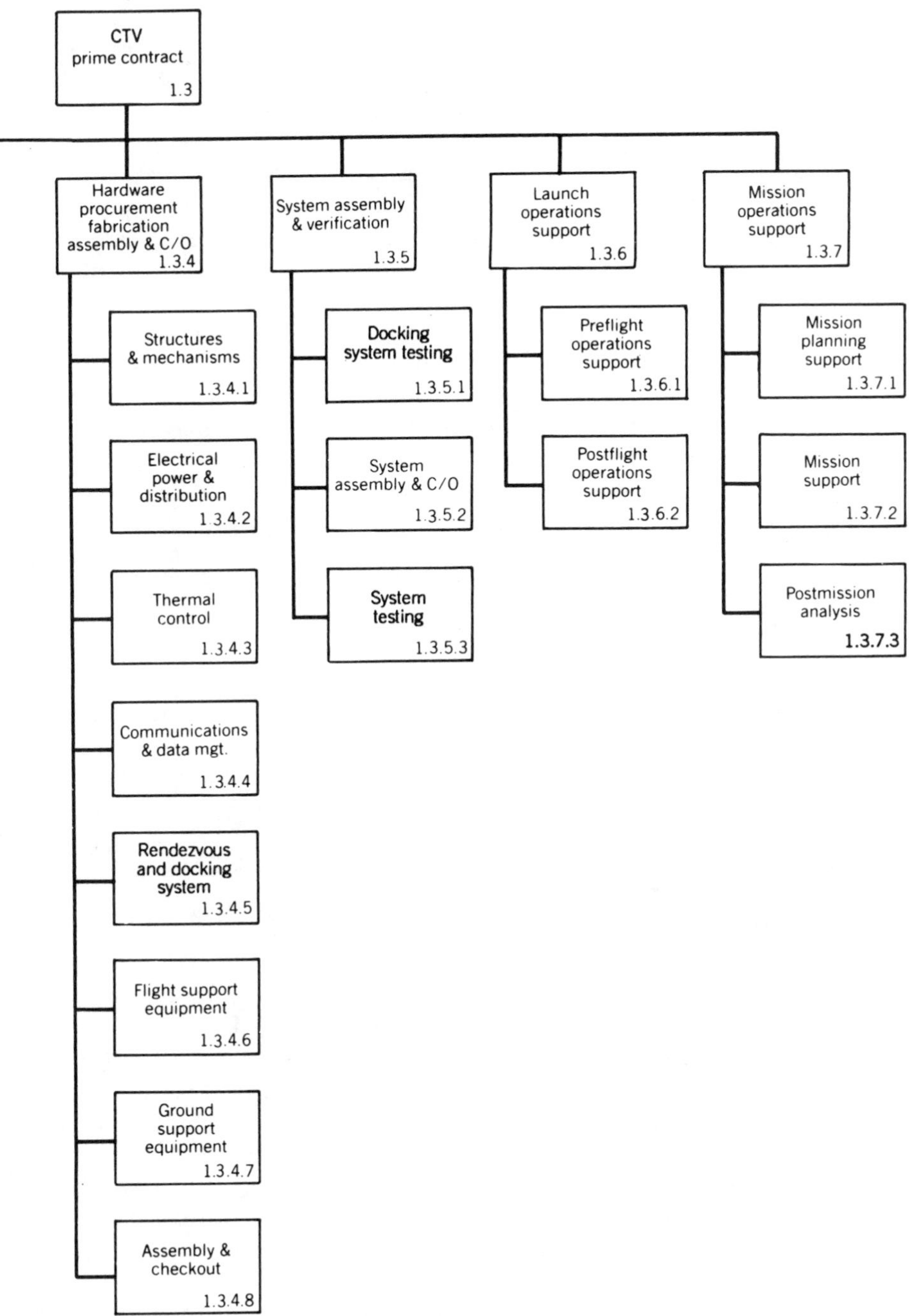

FIGURE 6.3 *(Continued)*

Award-fee provisions are usually monitored and controlled by a customer performance evaluation board, which reviews the contractor's work at uniform intervals during contract performance. The proposed resources allocated for the project must include sufficient labor hours, materials, and travel costs to participate in these reviews, to prepare self-evaluation reports required by the customer, and to make corrections in the next performance evaluation period in the event that it becomes evident that performance improvement is necessary to earn a higher award fee. The customer's determination of the percentage of award fee to be approved during a given contract period is principally subjective in nature, but can be a result of qualitative evaluations used to glean a qualitative composite measure of performance excellence. Except in cases of extreme or obvious inequities, it is difficult to change the award-fee decision of the performance evaluation board. Therefore, fee expectations stated in the proposal should be based on realistic anticipations of performance. The award fee is usually paid on a provisional basis based on the percentage of completion of work as determined by the contracting officer. Award-fee structures usually include a base fee that is paid regardless of the level of contract performance excellence and an award fee that can range from zero to a prestated maximum figure. A complete understanding of the award-fee structure is necessary before starting the proposal preparation process; therefore, it is necessary to fully understand and evaluate this part of the request for proposal before starting to develop detailed text, schedule, or costing information for the proposal.

Technical Direction and Management. Requests for proposals contain, in their contract document or contract schedule, specific information concerning how the proposer will receive technical direction and management under the proposed contract. In government contracts, technical direction is a function that is separated from the contracting-officer function, and the contract document spells out in detail just what direction the proposing company can receive from the technical manager and what direction it can receive from the contracting officer. The contracting officer is the person held legally responsible for the successful completion of the work. The technical manager can provide direction *within* the contract scope of work, while the contracting officer must approve changes *outside* the scope of work. *Technical direction* is defined as a directive that approves approaches, solutions, designs, or refinements; fills in details; or otherwise furnishes guidance to the contractor. Technical direction includes a process of conducting inquiries, requesting studies, or transmitting information or advice by the technical manager, regarding matters within the definitions and requirements of the statement of work. Any company proposing to a government agency should find out as much as possible about the personality, policies, attitudes, and management techniques of the person or persons who are most likely to provide technical direction of the work, because the actions of the technical manager and his or her staff can lead to expenditure of more work than had been anticipated, proposed, or negotiated at the outset of the project. Overruns can result from excessive technical direction or involvement from the customer, unless the technical manager has the interests of the cost control and harmonious working relationships at heart. Conversely, a cooperative and understanding technical manager can make it easier rather than harder to meet program commitments by exercising flexibility and understanding in developing and approving program tradeoffs.

In addition to specifying duties of the customer's technical manager and contracting officer, most government customers and some industrial customers require that the proposer designate one person to act as the project manager and that this person be delegated complete authority to decide all technical matters related to the contract. This project manager is the counterpart of the customer's technical manager and is usually expected to stay on as project manager throughout the duration of the project. The contract document often requires not only that the project manager and other key personnel be named, but that the proposer's company certify that these key personnel will stay with the project unless approval for a change is granted by the customer. This contract article is designed to discourage the bidding of a job, ostensibly to be managed by experienced personnel, who are subsequently replaced with a less experienced staff.

Contract Changes. Because contract changes can be troublesome areas that cause overruns and program slippages, the contract document usually contains considerable verbiage that addresses how changes will be handled. The main changes clause of the contract is in the "general provisions," to be discussed later. The changes clause in a government contract allows the contracting officer to make changes within the general scope of the contract. If the change results in an increase or decrease in the cost or time required to do the job, then an "equitable adjustment" is made in the estimated cost, delivery schedule, or other contract provisions. Defining and agreeing on what is an "equitable adjustment" may cause an adversarial relationship between the customer and supplier unless the work scope and the change are sufficiently defined. Almost any work activity that is changed while in process will cost more to accomplish than initially was planned, unless the change involves the removal of a large part of the work activity itself.

The contractor usually has 60 days to file a claim of increased costs or slipped schedule due to the change. Occasionally, for very large contracts, both parties will agree in advance to specify that a certain dollar value increase or decrease will not give rise to an "equitable adjustment." The purpose of this agreement is to allow small changes to take place that balance out to a zero or minimal change in total contract costs or delivery dates. Also, articles are included in the contract schedule that permit no-cost engineering change proposals to be submitted by the contractor and approved by the customer's technical manager. Other articles include information and potential contract clauses on the notification procedure for changes (how the customer is notified and responds to contract changes) and the method of tracking and accounting for contract changes. These contract schedule articles are also important to read thoroughly and understand before initiating detailed proposal preparation activities. They may affect how much is put into the category of allowance for cost growth when making the proposal cost estimate and establishing a contract bid price.

Miscellaneous Articles. A number of miscellaneous articles are also given in the contract schedule included in the request for proposal. These are listed below along with a brief phrase telling what each is designed to accomplish.

Total Sum Allotted. Specifies the limitation of the customer's financial obligation for the work.

Place of Performance. Specifies where the work will be performed.

Inspection, Acceptance, and Free on Board (FOB) Point. Specifies the location of interim inspection and interim acceptance of the services and/or deliverable items called for under the contract.

Reports Distribution. Specifies the number of copies required and the distribution location for all reports.

Packing, Packaging, Identification, Marking, and Preparation for Shipment. Specifies the method of shipment and calls out specifications for packing, packaging, identification, marking, and preparation for shipment.

Designation of Special Representatives. Designates special representatives of the customer who are charged with the responsibility for special areas, such as patents and new-technology reporting.

Information Releases and Publications. Specifies customer review of material to be presented orally or in written form in the public or professional media.

International System of Units. Specifies whether or not the international system of units (SI) will be used in all documentation under the contract. (Watch out for this one—it can be expensive if it is not expected.)

Rights to Proposal Data. Specifies which pages in the proposal the customer cannot duplicate, disclose, or have others do so, for any purpose.

General Provisions

The general provisions of the proposed contract, often called the "boilerplate," are the detailes that define all of the legal implications of going into business with a specific customer. In a large government request for proposal, these details can occupy 50 to 100 pages (they average 70 pages), many of which have very fine print and a very high word count per page. One agency has 41 general provisions with an added 39 special provisions in each contract. Although those who are experienced in working with the government have become accustomed to living under these provisions, which were mostly imposed because of enactment into legislation by Congress but which include some agency-imposed provisions, others who are new to large-scale contracting for industry and government must learn the significance (or insignificance) of each of these provisions before starting in earnest to prepare a proposal.

Most of the general provisions, at least for government contracts, are reflected as required "representations and certifications." discussed earlier. Many have little or no effect on a company that is doing business in an honest and straightforward manner. Neither space nor prudence permits the duplication of the general provisions in this book. Copies may be obtained by writing the customer or agency to be contacted for new business. The legal department of any proposing company should thoroughly read and understand these general provisions before proposal preparation starts, however.

OTHER TECHNICAL AND PROGRAMMATIC DATA INCLUDED IN THE REQUEST FOR PROPOSAL

In addition to the above information, many requests for proposals include detailed technical reports, drawings, specifications, and sample plans for the work activities or work outputs being proposed. This is often done to assure that all bidders receive the same information on which to base their proposals, because much of this information may have been provided by the bidders themselves, or it is in the public domain or in business or scientific literature openly available to the public. These documents provide an in-depth insight into at least the customer's understanding of the complexity, content, and timing of the work to be proposed and, therefore, are helpful in preparing the technical portions of the proposal.

OTHER TYPES AND SIZES OF REQUESTS FOR PROPOSALS

The request for proposal just described is perhaps one of the most complex and demanding, since it is for a very large procurement, probably for the acquisition of a major system, product, or facility or major subsystem thereof. There are other sizes and types of proposals—as many as there are items to be procured. The following adjective descriptions (used earlier in part) are assigned to classify requests for proposal by size:

Adjective	*Number of Pages*
Very small size	0–75 pages
Small size	75–150 pages
Medium size	150–500 pages
Large size	500–750 pages
Very large size	750–1,000 pages
Giant RFP	over 1,000 pages

The contents of the request for proposal will also vary with the subject matter of the procurement. Many of the request for proposal elements described in the first part of this chapter will disappear and others will be added when proposals for service-type activities are requested. Content changes and is dependent on whether a process, product, project, service, or combination of these is being requested.

Typical types of requests for proposals other than that discussed earlier in this chapter are described below.

Standard Products or Supplies

Requests for proposals are more often called requests for quotations (RFQ) where a price and delivery date for standard products or supplies are concerned. These requests for quotations are classified as "very small" or "small" and usually only contain a standard part number or series of part numbers, a desired delivery date, and any special packaging, shipping, and handling conditions. The more the product or supply item deviates from standard, the more detailed the request for quotation will be, because special accessories, features, and associated services required by the customer must also be described. This type of request for quotation seldom exceeds 15 or 20 pages, plus a letter of transmittal, and normally includes a sketch or drawing of the completed item; a parts list or work element structure dictionary; and delivery dates, quantities, and conditions.

General Support Service Contracts

Requests for proposals for general support service contracts usually fall into the "medium" size category. This type of RFP includes a description, map, or plan of the facility or equipment to be serviced and other elements that appear in the larger requests for proposals such as general background information, proposal format and content requirements, evaluation of proposals, reporting instructions, and a detailed statement of work. If the general support service contract is to be a government procurement, the request for proposal will also include the socioeconomic provisions such as equal opportunity provisions, buy-American act provisions, clean air and water act compliance, small business utilization reporting, and so forth. General support services contracts for the government are often directed to small businesses or small minority-owned businesses, because they do not require a large financial investment or business base to successfully accomplish the work.

Usually required in proposals responding to these requests for proposal are a management plan and organization; a staffing plan giving staff numbers, certifications, sources, availability, and flexibility; and a section on key personnel. A cost proposal is required, as is a section on experience and past performance. Since contracts resulting from these requests for proposal can consist of time-and-materials contracts or mission-type support contracts, the method of costing and the amount of cost detail required varies considerably. A minimum requirement is a cost breakdown by calendar period and element of cost: labor, materials, overhead, general and administrative costs (G & A), and fee. As in larger requests for proposals, the relative importance of the various evaluation factors and criteria are also described, to allow the proposer to place proper emphasis on the various aspects of the procurement. The method of contractor evaluation and fee award, as well as the method of technical supervision and direction by the customer, is also provided.

Base Maintenance Support Services for a Large Plant or Facility

Requests for proposals for activities such as base maintenance and support services for a large plant or facility will normally fall into the "large" category. This type of contact differs from the general support services RFP in that a large number of activities are required; more documentation, inspection, and evaluation of results is required; and more personnel and equipment are needed. The large RFP size is brought about by the need for more detailed descriptions of the task or tasks involved. In general, the complexity of the request for proposal is in proportion to the complexity of the job itself, which is, in turn, associated with the complexity of the plant or facility being supported. Checklists, procedures, property lists, facilities descriptions, and past maintenance and operation requirements normally form a part of this type of request for proposal. Most of the contract boilerplate, general provisions, evaluation guidelines, proposal formulation instructions, and scope of work provisions that are in a major procurement will also show up in the large request for a base maintenance support services proposal.

A FINAL COMMENT ON REQUEST FOR PROPOSAL QUALITY

One characteristic that is noticeable in many large requests for proposals, particularly some of those received from government departments or agencies, is the absence of consistency in the quality of writing, publication, and production. This characteristic is understandable because requests for proposals, like proposals, are usually put together in an inordinately short time frame with a limited staff. Inputs are sometimes received from various organizations, in the process of the RFP preparation, and little time is available for correlation, coordination, and synthesis of inputs into a cohesive, readable, and attractive document. It will be observed in some requests for proposals issued in the past that page numbering is difficult to follow, various styles and sizes of type are used, seemingly at random, format is not consistent, and editorial work has not been thoroughly accomplished. These characteristics in a request for proposal do not give license to the same characteristics in the responding proposal.

Improvements are constantly being made to the request for proposal preparation process. Industrial and government organizations are turning to the use of word processors to speed the preparation of requests for proposals and to improve their readability, ease of reference, and attractiveness. Progress has been made since the writing of the first edition of this book in 1981/1982. At this writing, however, much still remains to be accomplished in this area. Some day all requests for proposals will be equal in quality to that of the proposals they expect in response. Nevertheless, the burden will always lie with the potential bidder to respond with a high-quality, editorially excellent, attractive, neat, readable, easily referenced, and convincing proposal no matter what the condition of the RFP.

7

THE TECHNICAL PROPOSAL

Through wisdom is a house built; and by understanding it is established; and by knowledge shall the chambers be filled with all precious and pleasant riches.

—Proverbs 24:3, 4

The most important part of a proposal is the technical proposal volume. It is the major basis for performing a good work activity or for providing a good work output. The technical proposal tells exactly how, when, and sometimes why the work will be performed in a certain manner, hence, the technical proposal must describe the work in detail. It must provide discussion, references, drawings, specifications, schedules, and documented evidence to show how and when the various phases of the work are going to be completed. It must contain enough substance to demonstrate competence and technical expertise in the field being proposed, it must demonstrate understanding of the requirement, and it must provide assurance to the customer that the job will be completed in a manner that meets all technical requirements.

As discussed in preceding chapters, the objectives, ground rules, constraints, and requirements of the work must be spelled out in detail to form the basis for a good proposal. A thorough knowledge of the requirements of the request for proposal and the proposer company's capability and approach to the work activity are essential. Cooperation and coordination of the entire proposal team is required to produce the technical volume. The members of the proposal team preparing the technical volume are the most likely source of a detailed definition of the work required by the request for proposal. These work descriptions usually take the form of detailed specifications, sketches, drawings, materials lists, and parts lists. More detailed designs will invariably produce more accurate cost estimates, and the amount of detail itself produces a greater awareness and visibility of potential inconsistencies, omissions, duplications, and overlaps.

It is important for the technical proposal team to be aware of project ground rules concerning production rate; production quantity; and timing of initiation, production, and completion of the job before starting the technical volume. Factors such as raw materials availability, labor skills required, and equipment utilization often force a work activity to conform to a specific time period. The technical team must, therefore, obtain or establish the optimum time schedule early in the proposal preparation process, verify key milestone dates, and see that the overall work schedule is subdivided into identifiable increments that can be placed on a calendar time scale. This time schedule must coincide with that required by the request for proposal.

The technical proposal is the basis for the organizational and management arrangements spelled out in the organization and management volume and for the resources estimates upon which the cost or price proposal or cost volume is based. The technical proposal volume should be prepared in advance of the organization and management and cost or price volumes. Since time for proposal preparation is usually short, however, the three volumes are often prepared concurrently, with the technical proposal feeding periodically into the other two volumes.

The essence of a successful technical proposal is responsiveness. This means far more than merely parroting the words contained in the request for proposal or request for quotation. It demands that the proposer comprehend and accurately describe the requirements and the proposed solution.

This dimension of responsiveness is seldom appreciated and frequently is neglected in the preparation of the technical proposal. The proposal manager must accept responsibility for: (1) the development of complete understanding in the mind of the customer, and (2) establishing the fact that the bidder also has a complete understanding. To accomplish this, he or she directs the team in such a manner that they deliberately share the responsibility and structure their output to foster customer comprehension and technical credibility. A practical, simple device for ensuring customer understanding is to have each member of the preparation team write out a brief comment as to what he or she, the writer, would like to have the listener express after reading the section. In large proposals, this description is accompanied by a sketch or diagram and takes the form of a storyboard. This will tend to structure the writing of the technical proposal toward the intended result.

The intended result of the technical proposal is, of course, the acceptance and implementation of the proposed solution. Sales in technical and industrial markets are made by supplying pertinent and usable information upon which the customer can base plans or future actions. Pertinent and useful information can generally be reduced to a few simple terms or relationships. If these relationships are clearly defined and then firmly established in the reader's mind, the technical proposal volume will provide a solid basis for the remainder of the proposal.

THE STATEMENT OF WORK: BASIS FOR THE TECHNICAL PROPOSAL

The basis for the technical proposal is the proposed contract statement of work (SOW), which is normally provided in the request for proposal. The contract statement of work should be correlated as closely as possible with the work element structure (1) to organize the work and divide it among the available performers and

(2) to estimate and collect the resources required to do the job. (See Chapter 8, "The Organization and Management Proposal," for details on the development of the work element structure.) If feasible, the statement of work and work element structures should be closely correlated with schedule elements. In some proposals it is feasible to correlate the statement of work, work elements, and schedule elements on a one-to-one basis. This correlation will permit a better understanding of the proposal by the evaluator and will greatly enhance the process of developing resource estimates of the technical tasks to be performed.

THREE BASIC FUNCTIONS OF THE TECHNICAL PROPOSAL

Technical proposal preparation is accomplished within the framework of the proposed contract work statement and its work element structure and schedule. The technical proposal performs three basic functions: (1) it describes the work activity or work output; (2) it describes the means of accomplishing the work; and (3) it demonstrates an overall understanding of the job in relation to requirements.

Description of the Work

The work to be accomplished is described through the effective presentation of sketches and drawings, photographs, flow diagrams and schematics, written specifications, materials lists or drawing trees, and development plans. All of these methods of presentation are effective in presenting technical data, but their use must be orchestrated by skillful mixing and combination of pictorial, graphic, photographic, and text material to provide an understandable, readable, and easily evaluated technical proposal. It is in the combination of these presentation techniques or methods that the proposal comes the closest to resembling a sales brochure. Although elaborate and expensive graphics and photography are not required or desired, judicious use of these presentation techniques can do much to make a proposal more understandable, attractive, and technically credible.

Sketches and Drawings. The sketches and drawings in a proposal can range from a simple reproduction of a conceptual sketch by a principal investigator to an elaborate drawing tree or a photographlike airbrush rendering of the activity or product being proposed. Rough sketches are seldom acceptable for use in a proposal unless they are exact copies of an original scientist's notes on an important discovery or invention. In these instances they can be used to lend credibility to the originality of design or concept. The most acceptable and most cost-effective approach to technical proposal illustrative material is to include reduced photographically or manually enhanced copies of actual engineering or architectural drawings, or of drawings and graphics generated by computer-aided design and graphics software. The use of computer-aided design and graphics techniques will speed the proposal preparation process considerably, because computer-aided design equipment is capable of producing isometric drawings, cutaway views, exploded views, and perspective drawings. Through computer-aided techniques, the legends, dimensions, and notes on computer-generated drawings can be increased in size to permit legibility when the drawing is reduced to standard proposal page or foldout size.

Although special isometric, cutaway, and exploded views require highly competent engineering technicians, the use of selected special views will do much to enhance the evaluator's understanding and comprehension of a proposed product or project configuration. If engineering or architectural drawings are included in a proposal, only the top-level drawings—backed up by sufficiently detailed drawings to illustrate that in-depth design has been completed—may be desired. For large architectural projects and fixed-price manufacturing activities, however, it is desirable to present to the customer—or at least agree with the customer on—all drawings down to the lowest level of detail available.

Photographs. Although large numbers of photographs should be avoided in the technical proposal volume, there is no substitute for a photograph or three-dimensional computer graphic rendering when it is necessary to demonstrate that a given piece of equipment, model, mockup, component, or facility exists. Photographs do much to show the size, shape, and configuration of physical items that are a part of or will be engaged in the work activity or output. An effective way to use a photograph of a partially completed or prototype item is to combine photography with a skillful airbrush technique to develop a realistic picture of the completed product. Although this technique should not be used to mislead the customer into thinking an actual full-scale production item has already been completed, it will help the customer visualize the appearance of the final product and will improve credibility by offering a picture very close to that which will be the desired end product. Photographs are most commonly used in technical proposals to show laboratory breadboards, prototype models, mockups, and test equipment identical to or similar to that which will be used in the proposed project.

Flow Diagrams and Schematics. If relatively simple and straightforward in nature, line or shaded drawings or graphic renderings of flow diagrams and schematics are effective tools to be used as necessary to illustrate a technical proposal. Patterns can be used to indicate various fluids or pressures in hydraulic or pneumatic devices, and sequential flow diagrams can show flow or pressures in various modes of operation. Electrical and electronic schematics can be presented either in block diagram form or in wiring diagram form, showing each electrical or electronic component and its interrelationship with the others. Color coding is often useful in flow diagrams and schematics. It should be kept in mind that maintenance-manual simplicity is preferred when presenting complex material in a proposal. The technical proposal should strive to educate rather than to impress or confuse the reader with overly complex technical information. Flow diagrams and schematics must always be accompanied by a description of the symbols, legends, and conventions used in the construction of the illustration. Patterned or colored flow diagrams will show system components, system operational modes, fluid flow and valve action. Block diagrams can be used to show the interrelationship of major components and the number and complexity of operations. For service and technical support proposals, flow diagrams of processes, sequences of activities, and support functions help to display the proposer's understanding of the requirements.

Typical Specifications. The specifications of a work activity or a work output are written statements that depict its principal technical characteristics. Hence, reading and understanding the specifications are the most important parts of any new proposed technical activity. The accompanying lists of words ("Work Output Specifications" and "Work Activity Specifications") are typical characteristics that are specified for a work output and a work activity.

Work Output Specifications

Acceleration	Load-carrying capability
Braking	Maintainability
Capacity	Power requirements
Cleanliness	Range
Contamination resistance	Reliability
Cornering capability	Square feet
Density	Take-off distance
Depth	Shock resistance
Electrical output	Speed
Energy usage	Temperature resistance
Fuel economy	Tolerances
Height	Traction
Human factors	Vibration resistance
Humidity resistance	Volume
Interface requirements	Weather resistance
Landing distance	Weight
Length	

Work Activity Specifications

Design reviews	Number of documents
Labor hours	Number of trips
Maintenance frequency	Output rate
Mixture of skills	Number of problem reports

Since specifications are usually set forth in requests for proposals, it is important to know how to treat these specifications in the work description.

Keys to Meeting Specifications. Those experienced in proposal preparation insist that the most common pitfall in embarking on a new activity is failure to read and understand the specifications. There are five key steps to meeting the specifications for a new work activity or work output, as described below:

1. *Assess Relative Importance*. In reviewing the specifications for any new work activity or work output, the proposer should develop an assessment of the relative importance of each of the specified areas. The relative importance of meeting each area of specified performance in a technical activity is normally not stated clearly by the requestor; therefore, the proposer must make a judgment about which of the specified characteristics are most important and which are the least important. Invariably instances will be encountered when the specified requirements cannot be met and trades must be made to determine where a relaxation in requirements will be the least detrimental to the final output and the least detrimental to the proposer's being chosen to do the work. Hence, any compromises that must be made in design will be those that will be the least detrimental to the overall project. Usually some trends can be detected in the proposal request, but thorough technical homework is required to fully understand the implications of compromises made in meeting the specifications, both from a potential work performance standpoint and from a contract selection standpoint.

2. *Point Out Where Requirements Are Exceeded*. As part of the assessment of relative importance of the specification requirements, it also should have been determined if it is to the proposer's and the project's advantage to exceed certain specifications. If the work activity or output being proposed exceeds the specifications in these areas or in any area, it is wise to point out clearly and emphatically that the proposal exceeds specification requirements and indicate in which areas it does so. One must be careful, however, that designers or planners have not "gold plated" the desired output, because this may be viewed by evaluators as a lack of cost consciousness or as taking excessive technical risks. Exceeding performance requirements may be of such importance to the project that it should not only be discussed in the technical proposal, but it should be mentioned in the letter of transmittal and the proposal executive summary as well.

3. *Explain Specification Deviations*. If the proposed work activity or work output does not meet the required specifications in any area, the reason for this should be explained in the technical proposal. Perhaps a better solution to the problem, or an innovative approach that eliminates the need to meet certain requirements spelled out in the request for proposal has been developed. If so, deviation from specifications will be considered by the evaluator to be a strength instead of a weakness. Perhaps the stated requirement was unrealistic or unreasonable to begin with. This is another reason that a clear technical description of the reason for deviation will be judged as a strength instead of a weakness. At any rate, a well-written explanation of the deviation will be a convincing addition to the measure of credibility that the evaluators attribute to the technical proposal.

4. *Summarize Specification Adherence Information*. Information concerning the qualities or performance of the proposed work activity or output that do or do not meet those levels specified in the request for proposal should be summarized in a clearly visible format. The proposer should not depend on the proposal evaluator to ferret out this information. It is not necessary to wave any red flags that say the job cannot be done because specified requirements have not been met, but the

customer should be given an honest look at the capabilities being proposed in relation to what has been requested. A summary of the technical and performance specifications of the project in tabular format with an indication of which requirements were exceeded and which not met in the proposal is a handy way to summarize this information. It should not be done, however, before calling proper attention to the benefits of exceeding or not meeting requirements, or else the reader may lose interest before evaluating the full advantages of the proposed plan or design.

Types of Specifications. Specifications fall into five general categories: (1) product specifications, (2) qualification specifications, (3) acceptance specifications, (4) component specifications, and (5) process specifications. It is important for the proposer to recognize the differences and the interrelationships among these five categories of specifications.

1. *Product Specifications.* Product specifications specify how the product is expected to perform when put to its intended use. These specifications are used in promoting or advertising the item, as performance has been proven or exceeded either by continued use or by laboratory tests. Product specifications should not exceed acceptance specifications or qualification specifications, described below.

2. *Qualification Specifications.* These are the specifications or requirements that a product must meet in the laboratory before it is placed into general use or marketed as an end item. The proposer must be acutely aware of the qualification requirements because a qualification test program may be required prior to completion of the proposed work activity. Since qualification requirements are usually more stringent than field use requirements to assure a degree or margin of safety when the product is put to use, qualification testing can be expensive and time-consuming to the producer. In many proposals, the supplier is required to describe the qualification test program that demonstrates how qualification requirements will be met.

3. *Acceptance Specifications.* Acceptance specifications are usually applied to a work output (product or project) to identify how the customer will expect the item to perform before he or she will accept it into his or her system for payment. Acceptance specifications are less stringent than qualification specifications, but more stringent than product performance specifications. Acceptance specifications state the degree of sampling or inspection, the nondestructive testing that will be performed on the item before acceptance, the tolerances in expected performance, and quality-assurance requirements.

4. *Component Specifications.* Component specifications require that components of the final product meet certain qualifications, acceptance, and product performance standards, as in the requirement for tire tread thickness or wear on a cargo truck to be purchased by a trucking company. Occasionally the customer will not only require that the product itself meet certain performance requirements, but also that the components of the product meet a different but compatible set of requirements. Component specifications are used by the supplier of the total product to designate required qualities of the product's subsystems and components.

5. *Process Specifications.* Process specifications are used within a company to specify methods or techniques for treating, handling, assembling, machining, welding, soldering, painting, or otherwise manufacturing a product. Occasionally, standardized process specifications are imposed on the supplier by the customer. In these instances, the supplier must either use the customer's specifications in testing for compliance or submit to a customer inspection and process-specification review to determine if the supplier's process specifications meet the customer's requirements.

Materials Lists and Drawing Trees. A vital part of the description of any new work output or work activity is the materials list or parts list. This listing is sometimes tied to the drawing tree or drawing system through the use of a compatible numbering system. The use of part numbers that correspond to the drawing number of that part can eliminate confusion in determining what part belongs to what subsystem. If the part numbering system and the drawing numbering system are both tied to the statement of work and work element structure, confusion brought about by the use of several different numbering systems and procedures can be eliminated.

The proposer should be prepared to include a parts list for any product or project supplied and should also be able to explain, in the technical proposal, the materials and parts quality-control procedures, the materials and parts tracking and numbering system, and the drawing-tree numbering system if it relates to the materials and parts numbers. These systems should be organized and documented for submission as part of the proposal or for its use in maintaining an efficient and effective materials and parts inventory, tracking, and reorder system.

The Means of Accomplishing the Work

Every proposer should do some detailed planning and scheduling of the work to be done before submitting a technical proposal; generally, such information is submitted to the customer as an important part of the proposal. The information on planning and scheduling submitted with the proposal is sometimes included in the proposal's technical volume and sometimes in the management volume.

The key to providing the best possible work output with a given amount of resources is the timing of work activities to result in as little waste, duplication, overlap, and redundancy of effort and materials as possible. In mass production, a delay in the delivery of a material or part into the process flow or assembly line can hold up the entire work activity. Delays in production can rapidly escalate product costs, because many overhead and direct labor costs are constant despite fluctuating work output levels. Too-early delivery of a material or part to a process or assembly line can also cause inefficiencies, because of the need for online storage and handling of the yet-to-be processed material. In multidisciplinary activities or large projects, the timing of application of each unit of resource, whether it be a unit of labor or material, is important because it usually affects another work activity. For example, labor hours cannot be expended on material until the material arrives; manufacturing labor cannot be expended before engineering labor is expended to design the hardware; assembly cannot be completed until the manufacturing is complete; and so on. These interrelationships of resource elements make timing of work activities a most important factor in the overall development of a competitive technical proposal.

Delivery or Availability Keyed to Need Dates. The most important factor to consider in formulating an estimated schedule of any work activity or work output is to provide the work on the date or dates and at the rate or rates required by the market. The technical scheduler should have on hand the results of the marketing or planning analysis to show: (1) the goals and objectives of the work activity, (2) the action plan of supplying the work output, (3) the requirements of the delivered product or service, and (4) the work elements that make up the overall work activity. The marketing or planning analysis will show future delivery dates and rates that project the most probable future market needs. The technical scheduler will then use the future date of delivery of the work output as an end point and work back in time to the present to develop a detailed milestone schedule.

Developing a Schedule. Schedule elements are time-related groupings of work activities that are placed in sequence to accomplish an overall desired objective. Schedule elements for a process could be represented by very small time periods (minutes, hours, or days). The scheduling of a process covers the time the raw materials take during each step to travel through the process. The schedule for manufacturing a product or delivery of a service is, likewise, a time flow of the various components or actions into a completed item or activity.

A project contains distinct schedule elements called *milestones*. These milestones are encountered in one form or another in almost all projects: (1) study and analysis, (2) design, (3) procurement of raw materials and purchased parts, (4) fabrication or acquisition of components and subsystems, (5) assembly of the components and subsystems, (6) testing of the combined system to qualify the unit for operation in its intended environment, (7) acceptance testing, preparation, packaging, shipping, and delivery of the item, and (8) operation of the item.

Techniques Used in Schedule Planning. There are a number of analytical techniques used in developing an overall schedule of a work activity that help to assure the correct allocation and sequencing of schedule elements: precedence and dependency networks; arrow diagrams; critical path bar charts; the program evaluation and review technique (PERT), a tool pioneered by the Navy; and program evaluation procedure (PEP), the Air Force version. These techniques use graphic and mathematical methods to develop the best proposed schedule based on sequencing in such a way that each activity is performed only when the required predecessor activities are accomplished. A simple example of how these techniques work is shown in Table 7.1. Eight schedule elements have been chosen; the length of each schedule activity has been designated; and a relationship has been established between each activity and its predecessor activity. The resulting bar chart that is plotted after the analysis is shown on Figure 7.1.

TABLE 7.1. Schedule Relationships

Schedule Element	Title of Schedule Element	Time Required for Completion (Months)	Percent Completion Required
A	Study and Analysis	6	33⅓%
B	Design	8	50%
C	Procurement	8	50%
D	Fabrication	12	66⅔%
E	Assembly	12	100% plus 2 months
F	Testing	8	100%
G	Delivery	4	100% plus 4 months
H	Operation	36	100%

Notice several things about the precedence relationships: (1) some activities can be started before the preceding activities are completed, (2) some activities must be fully completed before their follow-on activities can be started, (3) some activities cannot be started until a given number of months after the 100% completion date of a preceding activity. Once these schedule interrelationships are established, a total program schedule can be developed by starting either from a selected beginning point and working forward in time until the completion date is reached, or by starting from a desired completion date and working backward in time to derive the required schedule starting date. If both the start date and completion date are given, the length of schedule elements and their interrelationships must be established through an iterative process to develop a schedule that accomplishes the job in the required time. If all schedule activities are started as soon as their prerequisites are met, the result is the shortest possible time schedule to perform the work.

Most complex work activities have multiple tasks that must be accomplished contemporaneously because they are interdependent. The sequence of interdependent tasks that take the longest time is called the *critical path*. The *schedule critical path* is developed by connecting all of the activities on the critical path. Construction of a schedule such as that shown in Figure 7.1 brings to light a number of questions. The first of these is, "How is the length of each activity established?" This question reflects on the cost estimate that will be included in the cost volume, since many costs are incurred simply by the passage of time. Costs of an existing work force, overhead (insurance, rental, and utilities), and materials handling and storage continue to pile up in an organization whether there is a productive output or not. Hence it is important to develop the shortest possible overall schedule to accomplish a job and to execute each schedule element in the shortest time and in the most efficient method possible. The length of each schedule activity is established by an analysis of that activity and the human and material resources available and required to accomplish it. The labor and materials estimating techniques described in Chapter 8 are used extensively by the estimator in establishing the length of calendar time

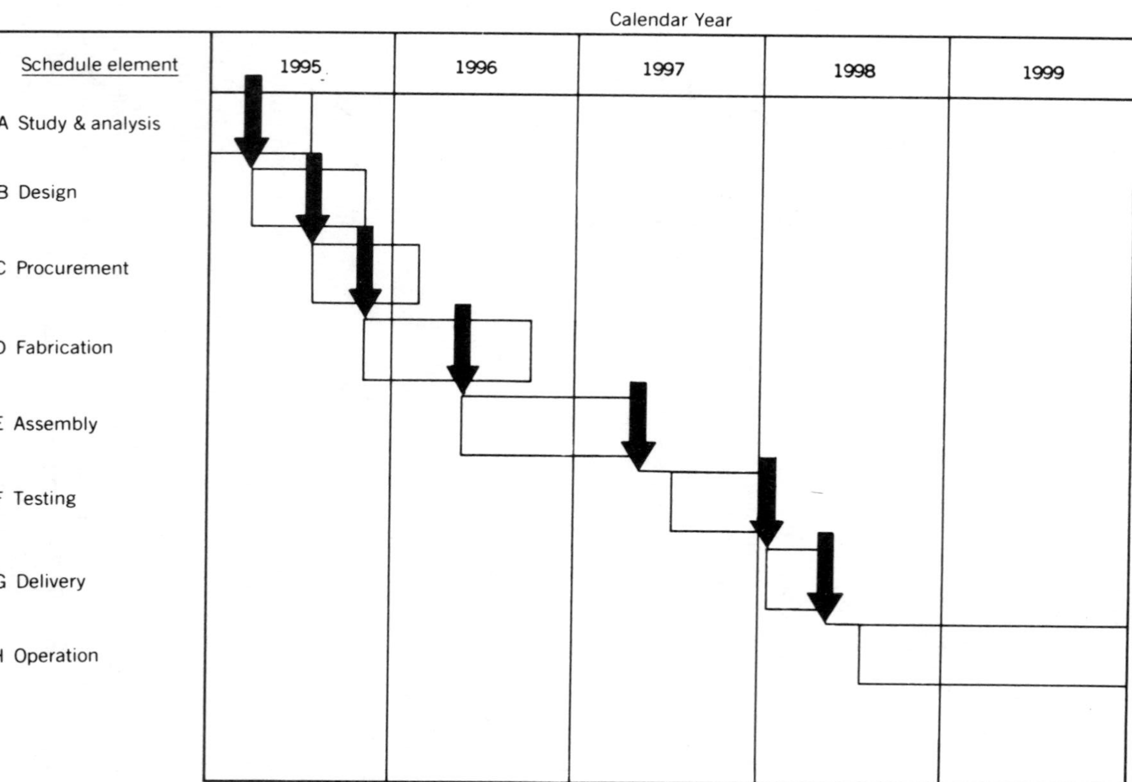

FIGURE 7.1 Scheduling a project.

required to accomplish a schedule activity as well as the labor hours and materials required for its completion.

A second question is, "What can be done if there are other influences on the schedule, such as availability of facilities, equipment, and labor?" This is a factor that arises in most proposal situations. There are definite schedule interactions in any multiple-output organization that must be considered in planning a single work activity. Overall corporate planning must take into account these schedule interactions in their own critical-path chart to assure that facilities, manpower, and funds are available to accomplish all work activities in an effective and efficient manner. A final question is, "How is a credible 'percent complete' figure established for each preceding work activity?" This is accomplished by breaking each activity into subactivities. For instance, *design* can be subdivided into *conceptual design*, *preliminary design*, and *final design*. If the start of the procurement activity is to be keyed to the completion of preliminary design, then the time that the preliminary design is complete determines the percentage of time and corresponding design activity that must be completed prior to the initiation of procurement. Scheduling activities performed during technical proposal preparation will have a significant bearing on the cost-effectiveness of the work and the competitive ranking of the resulting cost proposal. (See the section on "Automation in Proposal Pricing" in Chapter 11, which describes computer-aided scheduling features of several typical software packages.)

Schedule and Skill Interactions. Four factors established by the technical proposal team and by company management determine the optimum time-phasing of skills for any given work activity: (1) the attrition rate or turnover rate for each skill category, (2) the overall company or shop growth rate in numbers of personnel, (3) the merit salary increases (above inflation or cost-of-living increases), and (4) the initial and final mixture of skill categories and skill levels. A company with a high attrition rate or a high overall growth rate will have ample opportunity to add skills at lower salary levels to keep its labor costs low despite inflation, and it can provide room for advancement of its total work force through merit salary increases and promotions. A relatively static organization with a stable work force and continuous merit increases and promotions will find it difficult to stay competitive because of salary creep and skill-level creep. The only solution to staying competitive in this environment is to seek out other complex and difficult tasks that can be accomplished by a highly skilled, competent, and mature work force.

Any organization that is preparing a proposal must observe and consider the interactions of the job being estimated with other jobs that are in various stages of completion. Optimum planning and scheduling of work activities within an organization will result in full utilization of all available technical personnel. As the more difficult early phases of a project are completed, it is desirable to offload the experienced, highly paid individuals to other projects that are just entering their early phases. This macroscopic view of an organization's workload must continue to be considered when planning, scheduling, and estimating technical work activities.

Treatment of Long Lead-Time Items. Because of reduced raw materials availability and long transportation times, it is often necessary to place the order for materials, parts, and supplies long before the appropriate preceding activity is completed. Often a company must take a calculated risk that the materials, parts, or supplies possess the right composition, shape, size, and performance to meet the job specifications suitably; it has to order these items far in advance of expected delivery just to be sure that the subsequent milestones can be met. It usually pays to have a stock of scarce materials and parts on hand to avoid undue delays. The proposed work schedule should take into account the use of existing material stocks as well as the lead times required for their procurement and replenishment.

Make-or-Buy Decisions. When formulating a technical proposal, it is necessary to determine which items will be built in house (made) and which items will be purchased or subcontracted (bought). The best way to arrive at the make-or-buy structure is to (1) do a cursory analysis of the in-house workload, (2) compare vendor capabilities with equivalent in-house manufacturing capabilities, and (3) select the most attractive alternative. The amount of work a company subcontracts is a function of its expertise and skill in the required disciplines, its internal workload, and its overall company policy regarding subcontracted work. Generally, it is not effective to subcontract more than 60% of an organization's work, because the amount of technical management and direction for guiding and instructing subcontractors exceeds that required for that part of the project that remains in house.

Establishing Schedule Elements. If the overall milestone schedule for a work activity is considered as a level-one schedule, the schedule elements listed earlier in this chapter (study and analysis, design, procurement, fabrication, assembly, testing, delivery, and operation) can be considered as level-two schedule elements. Schedule elements can be subdivided into subelements just as work elements can be subdivided into subelements. These level-two schedule elements can be further subdivided and described as follows:

Study and Analysis. Establishing even the simplest operation will require some study, planning, analysis, scheduling, and estimating before the activity is started. Often, the careful planning and scheduling that has been done during the proposal preparation process must be redone once the contract is negotiated, because ground rules have changed, personnel and equipment may have been updated, and resources available may have changed. It is during this initial phase of a schedule that the proposer studies the requirements of the work output, analyzes alternative approaches for accomplishing the objectives of the work, and selects the most effective approach. This study-and-analysis phase includes definition of the technical, schedule, and cost aspects of the work and usually includes development of a preliminary concept of the process, product, project, or service. The preliminary concept could be represented by a block or flow diagram, sketches or artist's concepts, listings of requirements, preliminary specifications, and process or service plans. Each of these schedule subelements (which could also be duplicated as work elements or cost elements) represents a level-three schedule element and can be depicted on a bar chart.

Design. The second element of a schedule is usually the design of the process, product, project, or service. Design is usually subdivided into conceptual design, preliminary design, and final design. In large projects, these subelements are distinctly separated and are usually interposed with design reviews at the end of each design phase. In smaller projects, the subdivision is less clear, yet present. Conceptual design usually starts during the study and analysis phase, and completion of preliminary design is usually when orders can be placed for long-lead-time raw materials. Initiation of fabrication activities (if it is a product or project that is being estimated) must usually await the completion of the detailed design phase.

Fabrication, Construction, and Assembly. Fabrication, construction, manufacturing, and assembly of tools, parts, components, structures, and subsystems must start in time to have the appropriate hardware elements ready for testing and subsequent delivery. Part of the scheduler's task is to determine the optimum time to initiate fabrication, construction, or manufacturing and assembly of each hardware or facility item. The scheduler must also plan the best use of company facilities, equipment, and personnel in establishing a fabrication schedule.

The key to fabrication planning is the development of a process plan or operations sheet for each part. This sheet depicts the sequence for all fabrication operations. It shows the materials, machines, and functions required for each step of the fabrication process and will be used later as a basis for the development of the resources estimates of the cost volume. Typical functions included in the fabrication process are annealing, coating, coil winding, cutting, deburring, drilling, encapsulating, forming, heat treating, grinding, notching, milling, printing, plating, processing, punching, riveting, sawing, silk-screening, soldering, turning, welding, and wiring.

In the chemical process industries, typical operations that would be performed are atomization, baking, blending, coagulating, condensing, cleaning, diluting, distilling, evaporating, fermenting, filtering, freeze-drying, gasifying, polymerizing, precipitating, pumping, purifying, separating, and settling. The technical proposal should be developed with an awareness of the schedule impacts and implications of each of these activities.

In either the processing of a substance, the fabrication of a product, the conduct of a project, or the delivery of a service, each appropriate work operation or work function must be listed in sequence in the technical proposal or the technical proposal backup material to serve as a basis for resource estimates to be included in the cost proposal or cost volume. Operations sheets for any work activity should consider the flow of materials and parts into the fabrication process, the need for standard and special tooling, and the adaptability of the process or operations sheet to the later application of "standard time" data. Stand and labor times are pre-established average seconds, minutes, hours, or days required to perform a specific task.

Scheduling of the assembly process for fabricated items or parts is similar to the scheduling of the fabrication process. Operations or planning sheets should be developed for each assembly. Where multiple outputs are planned, it is wise to build a sample unit to verify the operations and standard times required to assemble the end item. In the planning of the assembly process, the assembly sequence should be itemized in detail, noting the location and method of attaching each part or subassembly and the requirement for design and construction of special jigs, fixtures, tools, and assembly aids. Time cycles of final assembly-line workstations should be balanced to provide economical, effective, and efficient flow of the product through each assembly operation.

Testing and Inspection. The quality and acceptability of any work output not only will depend on the care and precision with which the total job is performed, but also will be assured and verified through a testing and inspection activity. Simple products, processes, or services may require only a sampling inspection to verify a high-quality work output. The more complex work activities, such as complex, high-technology products or projects, will require the planning of several steps or phases in the testing and inspection activity. As mentioned earlier, the three key steps in the testing process are (1) performance or development testing, (2) qualification testing, and (3) acceptance testing. Performance or development testing is done during the early phases of many new products' or projects' life cycles to prove that the items will fulfill their intended function. Often this development testing is repetitive and interspersed with design and manufacturing changes that make the item work better, perform more efficiently, require less maintenance, and/or cost less to produce. Once the iterative cycle of testing and design is completed, the item is subject to a qualification test program. The final product design is subjected to a prespecified test sequence in which various conditions are imposed on the item to prove that it will work in a wide variety of environments and operational conditions. Qualification testing usually subjects the item to more severe environments than does the next step, acceptance testing.

Once the qualification tests are completed, the end item production is started. As units are completed, they are subjected to acceptance tests to verify that each unit will perform as advertised. Where qualification testing qualifies the design of an item and proves that it will operate under more severe conditions than those expected to be encountered normally, acceptance testing proves that a particular end item will perform under nominal conditions.

In planning any of the categories of testing, it is important first to determine the best overall testing method to demonstrate performance capabilities. It is in test planning that cost targets come strongly into play because the actual verification of many design parameters or goals is time-consuming and expensive. Examples are lifetime testing and reliability testing. It is usually not practical or cost-effective to test an item for two years to demonstrate a two-year lifetime. Methods of accelerated testing and sampling must be proposed to make this type of testing practicable.

Delivery and Operation. An often overlooked schedule element during proposal preparation is the delivery schedule. The importance of packaging; preserving; shipping; storage; handling; and supplying of parts, materials, and end items has been recognized recently by the emergence of a whole new profession called logistics, which is devoted to these activities. There have been many instances of delay, damage, and equipment malfunction brought about by a failure to consider logistics during the delivery and operation phase of a work activity. The best cure for these problems is to obtain the advice and expertise of a qualified logistician when developing the delivery and operations proposal for a work activity. Using the appropriate expertise in this area will cause the proposer to avoid the possibility of unforeseen delays brought about by inadequate consideration of the logistic aspects of the job.

Quality Control and Reliability Functions. Most customers for high-technology, complex, or large work activities will either impose a series of requirements to assure product quality control and reliability, or will expect the proposer to originate and describe his or her own systems, procedures, and methods of performing these functions. Frequently, the respondent to a request for proposal will be asked to submit formal or informal preliminary documentation that describes quality control (sometimes called *quality assurance*) and reliability policies, procedures, and practices to be used in the performance of the proposed work. These documents may take the form of product-assurance plans, critical parts control plans, nonconformance review procedures, failure-mode and effect-analysis plans, contamination control plans, maintainability plans, and others, depending on who the customer is and what the project's requirements are.

Product-Assurance Plan. A brief synopsis of a product-assurance plan is usually required to be submitted with proposals as part of the technical or management volume or as an appendix to the proposal, and a final product-assurance plan is submitted before the work starts. This plan describes the organization and method for implementing the product-assurance program. Customers will be looking for the degree of receiving, in-process, and final inspection; the method of identification, treatment, reporting, and correction of nonconformances; the amount and type of inspection and control exercised at vendor and subcontractor plants; and the degree of autonomy and independence of the quality-control and inspection functions.

Parts Control Plan. The method of identification, inspection, and control of critical or prequalified parts or materials will usually be described in the technical proposal volume and then submitted as a plan before the work begins. Some customers have approved only certain vendors for parts and materials that go into their products, the approval having been given after a thorough familiarization with the vendor's own product-assurance program and/or qualification of the parts through extensive testing or field use. Identification of parts that will be acquired from qualified vendors, parts selection criteria, parts specifications, parts qualification procedures, parts failure analysis, parts traceability, parts handling and storage, and time cycle and age control are subjects covered in the parts quality-assurance program. Critical parts or subsystems are identified with various levels of criticality based on the potential impact of failure of the part or subsystem on mission success or end-item performance.

Nonconformance and Resolution Reports. The technical proposal should include a description of the method of detecting, reporting, and correcting nonconformances that are identified in the quality-assurance, reliability, testing, and/or verification programs. This description should include: (1) an indication of which nonconformance and resolution reports will be submitted to the customer for information; (2) a description of the content and format of the reports; and (3) a description of the method of following up on open discrepancies. If the customer requires any special reports of nonconformances that would indicate a safety or mission hazard on other concurrent programs or projects, the content of these reports should also be described in this section of the proposal.

Failure Mode and Effect Analysis. Various forms of hazard analysis, safety analysis, and failure mode and effect analysis are required by customers who are purchasing complex, high-technology hardware or software. The methods, techniques, and procedures that will be used in conducting and reporting these studies should be described in the technical proposal. The customer is usually interested in receiving assurance that failures that could result in fire, explosions, structural damage, personal injury, leakage, shock or mission malfunction are thoroughly investigated as to their potential probability of occurrence and their most probable consequences. It is in this section of the proposal that any preliminary information relative to these hazards is described along with the method to be used in continuing to identify and report them.

The Contamination Control Plan and Others. The contamination-control plan is only one of many other plans that may conceivably be required by a potential customer. In complex medical, space, or military applications, cleanliness and contamination-control techniques and specifications have been developed to reduce the potential effect of foreign particles and substances on the operators or users of delicate devices as well as their effect on the device itself. Plans for controlling or reducing dust, dirt, moisture, foreign organisms, and other particles during the manufacture, testing, and packaging of the item or items for shipment are described in the technical proposal volume.

Maintainability Plan and Other Qualities. Although "maintainability" is not always included under the heading of quality assurance and reliability, it is included in this section as an example of other qualities that are often addressed in the technical proposal, which may include producibility, manufacturability, operability, etc. Maintainability is one of the more important ones, as it must be considered in the design of the item to be delivered and therefore must be considered at the outset of the project. Ability to maintain the item easily during its lifetime is a major consideration for purchasers, who are considering the life cycle cost of an acquisition as an important factor in their purchase decision. As part of the technical proposal, the proposer should describe his or her approach to assuring that product maintainability is designed and built into the item. The proposer's approach to the other qualities should also be described in the technical proposal.

Systems Engineering. Most complex, high-technology projects employ a discipline known as systems engineering, which deals with the application of scientific, engineering, and engineering management skills to the planning and control of a totally integrated project. A multielement discipline, systems engineering includes: (1) transformation of an operational need into a technical description of performance parameters and a configuration through an iterative process of definition, synthesis, analysis, design, test, and evaluation; (2) integration of related technical parameters in order to assure compatibility of all physical, functional, and program interfaces in a manner that optimizes the total project definition and design; and (3) integration of the necessary engineering disciplines (e.g., maintainability, reliability, producibility) into the total engineering effort. The technical proposal should describe the anticipated systems-engineering effort required to deliver the proposed work and should describe how the systems-engineering effort will be carried out. Systems engineering includes such disciplinary elements as system and subsystem analysis, dynamic analysis, thermal analysis, electrical power and energy analysis and control, mass properties analysis and control, materials and processes compatibility analysis, and establishing and controlling instrumentation lists (see Blanchard, 1991).

The reader should take special note at this point in preparing a proposal for continuing engineering activities, that there is an important new concept that is emerging as a highly effective competitive procedure. It is called *Concurrent Engineering*. Concurrent engineering, made increasingly effective by emerging improvements and innovations in computer-aided engineering and electronic communications, includes the early introduction of all engineering disciplines at the very beginning of the engineering process. Through the use of recently developed software and high-speed computers, concurrent engineering can start as early as the initial formulation of engineering requirements, and proceed throughout an engineering or scientific project through to completion. Use of the most recent automated tools and methods in designing, simulating, testing, manufacturing, and delivering high-technology hardware and software, coupled with an innovative and aggressive application of concurrent engineering principles will result in your firm having a dramatic competitive advantage over firms who do not use these tools and approaches. (See Hartley, 1991).

Usually, all of these disciplines are summarized in a systems engineering plan, which is submitted in condensed or preliminary form in the proposal and then fully developed and implemented as the work commences. The proposal should address each of the elements of the systems-engineering activity and describe how and by whom they will be accomplished.

Systems and Subsystems Analysis. Systems and subsystems analysis interrelates and integrates all of the hardware, software, and supporting equipment of the system from an analytical standpoint to provide assurance that the various parts of a system will work together as a whole. Interfaces are identified and analyzed to assure that any adjacent parts, components, subsystems, or elements fit and work well together and that they do not adversely affect each others' performance. Systems and subsystems analysis includes systems trades, stress analysis, fracture mechanics analysis, dynamic analysis, mechanisms analysis, thermal analysis, electrical analysis, data transfer analysis, radiation and magnetic field analysis, maintenance analysis, and operations analysis. Other special analytical studies may be required for specific types of systems, such as venting analysis, transportation analysis, environmental impact analysis, and so forth. The proposal should include a description these analytical disciplinary elements as they occur in the systems engineering, activity, the approach to be taken in the analysis, the expected or targeted results, and the methods proposed for verifying the analysis and implementing the results in the project.

Dynamic Analysis. The dynamic analysis of the total system may consist of analog, digital, or manual computation of the response and resistance to shock, vibration, and stress under various modes of nominal and unusual operation. Thermal and electrical characteristics of the system as well as mechanical characteristics may also be subjected to dynamic analysis. The proposal should describe the techniques and methods to be used, reference the mathematical or computer models to be employed, and describe the method of verification and correlation with actual test or operational results.

Thermal Analysis. If static, dynamic, or steady-state thermal analysis is to be performed to establish thermal interfaces or performance, the mathematical models should be named or the concept to be used in their development should be described.

Electrical Power and Energy Status and Control. The systems engineering discipline is usually charged with the job of establishing and tracking electrical power and energy usage budgets. Where multiple power sources or multiple power consumption points exist in a system, trends above power and energy usage budgets are recognized and recorded so that power allocation can be readjusted accordingly. Power and energy utilization timelines are developed to assure that power supplies are commensurate with consumption requirements during system operation. If the system is sufficiently complex to warrant an electrical power budget control activity, the technical proposal should describe anticipated power budgets; provide electrical diagrams, schematics and lists; and describe how the power budget will be tracked and adjusted.

Mass Properties (Weight) Control. If weight or mass is a critically important specification requirement, the technical proposal should recognize this and provide a plan for keeping close track of the projected weight of the system and its components as the work proceeds into final design and production. Some systems are sufficiently complex to require computerized tracking of the projected design weight of each component as the design proceeds, updating this weight as prototypes of the parts are built, and adjusting the total weight budget or allocations as appropriate. If such a system is to be used, it should be so stated and described in the proposal.

Materials and Process Control Plans. Often, the materials to be used in the makeup of high-technology systems are sophisticated or newly developed, and their interaction with other materials or substances in the system is not well-known or defined at the time the system is designed. The systems engineering function may need to establish a method of keeping track of the various types of materials to be used in the system to assure compatibility. This materials control plan is also used for tracking rare or scarce materials and helps assure material availability. Any material and process control plans that are specially established, or existing procedures that are used, should be described in the proposal.

Instrumentation Lists. Where there are many channels of instrumentation or measurement required for a complex system, a budget and tracking system may also be established under the systems engineering function to track, establish, and approve instrumentation measurements. Such information as measurement number, name, location, purpose, transmission mode, range of measurement, calibration requirement, bandwidth, etc., will be recorded on a tracking system, and measurements will be given a priority within an overall instrumentation budget. The proposal should describe such a procedure if it is to be used in the proposed work or project.

In addition to the above functions, the technical proposal should also describe other activities that are usually performed under the systems-engineering discipline. Among these are the development of an engineering requirements document early in the project and the development of a system description handbook during the final phases of project completion. If the project requires maintenance or operations handbooks or documentation, the technical content, format, and purpose of each of these documents should be described in the technical proposal along with the method of compiling, reviewing, and assuring the technical accuracy of these documents.

Configuration Management. When complex systems, products, or projects are undertaken, they invariably generate a large number of interfaces, not only in the hardware or software itself, but between various organizations that are contributing to the work. The abundant interfaces that grow out of these projects must be documented and controlled so each performer will know what the interface is with the adjoining hardware, software, or organization. The discipline that has developed from the need for systematic definition and control of interfaces is termed *configuration management*; it is important to address this subject in either the technical volume or the organization and management volume of the proposal.

Configuration management is treated as part of the technical volume because of its close kinship with systems engineering. In reading a description of the configuration management function proposed to accompany a new enterprise, the customer may want to know the details of how the function will be performed or the customer may only want to know that the proposer employs a disciplined method to assure hardware and software interface compatibility (and organizational interface compatibility, where appropriate). The more inquisitive customer will want to know the organizational setup of the configuration management function and how it relates to the project management organization. The customer is interested in the supplier's project manager having maximum control over the end item's configuration. Other things the customer will want to know are: (1) the methods and documentation (specifications, drawing practices, engineering release systems, etc.) that will be used to establish the system requirements and system baseline; (2) the policies and procedures to be used in controlling changes to the baseline; (3) the configuration accounting system, including the method for making and approving configuration changes; (4) the methods to be used in controlling changes to vendor- or subcontractor-supplied items; (5) the major milestones for implementation of configuration management; and (6) the plans that will be (or have been) established for conducting and supporting appropriate configuration management and design reviews.

The configuration management function has charge of the all-important systems specifications and interface drawings, so the customer will want to know how often and by what methods these documents will be updated and if the customer will have a say in the approval of changes that may affect external interfaces. If interface control documents (ICDs) are to be used, the method of initiation, control, approval, and issuance of these documents will be important. The customer will also want to know if a configuration control board has been or will be established and who the members are or will be, by title and function.

Test Planning and Verification. The test planning and verification program is a vital part of any project in which a significant amount of new development work is a part of the activity being proposed. An assembly and verification plan, test plan, or development plan should spell out all of the tests to be performed, citing their objectives, conditions, and test limits for subsystem tests, system tests, development tests, qualification tests, and acceptance tests to be conducted during the program. The test plan includes all or a part of the following, depending on the complexity of the project and the desires of the customer:

1. A description of the organization, policies, methods, and controls to be implemented. Includes general test requirements, test levels, and durations.
2. Descriptions of the verifications to be performed, including prerequisites, constraints, test objectives, and methods.
3. Test-item configuration identification and quantities to be tested.
4. A detailed time-correlated sequence of verification operations from component level through subsystems' and systems' final acceptance and integration.
5. Description, planned usage, and scheduling of the support equipment, facilities, and tooling necessary to execute the verification activity.

The verification plan is usually accompanied by a test-and-checkout requirements and specifications document that is prepared after the program work is initiated. This document need not be included in the proposal but its existence and content should be pointed out to the customer at the time of proposal submittal as part of the prepared supporting material on verification testing.

The document will identify each test requirement, specification, and constraint applicable to the various functional and environmental tests required for qualification and/or acceptance during the buildup, subsystem, and system test activities. Specifications include allowable tolerance for standards or judgment to be used in determining acceptable performance. Test types, levels, and durations are included. Since qualification test levels are generally more stringent than systems acceptance test levels, the qualification test requirements include test level margins and factors of safety.

On completion of verification tests, most customers will want to know that they will receive a detailed test report.

Demonstrating an Understanding of the Requirements

When preparing a technical proposal, the question always arises as to how one can best demonstrate an understanding of the requirements. This can be done in several ways. First, it is important to demonstrate that one is fully knowledgeable about the work activity or work being proposed. Knowledgeability will become evident to the evaluator as he or she reads through the description of the work and the description of the means of accomplishing the work. In this review, the evaluator will look for evidence that the firm has a complete understanding of the theory of operation, the disciplines needed to accomplish the job, and the facilities and equipment needed to perform the job efficiently and effectively. The proposing company can convey the fact that it has a thorough understanding of the tasks to be performed by comparing these tasks to those that have been performed successfully by the company in the past, describing any differences, improvements, or similarities. An understanding of the total field of effort can be conveyed by mentioning or discussing what others in the field have done or are doing. An awareness of interfaces with other mating hardware, other activities, and other organizations involved in the project should be demonstrated.

Second, the orderly and logical presentation of the overall schedule and the depth of penetration into the schedule will help to show that there is a full understanding of the scheduling aspects of the job. Typical detailed scheduling activities down to a lower level in the project can be presented in the technical proposal to demonstrate the fact that this planning has been done and that the schedule interactions and interrelationships are understood. Although it is not always necessary to show all detailed schedules, those detailed schedules for the key or critical areas of the work should be included.

Third, an understanding of the job can be demonstrated by accurate and realistic estimates of the resources required to do the job. Although the cost estimates are covered in the cost volume, the magnitude of and rationale for these estimates will be an indicator of the technical knowledge of the magnitude of the job and the relative importance of its various parts. It is for this reason that a close coordination and team relationship exist between the technical and cost proposal teams. Since the technical person who is going to perform the work will be making the original labor-hour or material estimates, the size and distribution of these estimates reflect the understanding of the technical team.

DESCRIPTION OF CUSTOMER INVOLVEMENT

An important part of the technical proposal is the description of customer involvement in the work and in the review process. This description can be in a separate section of the technical proposal or it can be interspersed within the various technical work descriptions. The customer will want to know when and what type of technical reviews will be held, the types and frequencies of technical reports, and the means of day-to-day communication between technical counterparts within the supplier's organization and the seller's organization. A description of all technical documentation

I Summary
II Introduction
III Proposed program schedule
  Statement of work
  System design
  Hardware/software configuration
  Optional equipment
  Operational functions—capabilities
  Production
  Reliability—quality control
  Related services
IV Deliverable items and services
  Reports and documentation
  Equipment
  Optional equipment
  Installation, test, checkout
  Training
V Facilities

FIGURE 7.2 Development proposal outline.

that will be submitted during the performance of the work, as well as formal and informal technical meetings, should be included in the technical proposal.

DISCUSSION OF PROBLEM AREAS

A frank and open discussion of problem areas is an asset rather than a liability to a technical proposal. The technical approach should discuss any problem areas that might be encountered, and their related solutions. These problem areas generally

I Introduction
II Hardware/software configuration
  Optional or additional equipment
III Deliverable items and services
  Production items
  Spares
  Quality control
  Testing
  Reliability
  Support equipment
  Related services
  Training
IV Facilities

FIGURE 7.3 Hardware/software proposal outline.

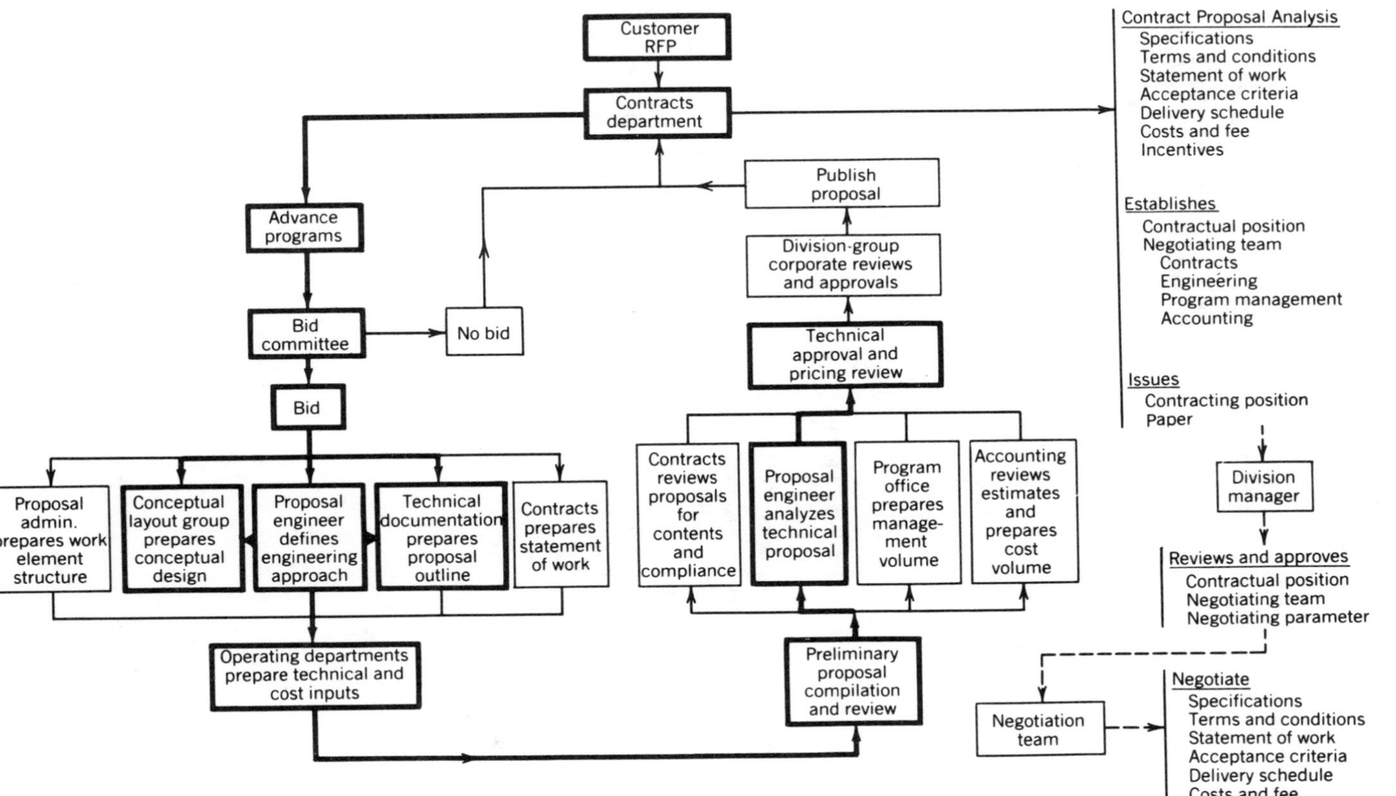

FIGURE 7.4 Proposal flowchart showing technical proposal activities (bold lines).

can be described in two ways. The first type of problem is comprised of difficulties that will have to be overcome in the performance of the job. Generally, the discussion of such problem areas addresses itself to a definition of the purpose, scope, and objectives of the program; a discussion of the strengths and weaknesses of alternative approaches; and a generalized description of the results that can be expected from the recommended solution. This section provides an opportunity for discussing the solutions that may be proposed by competitors and diplomatically but logically refuting them in advance.

The second type of problem includes those difficult points that would exist if they had not already been solved by the proposing firm. These solved problems represent a competitive advantage. Assertions that these problems have been solved must, however, be supported by factual evidence of prior performance. Showing the means to avoid pitfalls gives a proposer an opportunity to convince the customer that effort will be expended realistically and that technical accomplishment will be measured accurately.

TYPICAL TECHNICAL PROPOSAL OUTLINES

Since every requesting organization has its own desires about proposal outlines, it is difficult to develop a standardized outline. The request for proposal will usually specify a proposal outline. If not, the bidder is free to choose one. For reference purposes only, two typical technical proposal outlines are included as Figures 7.2 and 7.3. Figure 7.2 is a typical technical proposal outline for a development proposal and Figure 7.3 is a typical technical proposal outline for a hardware proposal. Figure 7.4 illustrates the technical proposal steps in an overall typical proposal flow chart.

8
THE ORGANIZATION AND MANAGEMENT PROPOSAL

And it shall come about that when the officers have finished speaking to the people, they shall appoint commanders of armies at the head of the people.

—Deuteronomy 20:9

In the organization and management volume, the evaluator will be searching for the real competence, personality, and character of the proposing company as well as any company or organization that will be supporting it in the work. The evaluator will be looking for sound, convincing, documented evidence that the organization will perform the work with diligence and dispatch, and that the persons, systems, and structures established for accomplishing the work have a high probability of success. The evidence that will be provided in the organization and management proposal regarding the company's management policies, company organization, project organization, key personnel, plant facilities and equipment, and past experiences and successes is designed to demonstrate the competence, willingness, and capability to manage the job and to keep it on schedule and within cost estimates. This volume of the proposal also contains a plan of how the company and its supporting contractors will apply organizational resources to carry out the proposed work.

The organization and management proposal addresses itself in detail to the structure of the company, the structure of the work, and the company's philosophy of managing the work. This discussion is important because it allows the customer to understand how the organization will exercise straight-line control in supporting the program. It defines the sources and channels of authority within the organization. It shows how the program manager can use authority and internal alliances to accomplish the program's objectives, and it reveals the rationale behind a company's performance of such management functions as planning, organizing, staffing, directing, and controlling.

THE WORK BREAKDOWN STRUCTURE

The work breakdown structure (WBS), whether specified by the customer or developed by the proposer, is an indispensable feature of a proposal for complex work activities or work outputs. Although it is usually described in the organization and management volume of the proposal, the other two volumes (technical and cost) share the use of the work breakdown structure to interrelate the technical, management, and cost aspects of the work.

The work breakdown structure is correlated with the company's organizational structure and with the proposed contract work statement. It serves as a framework for managing the work and for collecting, accumulating, organizing, and computing direct costs. It is also used for reporting technical progress and related resources throughout the lifetime of the work. There is considerable advantage in using the work breakdown structure and its accompanying task descriptions as the basis for organizing, scheduling, reporting, tracking, and managing the project. Hence, it is important to devote considerable attention to this phase of the overall proposal preparation process. A work breakdown structure is developed by dividing the work into its major elements, then breaking these elements into subelements and sub-subelements, and so on.

The purpose of developing the work breakdown structure is sixfold:

1. To provide assurance that all required work elements are included in the work output
2. To reduce the possibility of overlap, duplication, or redundancy of tasks
3. To furnish a convenient hierarchical structure for the accumulation of resource estimates
4. To gain greater overall visibility as well as depth of penetration into the makeup of the work
5. To provide a lower level breakout of smaller tasks that are easy to identify, staff, schedule, and estimate
6. To provide a work structure that can be correlated with the staff organizational structure to ensure that all tasks are assigned to groups that will perform the work

Hierarchical Relationship of a Work Breakdown Structure

A typical work breakdown structure is shown on Figure 8.1. Note that the relationship resembles a hierarchy where each activity has a higher activity, parallel activities, and lower activities. A basic principle of a work breakdown structure is that each work element is made up of the sum of the elements below it. No work element that has lower elements exceeds the sum of those lower elements. The lowest elements are described in detail at their own level and sum to higher levels. Many numbering systems are feasible and workable. The multiple decimal numbering system is one that has proven workable in a wide variety of situations. The level

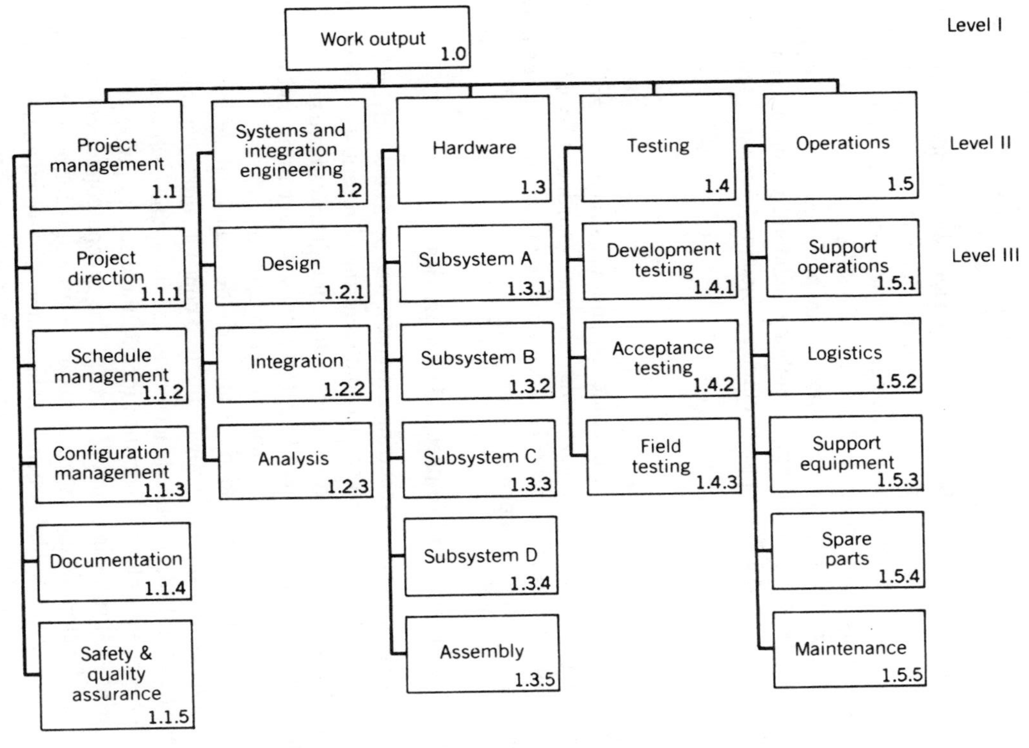

FIGURE 8.1 Typical work element structure.

is usually equal to the number of digits in the work element block. For example, the block numbered 1.1.3.2 is in level four because it contains four digits.

Functional Elements Described

When subdividing a work activity or work output into its elements, the major subdivisions can be either functional or physical elements. The second level in a work breakdown structure usually consists of a combination of functional and physical elements if a product or project is being proposed. For a process or service, all second-level activities could be functional. Functional elements of a production or project activity can include activities such as planning, project management, systems engineering and integration, testing, logistics, and operations. A process or service can include any of hundreds of functional elements. Typical examples of the widely dispersed functional elements that can be found in a work breakdown structure for a service are: advising, assembling, binding, cleaning, fabricating, inspecting, packaging, painting, programming, projecting, receiving, testing, and welding.

Physical Elements Described

The physical elements of a work output are the physical structures, hardware, and products or end items that are supplied to the customer. Starting at level two in the work breakdown structure, the physical elements can be broken down or subdivided into systems, subsystems, components, and parts.

Treatment of Recurring and Nonrecurring Activities

Most work consists of both nonrecurring activities: one-of-a-kind activities needed to produce an item or to provide a service and recurring or repetitive activities that must be performed to provide more than one output unit. The resource requirements (man-hours and materials) necessary to perform these nonrecurring and recurring activities reflect themselves in nonrecurring and recurring costs.

Separation of nonrecurring and recurring activities can be done in two ways through the use of the work breakdown structure concept. First, the two types of activities can be identified, separated, and accounted for within each work element. Resources for each task block would then include three sets of proposal estimates: (1) nonrecurring, (2) recurring, and (3) totals. The second convenient method of separation is to develop identical work breakdown structures for both nonrecurring and recurring activities. A third structure, which sums the two into a total, can also use the same basic work breakdown structure. If there are elements unique to each activity category, they can be added to the appropriate work breakdown structure.

Work Breakdown Structure Interrelationships

Considerable flexibility exists in the placement of both physical and functional elements in the work breakdown structure. Because of this, and because it is necessary to define clearly where one element leaves off and the other takes over, a detailed definition of each work activity block in the form of a work breakdown structure dictionary must be prepared and included in either the organization and management volume or in the technical volume. The dictionary describes exactly what is included in each work element and what is excluded; it defines where the interface is located between two work elements, and it defines where the assembly effort is located to assemble or install two interfacing units.

A good work breakdown structure dictionary will prevent overlaps, duplications, and omissions because detailed thought has been given to the interfaces and content of each work activity.

Skill Matrix in a Work Breakdown Structure. When a work breakdown structure is being constructed, it should be kept in mind that each work element will be performed by a person or group of people using one or more skills. There are two important facets of the labor for each work element: skill category and skill level. Skill categories vary widely and depend on the type of work being proposed. For a residential construction project, for example, typical skill categories would be bricklayer, building laborer, carpenter, electrician, painter, plasterer, or plumber. Other typical construction skill categories are structural steelworker, cement finisher, glazier, roofer, sheet-metal worker, pipe-fitter, excavation equipment operator, and general construction laborer. People in professions such as law, medicine, accounting, administration, project management, engineering, systems analysis, writing, and so forth are called on to do a wide variety of direct-labor activities. These people are included in professional skill categories.

Skill level, on the other hand, depicts the experience or salary level of an individual working within a given skill category. The skilled trades are often subdivided into skill levels and given names that depict their skill level; for example, carpenters are broken down into master carpenters, carpenters, journeymen carpenters, apprentice carpenters, and carpenter's helpers. Because skill categories and skill levels are designated for performing work within each work element, it is not necessary to establish separate work elements for performance of each skill. A work breakdown structure for home construction would not have an element designated "carpentry," because carpentry is a skill needed to perform one or more of the work elements (e.g., roof construction, wall construction). The skill mix is the proportion of each of several skill categories or skill levels that will be employed in performing the work.

Organizational Relationships to a Work Breakdown Structure. Frequently all or part of a work breakdown structure will have a direct counterpart in the performing organization. This interrelationship between the work breakdown structure and the organization of the company or project should be described in the organization and management volume of the proposal. In the proposal preparation process, early assignment of work elements to those who are going to be responsible for performing the work will motivate them to do a better job of proposing the work and will provide greater assurance of completion of the work within performance, schedule, and cost constraints because the functional organizations have set their own goals. Job performance and accounting for work accomplished versus funds spent can also be done more easily after the contract is awarded if an organizational element is held responsible for estimating the resources required for a specific work element in the work breakdown structure.

Treatment of Work Elements

Work elements are chosen for a work activity in a way that will make it easy to assign the prime responsibility for a work element to a segment of the organization that will do the work. The work elements conceivably could be identical to the schedule elements. This poses the difficulty that a work element for, say, procurement, would have very little interaction with other parts of the proposed work despite the fact that the procurement function is there to support multiple functions (more than one schedule element).

The interplay among the three factors of schedule, cost, and performance is paralleled by the interplay among schedule elements, cost elements, and work elements. This interplay among the three program factors can be maximized by what is known as a "matrix" of organization and management responsibilities. The choice of different schedule elements, cost elements, and work elements creates a matrix that forces equal attention on all three of the important program factors (schedule, cost, and performance). In a large high-technology project, it is highly desirable to define the elements of schedule, cost, and work separately and to have them interact with each other in the form of a three-dimensional matrix.

A work element list that has proven successful for a large number of projects is the following:

Project management

Systems engineering and integration

Subsystems development or acquisition

Assembly and verification

Operations support and logistics.

Notice that these work elements can be aligned easily with an organizational structure. For example, the project management activity can be accomplished by the project management organization; the systems engineering and integration can be performed by the systems engineering and integration organization; the subsystems development or acquisition can be performed by the subsystems development or procurement organizations; and so forth. Assignment of work elements to more than one organization is possible but is not good practice unless a lead organization is chosen. Choice of a lead organization for performing a work element provides greater management control of schedule, cost, and performance factors for that work element. Clearly explaining these interrelationships in the organization and management volume of the proposal will do much to lend credibility to the proposal.

THE PROJECT PLAN

The project plan should be defined in detail within the organization and management volume of the proposal. This plan discusses the means available (or to be created) within a company to achieve the objectives of the program. In general, the project plan describes the organizational base from which the program personnel and other resources will be drawn, the schedule to which they will be required to perform, and the manner in which funds and effort will be controlled to ensure accomplishment of program goals. The overall project plan is based on the technical schedule developed as part of the technical proposal. To win contracts (or even get a chance to bid) in today's competitive market, these descriptions must be based on existing capability—the inherent strength of the company. Organizational arrangements may be described under the name of several management structures, such as matrix management, project management, or divisional management. The inherent capabilities of the company, regardless of how they are described, must include the ability to create and staff the program organization.

In the organization and management proposal, established policies and procedures that constitute the proposer's means of direction should be summarized. These means should be condensed in the proposal to illustrate the way in which the management plan will be implemented. Items such as the work breakdown structure, master schedule, work authorization production orders, sales orders, purchase orders, and indeed all of the directive means that will be called upon by the program manager should be discussed and their application defined.

The organization and management proposal should address itself to the resources available to control effort within the organization. Effort is described in terms of cost, time, personnel, and technical achievement. Discussions about project controls are more fruitful if they concentrate on providing the customer with hard facts that prove that the proposer can respond to requirements and to newly discovered needs of the program, and that controls ensure traceability of decision-making. The customer needs assurance that the controls provide a means of identifying expenditure, effort, and other outputs of the plan. The planning discussion for the project should include the means of using information generated by these controls. These include detailed descriptions of reports, liaisons, and interchanges that will be accomplished for and with the customer. These measures naturally must include, at a minimum, all those reports and liaison efforts described in the request for proposal. Frequently, one can create a very favorable impression on the customer by describing additional means that will be used to inform the customer and to ensure that the customer's influence can be brought to bear as the program develops.

Financial, technical, and performance measurement reports normally directed to the project manager(s), line management, and customer's program office are, of course, vitally important to effective management of the program. As such, the plan for submitting them internally to the bidder's management and to the customer should be shown. This feedback of information provides data upon which decisions are made concerning the program.

It must be kept in mind that the objectives of employing a planning, scheduling, and progress tracking or performance measurement method during work performance are to achieve optimum skill and equipment use throughout the duration of the work, to level off peaks and valleys in workload, and to develop an optimum interface with other ongoing or proposed work. Optimum skill and equipment use will be a deciding factor in making a proposal cost-competitive and will provide evidence to evaluators that forethought has been given to achieving efficiency in performing the work. Leveling off peaks and valleys in workload will prevent the inefficiencies that could occur through the requirement for rapid fluctuations in the work force. Interphasing of the proposed work with other ongoing or proposed work will permit shared use of equipment, facilities, and personnel between work activities, minimize cross-project interference, and facilitate contingency planning based on the capture potential of the proposed work and of other pending or uncertain acquisitions.

Visibility

Planning, scheduling, and progress tracking or performance measurement must be proposed in a way that maximum visibility is provided to decision-makers in the performing organization during planning and performance of the work. Visibility can be provided in one or more ways as long as the material is presented in a manner that will provide "at-a-glance" status. An overly complex format will only confuse the observer. PERT (program evaluation and review technique) diagrams that have thousands of events and activities interrelated by a seemingly endless web of intricate interconnecting lines, although they may be effectively used for internal planning, do not impress customers or proposal evaluators. They are of little value in the proposal because they require detailed study to bring about an understanding of project planning and progress, even though they may be of value to planners. The most understandable and quickly observed format is the time-oriented bar chart (explained earlier in Chapter 7) that shows both events and activities on a calendar background. For simple scheduling situations, a listing of planned and actual dates is sufficient.

With the advent of video graphics display systems, real-time planning and scheduling status can be provided to managers and supervisors as well as to other select individuals involved in the planning and scheduling process. These systems are supplementing and replacing the more traditional reporting and display media (hard copy, charts, display panels, slides, and projected transparencies). The types of systems to be used for internal and external management visibility should be described in the organization and management proposal. The characteristic of visibility should also be used as appropriate to provide the customer with planning, scheduling, and progress information.

Competency in rescheduling work to correct for deviations from a plan is dependent upon the scheduling system characteristic that visibly compares actual progress or status with planned progress or status. Whether or not schedule status information will actually be reported periodically to the customer, the customer must be aware that the proposer has a means of tracking actual versus planned work and will constantly take actions to assure that the total job will be completed on time and within the resources proposed.

THE COMPANY'S ORGANIZATION AND TEAMING ARRANGEMENTS

The potential customer will want to know what specific organizational arrangements have been made to perform the work within the company or with teaming partners, what capabilities the company and its teaming partners possess, which of these capabilities will be committed to the proposed work, and how deeply the company's management is committed to a successful outcome. These subjects are normally addressed in the section on the company's organization and capabilities—specifically, in those related to the proposed work.

Relationship to Corporate Headquarters and Resources

Frequently, a source selection is influenced significantly by the reputation of the proposer in the appropriate field of work. If this reputation is not well-known, a considerable amount of detailed information may be required in the organization and management proposal. This information should include personnel, facilities, and equipment that are available at the corporate level to support the project and to assure that overall management and backup capability is available for the project organization. It should provide a discussion on how the project is expected to draw on corporate capabilities during day-to-day operation or during emergencies, and what degree of management and control will be exercised over the project by corporate entities.

An organizational chart of the corporation should be provided, and a description of how the proposed project relates to the corporate organization should be included. If there are one or more corporate division levels between the parent company and the project, these should be shown to clarify the position of the work within the overall company framework. Assurance of periodic personal attention by company officials to the work's performance, quality, and timeliness, should be provided.

Teaming Arrangements

In the past decade there has been a trend toward the teaming of two or more firms to perform large, complex tasks. Teaming arrangements with other firms must be well thought out and carefully planned because some evaluation teams in certain political, management, and technical environments consider teaming as a strength. Others consider the need to team with other firms as a weakness. The advantages of teaming with a well-qualified partner in a needed skill area can be significant. Through teaming, the strengths of two or more organizations can be combined to provide a unique and tailored mix of skills that could not be otherwise offered. In addition, the geographical location of one or more of the team members close to the customer's place of doing business, the location of one or more of the team members in a surplus-labor area, or the location of one or more team members near raw material sources and economical transportation systems could result in substantial cost and political advantages. On the other hand, the need for a company to team with others in order to collect the skills and expertise needed to accomplish the job could be considered to be a weakness by the evaluation team. Management relationships and project control become more difficult when two or more teams are participating simultaneously in a complex enterprise. Normally, customers desire clear and distinct lines of authority and control. Unless these clear and distinct lines of control can be demonstrated, along with rapid responsiveness to customer requirements, the advantages of teaming may be counteracted, and the company that teams with others may be at a disadvantage.

Whatever teaming arrangements are proposed, they must be presented in the organization and management proposal in a way that takes advantage of the benefits of teaming and minimizes the objections to teaming. Generally, if the proposed team members are not a "perfect fit," teaming should be discouraged. Straight-line control calls for the proposer to be in command of the project at all times, and if teaming dilutes this positive management factor, skills and experience should be obtained in other ways, e.g., hiring of skilled personnel contingent upon contract award.

Interdivisional Relationships

If parts of the work are proposed to be performed by other divisions within a company, the relationships of the project organization to these divisions should be described and the means of managing and transferring the work should be detailed. Information should be provided on the capabilities, organization, location, and resources available to these other company divisions. The method of approval, funding, and transmitting interdivisional work authorizations should be worked out in advance of the proposal submittal date, but need not be included in the proposal unless specifically required or requested in the request for proposal.

Other Internal Company or Intradivisional Support

Assuming that the work is to be accomplished in a matrix-type organization, company and division elements other than those in a direct-line project organization will be involved in supporting the proposed work. The organization and management proposal should show how the project or work manager is to direct, control, and manage the work of those elements that may be outside of his or her direct supervisory control. The degree of control that the project manager has over these company support elements in the areas of resources, schedules, and performance should be stated, and the method of control of these internal elements should be clearly established.

Relationships with Subcontractors

If major subcontractors are to be involved in the performance of the proposed work, the prime contractor's relationships with these subcontractors should be addressed. Existing or past working relationships, geographical proximity, and established lines of communication should be pointed out to show that a smooth and cooperative relationship will exist during the performance of the job. Since a complex subcontractor and sub-subcontractor arrangement can cause schedule delays and cost increases if not properly managed, these arrangements should be as simple and straightforward as possible. Since subcontractor responsiveness is a key factor in staying on schedule and within costs, evidence of past or current ties that improve or enhance teamwork between the prime contractor and subcontractor should be provided.

Indirect, General and Administrative, and Fee Rates

The company organization portion of the organization and management proposal is a convenient place to explain the company's policy on indirect, general and administrative costs, and fee or profit; and to describe the rationale and basis for establishment of factors, percentages, or amounts that will later be used in pricing the proposal in the cost volume. (Some companies place this rationale and discussion in the cost volume, but it is discussed here because it is conveniently tied to the company's organization, policies, and procedures.) As discussed later in this chapter, indirect costs are of significant concern both to the performer and to the customer because they are sometimes difficult to control. Since indirect costs are there only to support the direct activities of the company that represent marketable work activities or work outputs, any sudden reduction in the workload will cause a corresponding jump in indirect percentages or rate unless management action is taken to reduce or control the indirect costs. Continual monitoring of overhead expenditures within each department, division, overhead center, or profit center is necessary to keep overhead costs in line. This process should be described to the customer's satisfaction in the organization and management proposal.

An important feature of a sound indirect-cost management system should be the methods whereby the disciplines for control are established with the divisional or departmental manager. The divisional or departmental manager should be held responsible for coordinating requirements with other operating activities and setting the original indirect cost targets of the department. Once a departmental budget is agreed upon, he or she then has the continuing responsibility for justifying to top management variances of actual performance from budget objectives. This is a day-to-day responsibility and one that requires continuous interfacing with company operational elements. Variances from budget are not uncommon, but should not automatically be eliminated by routinely adjusted budgets. Variances should be thoroughly examined and investigated for cause before embarking on target revisions. If the cause can be corrected by improvement in the department operations, top management should enforce austerity on the operation causing the problem, to bring the indirect costs back in line. If the cause of variance is due to major business volume fluctuations, a management-directed overall revision in indirect cost budgets may be necessary.

The purpose of indirect cost variance analysis is to disclose the causes of indirect cost overruns so that corrective action can be taken. To direct that action to the source of the problem, it is necessary that each plantwide variance be traced to the specific cost centers where the variance originated. Clear indication, in the proposal, of how this process will be monitored will be convincing evidence to the customer that the provider is not going to allow the indirect costs to get out of hand.

The goal of department heads and supervisors at lower tiers of the organization is to minimize unfavorable variances in their respective indirect-cost-control areas. Hence, line supervisors should receive performance reports on labor and other significant costs at least weekly; daily labor reports might be justified in some instances. Prompt information will enable the supervisor to detect the occurrence of a variance early enough to correct a cost overrun that otherwise might become irreversible. Generally, company management will receive overhead reports monthly. Any longer period would not permit timely identification of problem areas, while a shorter period could be considered impracticable.

In normal circumstances, there should be no significant business volume variance from predictions over a short period, such as the budget year. Most companies can project production volume realistically from their backlogs and from statistical patterns on such new business as spare parts and components. Anticipated volume on new proposals for major programs or projects cannot be forecast accurately, but the impact of those potential work activities is often not severe during the first year regardless of a company's success or lack of success in winning them.

The principal impact on the projected volume for a short period comes from the sudden decrease in an ongoing major work activity through partial or total cancellation or deferment of delivery. When that action occurs, all projected costs, whether classified as variable or fixed, should be reevaluated. For example, the designation of some administrative staff or supervisory personnel as fixed costs at the previously projected volume range may have been a valid decision for the flexible budget within that range. However, no fixed costs, and especially no manpower costs classified as "fixed," should be immune from cost-reduction procedures when there is a substantial drop in volume.

The company should establish and maintain this indirect-cost-control procedure and should describe this procedure in the organization and management proposal volume when required by the request for proposal or when prior marketing feedback has indicated that this area may be of significant concern to the customer. The proposal discussion should include but not be limited to the following:

1. The proposal must indicate that top management recognizes the need for controlling and accurately estimating indirect costs. Without this top management support, the indirect cost estimate becomes meaningless.

2. The proposal should show that the indirect cost control and estimating function is placed high enough in the company organization to permit effective dealing with

all levels of management. It should be a part of the finance or comptroller organization, assigned sufficient authority to request and receive account accumulations compatible with an effective program, and to make reductions in indirect budgets where appropriate.

3. The proposal should show that company accounts are clearly segregated to categories that will easily identify the operating responsibility with respect to expenditures. Budgets should be established and then negotiated. Revisions should be made as changes in volume or overall business trends dictate. Revisions should not be made merely for cosmetic reasons. Each negotiation of the indirect cost budget should seek improvement in criteria and techniques for establishing this important element of the company's performance.

General and administrative costs should be treated in a like manner if requested by the request for proposal or if advance marketing information has shown that undue customer concern exists in the area of general and administrative expenses.

Exposure of both indirect cost and general and administrative costs is appropriate mainly for cost-reimbursable-type negotiated procurements. Many firm fixed price bid situations will not require exposure of these factors in a proposal; indeed, it may be detrimental to the company's competitive posture to do so.

In either the organization and management volume, the cost volume, or the letter of transmittal, it is prudent to provide rationale and supporting data for the fee structure to be used in performance of the work. In certain situations the fee structure is proposed by the customer in the request for proposal rather than by the proposer. In any case, the fee negotiations may make up a considerable part of the contract negotiations. Therefore, fee rationale must be carefully generated and made available within the company and exposed in the proposal only to the degree that would be necessary to support a strong company front and position relative to the establishment of a fair and equitable profit or fee for performing the work.

THE PROJECT ORGANIZATION

In proposals for smaller work activities, there may be no need to establish a project manager, project coordinator, or project office. Generally, however, even in small jobs there is one person designated in the company to see that the work is accomplished on time and within costs. This person may manage more than one job at a time, or may be exclusively assigned to the proposed work. The customer usually wants to know the identity, capability, authority, and responsibility of this person and how he or she will communicate and work with the customer in accomplishing the job. In larger projects, where much hinges on the abilities and authority of the project manager, much more information should be provided about the project manager and about other key project and company team members. The capabilities of key personnel will be revealed in a résumé or biography.

Inherent in the project organization description should be the relationship of the project manager with the company management and the relationship of the project office elements with the company divisional or line elements that support the project.

Most customers want to know that the project manager has direct access to company management and company resources. In other words, the project must be closely knit to the company structure in a way that company resources can be activated or put to work on the project on short notice to solve unexpected problems in management, schedules, personnel, or other resource commitments. The customer would also like to know that the company line organizations that are supporting the project, although not directly responsible to the project manager in a supervisory sense, will be cooperative and responsive to the needs of the project. Particularly important functions in any project organization will be the ones that report on and control technical changes, the ones that control costs, and the ones that review, revise, and establish schedules. The project organization, then, must have a good mix of business, financial, and technical skills. The proposal should show how these skills are interrelated and integrated to provide a cohesive, efficient, and effective management structure.

The project organization discussion should emphasize responsiveness, timely reporting, and effective communications and interfaces with the customer's organization. If there are counterparts in the proposer's and customer's organizations in specific disciplines, the method and degree of communication between these counterparts should be addressed. If extensive documentation is to be provided in the proposed contract, those functions in the project office that are responsible for preparing or coordinating each item of documentation should be singled out to provide the supplier as well as the customer the assurance that all vital areas of documentation are covered. The project organization portion of the organization and management proposal volume is also the place to discuss any special management, organizational, or programmatic ground rules, assumptions, or guidelines that will be followed during the management of the project. The overall philosophy of project management, as well as specific policies, should be spelled out and clarified here.

Specific areas that affect product quality, reliability, or cost-effectiveness should be discussed in the project organization section. Methods of inspection, quality assurance, reliability assurance, and other forms of work monitoring should be described as well as means and methods that the project office will use in taking corrective actions to overcome deficiencies. If the company uses techniques such as quality circles, total quality management, or special interdisciplinary teams to assure high productivity and innovative solutions to problems, the relationship and input of these systems to the project organization should be described. Other checks and balances, whether they be administrative policies, organizational arrangements, or special project procedures should be discussed along with their anticipated and demonstrated benefits. Some of the principal functions of the project organization which should be described in the organization and management volume are listed below.

Project Direction

The project direction function of the project management organization integrates all of the other project management functions while providing day-to-day management direction of the project. Project direction is usually accomplished by a project or program director or manager and is sometimes assisted by one or more deputy or assistant project managers or directors. In the project direction discussion of the organization and management volume, the proposer should describe how the project or program manager or director will integrate the work of the personnel and organizations that are supporting the project. This should include how task assignments are made, how internal review and tracking of progress is accomplished, how priorities are set, how decisions and conflicts are settled, and how appropriate corrective measures are determined and assigned to someone for action.

The participation of the project manager and the project staff in design reviews, management reviews, and project staff meetings should be described; and the frequency, format, attendance, and follow-up procedures for these meetings should be explained. If formal task assignments are made, the method of making these assignments as well as the means of tracking their progress and completion dates should be outlined. This discussion should include the procedure for establishing required completion dates and for the conduct of periodic reviews to assure that progress is on schedule for completion on these dates. The methods to be used by the project manager in reviewing and checking on progress and status of the work activity should be shown. If management control centers with schedule or network displays are to be used, these should be described along with the procedure for their periodic updating. The method used for setting priorities as the work progresses should be established and described, and the responsibilities for changing these priorities should be spelled out.

The discussion on project direction should explain how the project manager gets inputs relative to deficiencies and problems, who on the staff is available to help resolve these problems, and how corrective action is assigned to resolve them. The customer will want to be assured in this discussion that all anomalies will be addressed and corrected at the earliest possible time and in the most efficient manner. In this and the following sections on program office functions, any procedures, forms, systems, or policies that are to be used should be described to show what tools the project manager and project staff will have available to aid in making decisions among alternate paths and to assist the project manager's judgment in settling disputes and conflicts.

Project and Cost-Control Management

Project and cost-control management refers to those activities that assure the integrated planning, scheduling, budgeting, work authorization, and cost accumulation of all tasks performed during the project. It provides project performance planning, including preparation and maintenance of a project management plan, project schedules, resource status reports, and cost forecasting. Also included are the establishment of project performance criteria, the control of change parameters, and the analysis and summary of measured data. Continuous monitoring of all functional management disciplines is provided for central direction and control of the overall project, including timely resolution of problem areas to ensure that established schedules are met. Establishment, operation, and maintenance of a management information system is a portion of this element. Other task elements include interface with the customer, contract administration, and proposal administration. The following are specific areas included under project and cost-control management:

1. Updating of the work breakdown structure(s) dictionary
2. Preparation and maintenance of a cost allocation, tracking, and reporting system
3. Monitoring of costs versus budget allocations
4. Identification of technical performance measurement parameters and values, and technical achievement planning including preparation, submission, and maintenance of technical performance reports
5. Maintenance of surveillance of cost accounts in order to assure reasonably accurate accrued charges

Information Management (Documents and Data)

Information management refers to the overall management process and activities required to ensure proper control of documents and data. Services are included to identify, control, and monitor the preparation of and maintain status of the documentation for the project. Establishment, implementation, and maintenance of a data management plan and procedures are included. Monitoring and preparation of documentation and data required by task orders, agreements, or directives are included, as is the establishment, operation, and maintenance of a project level information file.

Procurement Management

Procurement management includes management and technical control of interdivisional work, subcontractors, and vendors. Tasks included are the providing of contractual direction to other divisions within a company, subcontractors, and vendors; authorizing subcontractor tooling and equipment; analyzing subcontractor reports; conducting subcontractor and vendor reviews; and onsite coordination and evaluation of procurements. Also included are the maintenance of records and submission of any required reports relating to the geographic dispersion of minority and small business participation in procurements.

Logistics Management

Logistics management includes spares management; inventory management; repair and overhaul policy; forecasts, and usage reports regarding propellants, gases, and fluids; warehousing and storage policy; and transportation analysis and planning.

The logistics management activity provides the effort required to implement, operate, and maintain a logistics function for support of the project. Included in logistics management are the preparation and maintenance of:

1. Systems support and logistics plans
2. Recommended spare parts lists

3. Maintenance analyses
4. Analysis of support requirements.

Safety Management

Safety management consists of the definition, direction, and monitoring of a safety program that will assure the development of a safe product, prevent accidents and incidents, and minimize hazards to personnel and property. Safety is an integral part of design, development, manufacturing, testing, handling, storage, and operation. Safety management is accomplished through training, analysis, safety program assessments, and preparation of a project hazard summary; development and implementation of procedures, controls, reviews, audits, safety analyses, and safety design; and a safety plan that covers the safety program and its implementation.

Other Management Areas

Other management areas that are sometimes assigned at the project organization level are configuration management, quality and reliability assurance, and the other engineering disciplines such as maintainability, producibility, etc.; (See Blanchard, 1991). Since these areas are considered to fall in the category of engineering management rather than project management, they are included in this book under the technical proposal volume (Chapter 7).

PROJECT LABOR RATES AND FACTORS

Either the organization and management proposal volume or the cost proposal volume should include a section providing backup and rationale for the current and projected labor rates used in the proposal for all skill categories and for skill levels within these categories. This information will support the pricing figures used in the cost volume. Any factors to be used for escalation or inflation should be described along with their method of application. If different overhead pools are applied to different categories of labor rates, these should be described. Labor rates for overtime, percentages of overtime, labor rate variations for different work shifts, and other labor rate factors should be backed up by suitable rationale, historical data or statistical analysis. Methods of keeping the labor rates and factors current throughout the job performance should also be described, along with any provisions that can or will be implemented for the control of these amounts.

THE MAKE-OR-BUY PLAN

The make-or-buy plan is a description of what parts of the work will be done within the company and what parts will be done by organizations other than the proposing company. The most convenient outline for the make-or-buy plan is the work breakdown structure outline. Each activity or product in the work breakdown structure can be assigned to a given organization in house (inside the proposing company) or out of house (outside the proposing company). When out-of-house organizations are used, the customer may want to know not only who these organizations are, but also how much of the work will be given to them. Use of a large subcontractor for a major portion of the work will arouse interest in the customer as to the subcontractor's company, project organizations, and key personnel. In some procurements a bidder may be asked to supply as much information on the subcontractor's company organization, project organization, key personnel, and management philosophy as is provided for the proposing company itself.

The make-or-buy plan will give the customer an overall view of the mixture of planned in-house and out-of-house effort. Available with this make-or-buy plan should be a description of the rationale or reasons for the identification of each work activity or output as either a "make" item or a "buy" item, and the reasons for selecting a specific subcontractor if it is a "buy" item. Although this information may or may not be required for submission to the customer, it is essential that it be in company records to satisfy management and to provide an information base for make-or-buy decisions on subsequent proposals. As mentioned in Chapter 3, the marketing decisions and bid/no-bid decisions may have been made based on a specific make-or-buy analysis. The results of this earlier analysis should be kept available during the proposal preparation, negotiation, and initial contract period to help substantiate earlier decisions and to serve as a basis for making changes in the make-or-buy structure if it becomes necessary to do so.

HIGH-IMPACT RÉSUMÉS FOR KEY PERSONNEL

Key personnel résumés are often not given the attention they deserve in the proposal preparation process. As we will see in Chapter 12, source evaluation boards usually assign a significant weight to the education, experience, and accomplishments of key personnel. The tendency of most proposal preparation teams is to draw its key personnel résumés from a supply of standard résumés at the last minute and include these résumés in the proposal with little or no modification. It is a serious strategic mistake to fail to interview the person in question to make sure that the résumé is up to date, that it carries with it all experience applicable to the proposed work, and that it has been purged of irrelevant or inapplicable information. High-impact résumés are required for all key personnel. The important areas to emphasize in this careful and deliberate rework and tailoring of key personnel résumés are discussed in the following paragraphs.

Education. The educational level of key personnel is often a scored criterion for high-technology, multidisciplinary, or complex work activities or work outputs. Merely listing the degree type and title, a common practice with even high-level engineering and scientific personnel, is not always enough to convey educational background. A paragraph that provides information on the academic level of achievement; the specific areas of study included; and the subject matter of any special theses, dissertations, papers, or publications that were a part of the educational process will enhance the proposal reader's in-depth knowledge of the individual's educational accomplishments. With advancing technology, the importance of continuing education and updating knowledge within a field has increased markedly. Accomplishments in postbaccalaureate areas should be listed in terms of subject, depth, and achievement level. Other forms of recognition by the academic community such as honorary degrees or teaching fellowships should also be included.

Experience Categories. Proposal evaluators usually divide experience into at least two categories: specialized experience and general experience. In evaluating an individual proposed for a key position in a major proposed contract or subcontract, they look not only at the length of specialized and generalized experience but at the depth of this experience in areas similar or identical to the proposed work. Depth of experience can only be adequately conveyed by a well-written, detailed description of the duties, responsibilities, and activities of the individual in previous positions. Specialized experience is either identical to or very closely related to that required for the proposed work. General experience is experience within the same field, discipline, or area of endeavor as the proposed work. Specialized experience rates highly with evaluators and is a significant factor in the selection of key scientific or engineering personnel. In the case of key management personnel, general management experience is often more important than experience in managing the specific work activity being proposed. A broad, rather than restricted, background is looked for in the résumés of the project managers and other key managers. Varied work experiences such as a term of employment with major competitors, the customer, or the "customer's customer" are usually regarded as important by the teams that evaluate organization and management proposals.

In describing specialized or generalized experience, it is desirable to state or quote specific evidence of the quality of the work that was produced during the tenure of employment. This can be done by citing specific achievements of goals at or before their required dates, cost reductions, innovative accomplishments, and special recognition by the company or the customer for specific outstanding work accomplishments.

Awards, Honors, and Other Outside Recognition. Awards, honors, and recognition by external organizations can add a convincing climax to a well-written résumé that matches the person to a specific job. Since the proposing company rather than the individual himself or herself is writing the proposal, there is no need to be overly modest when quoting actual results of outstanding individual performance. This information should be accompanied by evidence of continued efforts toward professional education and advancement.

Skill Mix Among Key Personnel. Not only should each résumé be updated and emphasis provided to adapt it to the specific task or tasks at hand, but the mix of skills and experience provided by all of the key personnel collectively should be a subject of discussion and positive reinforcement within the organization and in the management proposal. For example, key personnel in a multidisciplinary scientific project should not all have the same engineering backgrounds. Evaluators will look closely at this skill mix of key personnel to be sure it closely matches the discipline mix of the work activity or work output itself.

Staffing. The staffing section of the organization and management proposal should be used to correlate the organization with the functions of each position. The functions, as described in the position description, and the capabilities of the person tentatively assigned to this position should be clearly related. In most instances, four to eight people constitute the key individuals. Discussions of specific individuals should normally be limited to those key people. Their staff can then be described in more general terms.

PLANT FACILITIES AND EQUIPMENT

Another area of company capability too often taken for granted or described in an offhand manner in small or major proposals alike is the description of the company's facilities, testing and manufacturing equipment, tools, fixtures, utilities, and other capital assets. This, too, is a place where some updating, editing, and selectivity in presentation will be fruitful. It should be recognized that the customer/evaluator is principally interested in how well the job will be done, not how many facilities the proposing company has to perform all of its jobs. Highlighting the characteristics that make facilities specifically useful and attractive for performing the job at hand will economize on presentation space and present a more convincing argument that one indeed possesses and will fruitfully use the facilities, equipment, and other real assets required to do the job.

The customer is normally fully aware that a bidder may not have in existence all of the facilities necessary to perform the job. When this is the case, it should be stated frankly. Then the steps should be outlined that will be taken to obtain the facilities. In this way a seeming disadvantage may be turned into a positive advantage by demonstrating how well management solves its problems.

Specialized Equipment, Facilities, and Software Assets. Equipment, facilities, and software assets can be placed into two categories: specialized and general-purpose. Specialized assets are those directly and specifically applicable to performing the job. These assets could have been designed, built, programmed, or acquired for a previous job, or they may have been developed specifically for the proposed work as an aid to enhancing competitive posture. Specialized equipment includes special tooling; special electrical, electronic, or mechanical test equipment; special manufacturing equipment; and special microprocessors or computers. Since this type of equipment is itself expensive and time-consuming to design, build, test, and operate, on-hand availability of the equipment at the beginning of the contract activity could provide a significant competitive advantage. General software includes computer-aided design programs, database applications, scheduling software, estimating software, graphics programs, word-processing software, and desktop-publishing software. Special software includes company-generated simulation, analyses, computation, or management information systems.

General-purpose Equipment and Facilities. General test, manufacturing, tooling, or inspection equipment and facilities and related software need not be described in detail in the proposal, but can be included in a referenced document. Specific general-purpose equipment or facilities that will be used on the job can, however, be included in the organization and management proposal. The customer should be provided with the knowledge, however, that a proposing company is fully equipped not only to do the job but to take on almost any unforeseen emergency that would keep the job from getting done in a timely and cost-effective way. This should include the policy on providing backup equipment to be used in the event of primary equipment failure. If one does not have adequate facilities and equipment in-house to meet all circumstances, evidence should be provided to the customer that a subcontractor or vendor is readily available who can absorb negative slack caused by special problems. Methods of updating and maintaining equipment capabilities should also be described, and any planned equipment or facility modifications, acquisitions, or improvements should be disclosed.

Compatibility of Personnel with Equipment Skills. The customer must be convinced that the proposer not only has the proper specialized and general-purpose equipment and facilities to do the job, but that skilled personnel are available to operate and maintain the machines, test equipment, computers, and related facilities. Their skills must be continually updated and modified as new, more advanced equipment becomes available. The best evidence that can be provided here is the tabulation of the number of skilled operators and maintenance personnel available for the items scheduled for use in the project, and an indication how and on what schedule these persons will be applied to the job. The first part is provided in the organization and management proposal, and the second part is included in the cost proposal.

COMPANY'S PAST EXPERIENCE AND SUCCESSES

The organization and management proposal should do a thorough job of describing a company's past experiences and successes in (1) identical work (if any), (2) similar work, and (3) related work. In addition it should give as much information as possible on the record of performing quality jobs within cost targets and on schedule. An overall company management and financial report, condensed for the specific proposal, is a valuable constituent of the organization and management proposal. Figure 8.2 is an outline that can be used as a checklist for company data that may be requested to be included in the proposal. Needless to say, the disclosure need not be as extensive as that implied in Figure 8.2 unless RFP requirements so dictate.

I Company

Outline of company's history
Existing plants
Directors and share distribution
Listings of bankers, solicitors and auditors
Résumés of principals and key management personnel

II Products (existing and proposed)

Sales history (month, region, product–3 year)
Description (product brochures)
Patents (pending and issued)
Competitive advantages (e.g., design, service, modification, technology)
Research and development (description)

III Market

Size, share, location (existing and projected)
List of principal competitors
Potential competitor's reaction

IV Marketing Plan (first 3 years)

Sales estimates
Verification of sales estimates (e.g., letters of intent from potential customers)
Marketing organization (management, experience)
Advertising and sales promotion
Buyers, distributors, end users

V Manufacturing

Raw materials (sources, volumes, advantage)
Manufacturing processes and technologies
Production layout
Quality control
Plant (location, site requirements)
Buildings (plans, specifications, construction estimates or quotes)
Equipment (new and existing; cost estimates or quotes)
Employment (first 3-year projection by numbers, skill, wage and salary rates)

VI Financial

Three-year audited financial statements (including affiliates if a company is in existence or if shareholders own one or more other companies)

Capital requirements of project (estimate of land, building, equipment, working capital and start-up costs)

Sources of financing of capital requirements financial projections:

1. Pro forma *balance sheets* and *income statements* for first 3 years
2. Cash flow: monthly year 1; yearly for years 2 and 3
3. Return on equity/return on investment

FIGURE 8.2 Company data checklist for organization and management proposal.

VII Other information should also be included if appropriate, such as:

Appraisal reports (equipment)
Franchise agreements
Grant approvals or applications
Lease agreements
Feasibility studies
Timing of project (start-up)
Present or threatened litigation affecting the company or major principals

FIGURE 8.2 (*Continued*)

The most convincing evidence that can be provided about a company's ability to do a good job of management is an actual cost/performance/ schedule track record on a similar job. This type of documented evidence, accompanied by a letter of commendation or letter of recommendation from the previous customer, can be an essential ingredient in a winning proposal. Further, listings of the contracts held over the past several years, names of key personnel dealt with in the customers' organizations (for reference purposes), and a narrative evaluation of each contract's success will help the evaluator recognize competence and broad-based experience.

METHODS OF ACHIEVING EFFICIENCY, ECONOMY, AND EFFECTIVENESS

It is in the organization and management portion of the proposal that one has the greatest opportunity to show how a company can, through innovative, aggressive, and competent management, provide greater efficiency, economy, and effectiveness in performing the work than competitors.

In this part of the proposal, a proposing company should tell the customer how it plans to:

1. Attract high-quality personnel to accomplish the job
2. Combine skills in the right mixture to achieve optimum skill utilization
3. Time-phase work and account for real-time rescheduling of work if required to meet changing conditions
4. Motivate employees to provide a high quality work output
5. Adjust skill levels for economy as the work progresses
6. Evaluate progress in real time and correct errors quickly
7. Adopt new procedures, methods, and equipment as they become available
8. Keep employees well-trained and updated in their skills throughout the project
9. Provide management attention and control of indirect (overhead and burden) costs.

Employee and Staffing-Related Matters

Attracting and Retaining High-quality Personnel. The ability to attain excellence and to maintain this excellence in providing a work activity or work output will depend to a large degree on the quality of personnel a proposing company has attracted or plans to attract to do the work. The personnel management hiring and promotion policies of the company need not be described in detail in the proposal, but key points of this policy may be valuable in stating one's case. What is the policy on the encouragement of versatility and cross-training of personnel? Does the company have a policy of hiring at the bottom end of salary scale and promoting from within the company or of gaining supervisory and management personnel from the outside? Does the company plan to use "contingency hiring" (job offers contingent upon winning the contract) for the proposed task? What are the sources of new personnel planned for the job? How can the company be assured that the skill categories and skill levels needed can be located, hired, and retained?

Combining Skills in the Right Mixture to Achieve Optimum Skill Utilization. Are multiskilled employees to be used for more than one job? Will the mixture of skill categories and skill levels be changed as the job progresses? If so, how? These questions should be addressed to show the customer that the company is skillful in managing the human resources that are to be employed in doing the work.

Motivating Employees to Produce High-quality Work. Does the company have specific policies that motivate employees to do high-quality work? Are incentive awards, bonuses, and performance awards used? Are these awards sufficient in size to trigger substantial employee contributions or are they merely routine programs with little or no monetary or recognition benefits? What evidence exists of high employee morale, low turnover, low absenteeism, few employee grievances, or high productivity? What is the documented record of work quality over the past several years? Proposal recipients will be more convinced of the company's ability to achieve a high-quality work output if some of these specifics are provided.

Adjustment of Skill Levels as the Work Progresses. Are there plans or methods to shift more highly skilled workers onto the project if difficulties needing their services occur? Are means provided to reduce skill levels in later phases of the program by transfering lower-paid employees to the job when those with higher levels of skill are no longer needed? Are other existing or anticipated projects available that will absorb people with more highly paid skills when they are no longer needed for ongoing work under the proposed contract? The answers to these questions will give an indication of a company's flexibility and adaptability to changing conditions throughout the project or work activity.

Evaluating Progress and Making Corrections. What systems, methods, techniques, or policies not already described in other parts of the proposal are there for evaluating progress and making corrections in work output? How is a day-to-day account of actual progress versus estimates of cost, schedules, or performance kept? What methods are used to rapidly identify deviations from planned or expected performance and to quickly make corrections in work methods, materials, or procedures? Assurance of closely monitored work performance will demonstrate an adherence to the principles of high quality and timely performance.

Adoption of New Procedures, Methods, and Equipment. What is the company policy regarding upgrading and advancement of procedures, methods, equipment, and facilities? Is there an ongoing research activity funded with overhead or profit that continually seeks to improve the tools available to do the job? Does the company have a history of willingness to adopt new procedures, methods, and equipment as soon as they are proven workable? What is the company's record of accepting and adopting employee suggestions that enhance efficiency, economy, and effectiveness? A good record in these areas should be exposed and advanced as proof of a company atmosphere of continued improvement in management and capital assets.

Employee Training. Are internal or external training programs used to keep employees' skills up to date and up to quality? Is sufficient feedback given to give the employee a knowledge of shortcomings and areas of needed improvement? Does the company provide any other means for continued employee improvement?

Control of Indirect (Overhead and Burden) Costs

It is more than likely that at least half of every dollar that is paid to a company will be spent for overhead or overhead-related items. Because overhead is such a large budget item in any company's business, it must be constantly monitored and controlled to keep the company in a competitive posture for new work. As part of every proposal preparation process, a company should and must take a hard look at its overhead, analyzing its overhead expenditures and overhead elements in detail, and adjusting its business practices (if required) in order to propose its new work based on a competitive overhead. As maintained earlier in this chapter, the company's approach to analyzing and controlling overhead and burden costs should be described in the organization and management volume of the proposal. Because overhead costs are company-controlled, the proposal preparation team must be given support by the corporate, company, and/or division management in describing this large cost factor and in explaining how it will be kept under control. In the review, analysis, and control of overhead costs, the proposal team must recognize that there are some overhead cost elements that are difficult to control, and there are some that are more amenable to control.

Labor Burden. Of the three categories of overhead costs, the labor burden is the most difficult to control. Some labor burden elements, such as social security and unemployment compensation taxes, are established by law and therefore cannot be changed unless the law is changed. Other labor burden cost elements such as paid vacations, paid holidays, and sick leave are the subject of union–management

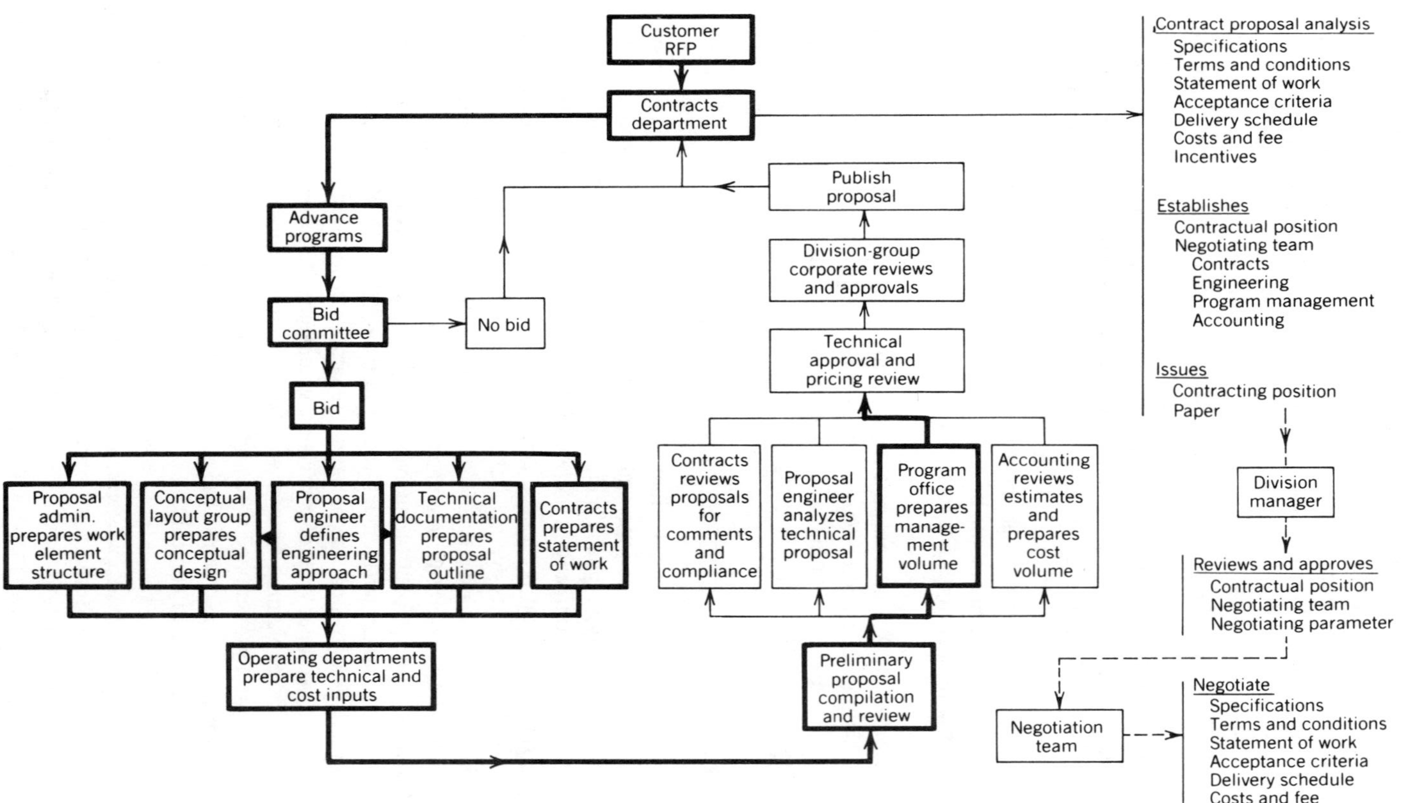

FIGURE 8.3 Proposal flow chart showing organization and management activities (bold lines).

agreements or employment contracts. Only in recent years has there been a trend toward reduction of some of these fringe benefits through labor negotiations, and these have been counterbalanced by job security provisions in union contracts that require employee retention even in times of decreasing workload. Some of this labor burden or fringe benefit amount consists of employer-granted benefits like bonuses, retirement plans, and profit-sharing plans designed to increase employee motivation and improve job performance. Reduction in these benefits is possible but may be accompanied by decreased employee morale and reduced worker efficiency. Most proposers, therefore, consider labor burden as a relatively fixed amount and are often obligated to project increases rather than decreases in labor burden due to the continuing competition for skilled professionals and skilled workers.

Material Burden. The material burden is more subject to cost control and cost reduction than the labor burden because high-technology systems can be used to reduce the costs of purchasing, inspecting, handling, storing, packaging, protecting, shipping, and disposal of materials. Although the material burden is sometimes the smallest of all of the three categories of overhead costs, significant gains in competitive posture can be made by describing how advanced-technology, cost-effective techniques will be applied in this area.

Overhead Costs. By far the largest part of a company's overhead consists of the wide range of activities listed under "Indirect/Overhead Costs." Heat, light, water, and rental are only a few of the costs the company must pay to stay in business. Companies of all sizes have been continually hit with increasing overhead costs. Despite the continual increase in overhead cost elements, much room exists for innovative ideas and methods of combining, reducing, and/or eliminating some of these costs. Some of the solutions stem from high-technology approaches to business and industrial management, but others are simply innovative management arrangements that use existing resources to their fullest rather than expending new resources. These subjects should be briefly discussed in the organization and management proposal volume to provide the customer an insight into the methods, procedures, and techniques that will be used to keep overhead costs under control.

In summary, the organization and management volume should show principally how the job will be "carried out with all diligence." It should reflect the capability, experience, and personality of the company and the team that is expected to do the job; it should offer persuasive arguments that the proposing company or firm can manage the job well and carry it through to a successful completion on schedule and within cost constraints.

ORGANIZATION AND MANAGEMENT PROPOSAL FLOW

Figure 8.3 shows the position of the organization and management activity in the overall proposal flow. Note that the organization and management volume uses inputs from all previous activities.

9

THE COST PROPOSAL

A false balance is an abomination to the Lord, but just weight is His delight.
—Proverbs 11:1

CREDIBILITY: A KEY OBJECTIVE

Whether a fixed price quote or a detailed cost proposal for a negotiated procurement is being submitted, a detailed cost and price estimate or volume will be required, either by the management in the first case or by the customer in the second. The most important characteristic of the proposal's cost volume is its credibility. Webster defines *credible* as "offering reasonable grounds for being believed; plausible." The major difficulty many proposers have encountered in establishing a proposal's credibility resides in the distinction between "cost" and "price." Price is the amount at which transactions take place in the market. It is a value set by a company officer with the appropriate authority. It is the dollar amount that he or she decides to propose to win a specific contract. The price of a work activity or work output is based on many factors. Two important ones are the estimated cost of producing the work and the desired profit. Other factors that help establish the price include competition, business plans, product line maturity, customer importance, and possibilities of follow-on work. In specific instances where it is desired to introduce a product or service into the marketplace or to capture a certain segment of the market, price can be less than estimated costs. But costs must be predicted or forecast with the highest degree of accuracy possible in order to determine the profit of the venture (or the amount of profit foregone).

It must be remembered that the profit of a venture is its price minus all of the expenses incurred in performing the work:

$$\text{profit} = \text{price} - \text{cost}.$$

The difficulty most often encountered in establishing price credibility is that some organizations have either deliberately or unknowingly quoted an unrealistic price in order to acquire the work. That is, they have stated a price that is less than the sum of the desired profit and estimated cost. This results in an undue suspicion by customers that a quoted price may not be achievable. To counteract this suspicion, it is necessary to take a systematic, deliberate approach to the establishment of credibility in the cost proposal. Assuming that a positive and quantifiable profit is desired, this means that both cost credibility and price credibility must be established in a proposal. Below we discuss some components of credibility.

Credibility Area 1: Matching the Work Content to the Resources Available. As skills and tools used in resource estimating have improved over the past several years, it has become evident that the resources (labor hours and materials) to do a job can be linked to the job in a methodical and systematic way. Cost-estimating techniques, methods, and procedures are now available to proposers that will permit them to establish a cost for each portion of the work which, with good management of each job task, will permit successful completion of the work within budget constraints. The proposer must first define the job in detail; second, estimate the resources required to do the job; third, add the cost estimate to the desired profit to determine the desired sales price; fourth, compare the desired sales price with the required market price; fifth, adjust the cost or profit to fit market price goals; and sixth, readjust the work content to fit any modifications to the cost. This last step, matching the work content to the resources available or established by the market price, is the step most often overlooked in pricing a proposal. The techniques and tools described in this chapter will allow this adjustment to be accomplished, thereby improving credibility.

Credibility Area 2: Appropriate Time Allocation of Resources. One fault or flaw in a proposal that could cause more damage to the potential of capturing the work than any other is failure to allocate resources (skills, materials, equipment, and funds) realistically to the various elements of the proposed work. Presumably, the customer will have at least some technical knowledge of at least certain portions of the work to be performed. If so, the evaluation of the credibility of the cost proposal will center on the proposer's understanding of the requirement as evidenced by careful and meticulous distribution and application of skills, materials, equipment, and funds to each element of the work. This allocation process will be scrutinized in two ways: first, in the time allocation of the various components of the overall job; second, in the distribution of these resources to each work element.

Credibility Area 3: Labor-Hour Estimates. The cost of labor, wages, and related fringe benefits is a large element in any cost proposal. Often this cost element exceeds any other single cost element. Many times, labor costs will exceed the costs of all other cost elements combined. Because of this, credibility in establishing labor-hour estimates and their associated labor rates is not only desirable but essential. The preparer of the proposal cost volume must be familiar with the various methods of estimating labor hours and must employ these methods to improve proposal credibility. Labor-loading methods, shop-loading methods, industrial standards, parametric estimating, learning curve methods, and "direct" labor estimates are summarized in this chapter.

Credibility Area 4: Traceability of Resource Estimates. Whether the proposer's own company management or an extend customer is evaluating the cost proposal, the evaluator will be interested not only in the bottom-line price, but in the derivation of this price. The cost proposal volume must clearly expose and explain the steps that were taken to build up the total cost and resulting price. The thread of methodology, from the basic man-hour estimates through the application of rates and factors and assembly of the total cost from its elements, must be observable and easily followed by an evaluator. An obscure or incompletely described methodology will often injure credibility as much as an incorrect estimate or mathematical error. Traceability of buildup of the resources required to do the job and description of the flow of resource element assembly into a final price are important factors in establishing proposal credibility.

Credibility Area 5: Supporting Data and Backup Material. Another word that encompasses the supporting data and backup material that accompany a cost proposal volume is *rationale*. Merely providing a large volume of statistics, historical costs, or written backup will not improve proposal credibility; but in-depth backup rationale, supplied in an organized manner and keyed to specific resource values, will firmly undergird proposal credibility. Skill in providing the best possible backup data within proposal page limits and volume size constraints will enhance selection potential.

Important things to be remembered in the reading and study of this chapter are (1) a cost estimate usually involves fitting the resources to the detailed work description and (2) a price estimate usually includes fitting the work to the resources available or the established competitive market price. Restating the earlier equation as:

$$\text{cost} = \text{price} - \text{profit}$$

shows that, when price or profit are established or adjusted, the cost must be established or adjusted by altering work content, delivery schedules, quantities, specifications, or skill levels used to match the resources allocated or available. (A "negative profit" in this equation would signify a loss.)

By far the most appropriate and most effective cost-estimating method to use for cost-proposal preparation is the industrial engineering type, labor-hour- and material-based cost estimate. Otherwise known as a "ground-up" cost estimate, this type of estimate will address all of the five areas needed for proposal cost credibility. Experience has shown that an in-depth task analysis of the work and estimating of work elements will create the most credible, supportable, usable, and accurate proposal cost estimate. This estimate can then be compared with an independently derived parametric cost estimate developed from one of many available cost models. This in-depth cost-estimating procedure usually consists of the following phases of activity:

1. Preparation of a complete list of all drawings, documents, publications, materials, and parts required to perform the job and analysis of these items to establish a make-or-buy decision on each
2. Detailed manufacturing or process planning, including a preliminary or conceptual design of each major tool or piece of special equipment, and a complete description and analysis of the manufacturing or process flow
3. Application of work standards and adjustment to account for expected performance against these standards
4. Definition of each administrative, engineering, manufacturing, assembly, testing, shipping, and support task by discipline and the use of standard industrial engineering methods, labor-loading techniques, judgment of skilled personnel, and historical experience to arrive at a detailed estimate
5. Application of standard catalog prices, recent purchase-order data, vendor quotations for materials and parts, and the competitive solicitation of quotations for subcontracts
6. Use of the latest available information on labor rates, travel costs, fringe benefits, overhead costs, and general and administrative expenses

Through the use of these techniques, sufficient detailed data can be accumulated to convince both the performing organization and the customer that the job can be performed within the estimated and proposed resources.

CONTENTS OF THE COST PROPOSAL

The cost volume of the proposal consists of eight sections:

1. The introduction—a brief comprehensive statement of the scope of work to be performed
2. The ground rules used for the formulation of the estimate
3. A program summary of the cost estimate, and the total program cost by work breakdown structure
4. The cost estimate by work package—a summary of costs at major work package levels and sublevels of the work breakdown structure, including task description and estimating rationale
5. The cost summary by elements of cost—sublevels of work breakdown structure showing functional cost elements such as direct labor by category, overhead, material; and other direct and indirect costs
6. Cost and pricing supporting information—the composition of rates and factors by projected fiscal years, economics, escalation, and sources of rates
7. Incremental costs—the computations for selected increments of procurement (where applicable)
8. Backup material, rationale, and supporting data.

All of these elements are not necessarily included in every cost proposal, since the secret to a competitive posture may thereby be revealed. However, in a cost-reimbursable contract, the agency charged with reviewing the bids and managing the contract will take full advantage of any detail presented to them by the bidder. In these instances the bidder will find that it will be advantageous to provide adequate estimate rationale in the cost proposal. Providing much more detail than the request for proposal asks for is not a good practice, because it may result in increased evaluation and negotiation times. However, to be competitive in the environment of negotiated versus advertised procurements, full disclosure, innovation, and cost efficiency are infinitely more successful than obscure pricing and estimating techniques and methods.

THE PROPOSAL-ESTIMATING PROCESS

The proposal-estimating process, used in the preparation of the cost volume, is composed of parallel and sequential steps that flow together and interact to culminate in a completed pricing structure. Figure 9.1 shows the anatomy of the proposal-estimating process. This figure graphically depicts how the various resource ingredients are synthesized from the basic labor-hour estimates and material quantity estimates. Labor-hour estimates of each skill required to accomplish the job are combined with the labor rates for these skills to derive labor-dollar estimates. In the meantime, material quantities and purchased parts are estimated in terms of the units by which they are measured or purchased, and these material quantities are combined with their costs per unit to develop direct material dollar estimates. Labor overhead or burden is applied to direct material costs. Then travel costs and other direct costs are added to produce total costs; general and administrative expenses and fee or profit are added to derive the proposal price.

The Process Plan

A key to successful proposal costing of manufacturing or construction activities is the process plan. A process plan is a listing of all operations that must be performed to manufacture a product or to complete a project, along with the labor hours required to perform each operation. The process plan is usually prepared by an experienced foreman, engineer, or technician who knows the company's equipment, personnel, and capabilities, or by a process-planning department chartered to do all of the process planning. The process planner envisions the equipment, work station, and environment; estimates the number of persons required; and estimates how long it will take to perform each step. From this information he or she derives the labor hours required. Process steps are numbered and space is left between operations listed to allow easy insertion or omission of operations or activities as the process is modified. A typical process plan is shown in Figure 9.2.

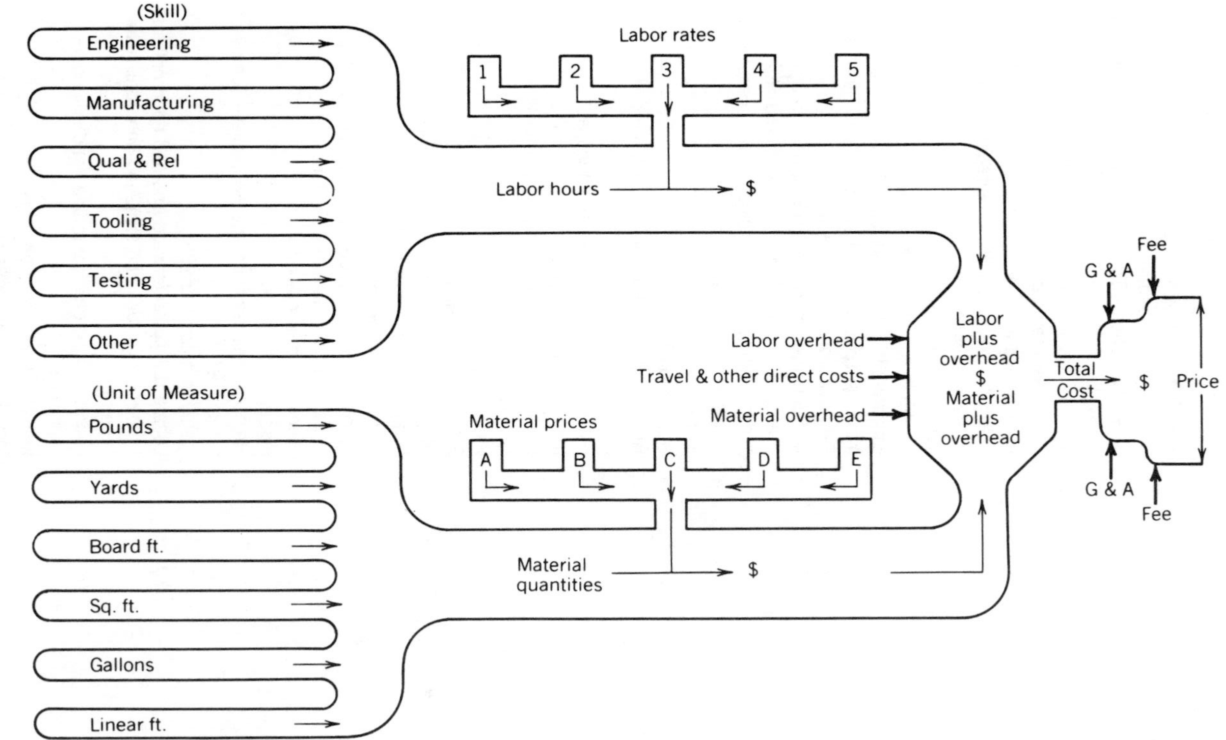

FIGURE 9.1 The anatomy of an estimate. G&A: general and administrative costs.

PROCESS PLAN

Drawing No. D21216 **Part No. 1D21254**

Title: Cylinder Assembly (Welded)

Operation Number	Labor Hours	Description
010	—	Receive and inspect material (skins and forgings)
020	24	Roll form skin segments
030	60	Mask and chem-mill recessed pattern in skins
040	—	Inspect
050	36	Trim to design dimension and prepare in welding skin segments into cylinders (two)
060	16	Locate segments on automatic seam welder tooling fixture and weld per specification (longitudinal weld)
070	2	Remove from automatic welding fixture
080	18	Shave welds on inside diameter
090	16	Establish trim lines (surface plate)
100	18	Install in special fixture and trim to length
110	8	Remove from special fixture
120	56	Install center mandrel—center ring, forward and aft sections (cylinders)—forward and aft mandrel—forward and aft rings—and complete special feature set up
130	—	Inspect
140	24	Butt weld (4 places)
150	8	Remove from special fixture and remove mandrels
160	59	Radiograph and dye penetrant inspect
170	—	Inspect dimensionally
180	6	Reinstall mandrels in preparation for final machining
190	14	Finish OD–aft
	10	Finish OD—center
	224	Finish OD—forward
200	40	Program for forward ring
220	30	Handwork (3 rings)
230	2	Reinstall cylinder assembly w/mandrels still in place, or on the special fixture
240	16	Clock and drill index holes
250	—	Inspect
260	8	Remove cylinder from special fixture—remove mandrel
270	1	Install in holding cradle
280	70	Locate drill jig on forward end and hand drill leak check vein (drill and tap), and hand drill hole pattern
290	64	Locate drill jig on aft ring and hand drill hole pattern
300	—	Inspect forward and aft rings
310	8	Install protective covers on each end of cylinder
320	—	Transfer to surface treat
340	24	Remove covers and alodine
350	—	Inspect
360	8	Reinstall protective covers and return to assembly area

Labor Costs

Labor rates that are applied to the basic labor-hour estimates are usually "composite" labor rates; that is, they represent an average of the rates within a given skill category. For example, the engineering skill category may include skill levels such as draftsmen, designers, engineering assistants, junior engineers, engineers, and senior engineers. The number and titles of engineering skills vary widely from company to company, but the use of a composite labor rate for the engineering skill category is common practice. The composite labor rate is derived by multiplying the labor rate for each skill by the percentage of man-hours of that skill required to do a given task and adding the results. For example, if six skills have various labor rates and percentages, the composite labor rate is computed as shown in Table 9.1. In this example in Table 8.1, the engineering composite rate would be:

$$(\$8.13 \times .07) + (\$10.31 \times .03) + (\$13.56 \times .10) + (\$19.15 \times .20)$$
$$+ (\$25.19 \times .50) + (\$32.10 \times .10) = \$21.87 \text{ per hour.}$$

Another common practice is to establish separate overhead or burden pools for each skill category. These burden pools carry the peripheral costs that are related to and are a function of the labor hours expended in that particular skill category. Though the mathematics consists only of multiplication and addition, multielement work activities or work outputs greatly increase the number of these mathematical computations. It becomes readily evident, therefore, that the computations required for a cost proposal are so complex that computer techniques for computation are essential for any cost proposal involving more than the simplest tasks.

Estimating Based on the Work Breakdown Structure. The entire cost estimate is developed using the framework of the work breakdown structure for collection, correlation, computation, and display of costs. Most companies have their own work-breakdown-structure-based automatic data-processing system that performs all of the computations needed to compile a complete multiskill, multiyear, and multielement estimate. Frequently the cost estimate will be developed to a lower level than that which will be included in the cost proposal. A customer may request pricing down to level three in the work breakdown structure; the company will elect to go down to level five in its own internal estimating process. This practice reduces the amount of costing material that must be provided in the cost proposal but causes lower-level, backup material to be generated, which can be used later in supporting negotiations. The work breakdown structure is the framework upon which the work elements, cost elements, and schedule elements are based and correlated.

The major cost elements of a proposal can vary considerably and depend on the work activity being estimated. Generally, however, the costs for any activity can

FIGURE 9.2 Process Plan. (Source: R.D. Stewart, *Cost Estimating*, New York: Wiley, 1982).

TABLE 9.1. Computing Composite Labor Rate

Engineering Skill	Labor Rate ($/Hour)	Percentage in the Task	Multiplier
Draftsman	$ 8.13	7	.07
Designer	$10.31	3	.30
Engineering assistant	$13.56	10	.10
Junior engineer	$19.15	20	.20
Engineer	$25.19	50	.50
Senior engineer	$32.10	10	.10
Total		100	

be subdivided into labor costs, labor overhead costs, subcontract costs, material and material overhead costs, travel costs, computer costs, other direct costs, general and administrative expenses, and profit or fee. Labor cost elements are derived by estimating labor hours for each skill category and each skill level and multiplying the labor rates for each skill by the hours for each. The proportions of each skill category and level required to do a total job and the related overhead materials and supplies will assist in the development of overhead rates for each major skill category. For example, an engineering overhead pool includes the overhead costs for a category that includes a predetermined mix of engineering skill levels. Along with an overhead pool, the composite engineering labor rate can be developed based on the same mix of skill levels. When composite labor and overhead rates are used, proposal estimates can be reduced to estimates of the hours required by several major skill categories, and labor costs can be computed using a composite labor and overhead rate for that skill category, such as: project management; engineering; manufacturing; tooling; quality, reliability, and safety; and testing. Another category, simply labeled "other," provides a means for costing labor hours that do not fall into any of the categories listed above. The seven skill categories are described in the following sections. (Other major composite skill categories can be developed to fit a company's wage, salary, and organizational structure.)

Labor Allowances

Standard times assume that workers are well trained and experienced in their jobs; that they apply themselves to the job 100% percent of the time; and that they never make mistakes, take breaks, lose efficiency, or deviate from the tasks for any reason. This, of course, is an unreasonable assumption, because there are many legitimate and numerous unplanned work interruptions that occur with regularity in any work activity. Therefore, labor allowances must be added to any proposal estimate that is made up of an accumulation of standard times. In most instances, these labor allowances can accumulate to a factor of 1.5 to 2.5. The total standard time for a given work activity, depending on the overall inherent efficiency of the shop, equipment, and personnel, will depend on the nature of the task.

Standard hours vary from actual measured labor hours because workers often deviate from the standard method or technique used or planned for a given operation. This deviation can be caused by a number of factors ranging from the training, motivation, or disposition of the operator to the use of faulty tools, fixtures, or machines. Sometimes shortages of materials or lack of adequate supervision are causes of deviations from standard values. These variances can add 5 to 20% to standard time values.

Another type of variance from standard time is called personal, fatigue, and delay (PFD) time. Personal times are for personal activities such as coffee breaks, trips to the restroom or water fountain, unforeseen interruptions, or emergency elephone calls. Fatigue time is allocated because of the inability of a worker to produce at the same pace all day. Operator efficiency decreases as the job time increases. Some delays are unavoidable, caused by the need for obtaining supervisory instructions, equipment breakdown, power outages, or operator illness. PFD time can add 10% to 20% to standard time values.

Although normal or routine equipment maintenance can be done during other than operating shifts, there is usually some operator-performed machine maintenance activity that must be performed during the machine duty cycle. These activities include adjusting tools, sharpening tools, and periodically cleaning and oiling machines. In electroplating and processing operations, the operator maintains solutions and compounds and handles and maintains racks and fixtures. Tooling and equipment maintenance can account for 5% to 12% of standard time values.

The overall direct labor hours derived from the application of the preceding allowance factors to standard times must be increased by additional amounts to account for normal rework and repair. Labor values must be allocated for rework of defective purchased materials, rework of in-process rejects, final test rejects, and addition of minor engineering changes. Units damaged on receipt or during handling must also be repaired. This factor can add 10% to 20% direct labor hours to those previously estimated.

For projects where design stability is poor, where production is initiated prior to final design release, and where field testing is being performed concurrently with production, an engineering change allowance should be added. Change allowances vary widely for different types of work activities. Even fairly well-defined projects, however, should contain a change allowance.

The labor hours required to produce an engineering prototype are greater than those required to produce the first production model. Reworks are more frequent, and work is performed from sketches or unreleased drawings rather than production drawings. An increase over first production unit labor of 15% to 25% should be included for each engineering prototype.

Project Management and Supervisory Labor. Project management and supervisory labor includes the overall administrative effort of planning, organizing, coordinating, directing, controlling, and approving the work activities required to accomplish the program objectives. The subelements of project management are project direction, cost-control management, logistics management, procurement management, configuration management, information management, and safety management.

Direct supervision costs will vary with the task and company organization. The cost estimator must carefully analyze the staffing plan prepared in the organization and management portion of the proposal to identify all direct management and supervisory personnel who will be charging their time directly to the project. Labor costs for all other management and administrative personnel will be accumulated under indirect costs or allocated as part of general and administrative expenses.

Engineering Labor. Engineering labor includes the study, analysis, design, development, evaluation, and redesign for specified subdivisions of work. This skill category includes the preparation of specifications, drawings, parts lists, wiring diagrams; technical coordination between engineering and manufacturing; vendor coordination; test planning and scheduling; analysis of test results; data reduction; and engineering report preparation.

Manufacturing Labor. Manufacturing labor, or manufacturing and assembly labor, includes such operations as fabrication, processing, subassembly, final assembly, reworking, modification, experimental production, and installation of parts and equipment. Included is the preparation and processing of material of any kind (metal, plastic, glass, cloth, etc.). Preparation and processing includes but is not limited to flashing operations, annealing, heat-treating, baking, refrigeration, anodizing, plating, painting and pretest, and production services. Fabrication (the construction of detail parts from raw materials) includes the hours expended in the cutting, molding, forming, stretching, and blanking operations performed on materials of any kind (metal, wood, plastic, cloth, tubing) to make individual parts. Experimental hours spent in construction of mockup models, test articles, testing, and reworking during the test program should be considered as direct manufacturing labor hours, as should machine setup time when performed by the operator of the machine.

Tooling Labor. Tooling labor includes planning, design, fabrication, assembly, installation, modification, and maintenance. It also includes rework of all tools, dies, jigs, fixtures, gauges, handling equipment, work platforms, test equipment, and special test equipment in support of the manufacturing process. This work includes effort expended in the determination of tool and equipment requirements, planning of fabrication and assembly operations, maintaining tool and equipment records, establishing make-or-buy plans and manufacturing plans on tooling components and equipment, scheduling and controlling tool and equipment orders, and programming and preparation of tapes for numerically controlled machine parts. It also includes preparation of templates and patterns.

Quality, Reliability, and Safety Labor. Quality, reliability, and safety labor includes the development of quality-assurance, reliability-assurance and safety-assurance plans; receiving inspection, in-process inspection, and final inspection (of raw materials, tools, parts, subassemblies, and assemblies); and reliability testing and failure-report reviewing. It includes the participation of quality, reliability, and safety engineers in design reviews and final acceptance reviews.

Testing Labor. Testing labor includes the work expended in the performance of tests on all components, assemblies, subsystems, and systems to determine and unify operational characteristics and compatibility with the overall system and its intended environment. Such tests include design feasibility tests, design verification tests, development tests, qualification tests, acceptance tests, reliability tests; also tests on parts, systems, and integrated systems to verify the suitability in meeting the criteria for intended usage. These tests are conducted on hardware or final designs that have been produced, inspected, and assembled by established methods. Skills expended in test planning and scheduling, data reduction, and report preparation are also included in this category.

ESTIMATING MANUFACTURING ACTIVITIES BY STANDARDS

The most common method of estimating the time and cost required for manufacturing activities is the industrial engineering approach whereby standards or target values are established for various operations. The term "standards" is used to indicate standard time data. All possible elements of work are measured, assigned a standard time for performance, and documented. When a particular job is to be estimated, all of the applicable standards for all related operations are added together to determine the total time. Then major adjustments are made to the estimate of time and resources by applying a "realization factor," which is derived by dividing actual experienced in performing previous similar work time by the standard time for a given series of manufacturing activities.

The use of standards produces more accurate and more easily justifiable estimates. Standards also promote consistency between estimates as well as among estimators. Where standards are used, personal experience is desirable or beneficial but not mandatory. Standards have been developed over a number of years through the use of time studies and synthesis of methods analysis. They are based on the level of efficiency that could be attained by a job shop producing up to 1,000 units of any specific work output. Standards are actually synoptical values of more detailed times. They are adaptations, extracts, or benchmark time values for each type of operation. The loss of accuracy occasioned by summarization and/or averaging is acceptable when the total time for a system is being developed. If standard values are used with judgment and interpolations for varying stock sizes, reasonably accurate results can be obtained. Sources of the standards, as well as the rationale for the application of these standards, should be included in the proposal.

Other Labor Costs. Other labor costs include categories of labor and skills not included in the foregoing six categories. These other labor costs can include direct supervision and management, direct clerical costs, and miscellaneous direct labor activities.

Documentation. In these times of high technology and sophisticated projects, products, and services, a large part of a company's resources is spent in formulating and writing specifications, reports, manuals, handbooks, engineering orders, and product descriptions. The complexity of the engineering activity and the specific document requirements are important determining factors in proposing the labor-hours required to prepare documentation.

The hours required for documentation will vary considerably depending on the complexity of the work output; however, average labor hours for origination and revision of documentation have been derived based on experience, and these figures can be used as average labor hours per page of documentation (see Tables 9.2 and 9.3). These tables are also handy in determining the amount of time to prepare a proposal, as a proposal is essentially a form of technical documentation.

Materials and Subcontract Costs

Where there are a large number of buy items (items or services to be procured from outside organizations), the subcontracts cost element is usually subdivided into major subcontracts and other subcontracts. Major contracts are usually established by setting up a cost value above which the contract is considered a "major subcontract." The reason that major subcontractors are singled out is that their performance has a significant effect on the conduct of the overall prime's work activity. Cost and financial reporting, schedule reporting, and performance reporting are usually required to a higher degree from major subcontractors than from other contractors. This greater depth of reporting and visibility into the work of major subcontractors gives the prime performing organization better management control of the work progress.

The "materials" cost element includes tangible raw materials, parts, tools, components, subsystems, and assemblies needed to perform the work. Service contracts are not usually covered under materials, but there are grey areas in the definitions of materials and subcontracts that result in some possible overlap. An organization's accounting system normally will specify clearly under which category each procurement falls. This relationship should also be described in the organization and management volume or the cost volume of the proposal.

TABLE 9.2. Labor Hours for New Documentation

Function	Manual Labor Hours per Page	Labor Hours per Computer-Generated Page
Research, liaison, technical writing, editing, and supervision	5.7	3.8
Typing and quality control	0.6	0.3
Illustrations	4.3	2.8
Engineering	0.7	0.7
Coordination	0.2	0.2
Total	11.5	7.8

TABLE 9.3. Labor Hours for Revised Documentation

Function	Manual Labor Hours per Page	Labor-Hours per Computer-Generated Page
Research, liaison, technical writing, editing, and supervision	4.00	2.66
Typing and quality control	0.60	0.16
Illustrations	0.75	0.18
Engineering	0.60	0.60
Coordination	0.20	0.20
Total	6.15	3.80

Spares and Spare Parts. In the proposal for a product or project, it is important to determine the degree of manufacture, distribution, stocking, warehousing, and sales of spare parts. In virtually every product manufactured today, there are certain subsystems, components, or parts that have a limited lifetime. Items such as batteries, gaskets and seals, drive belts, and illumination devices invariably have to be replaced before the useful lifetime of the overall product is expended. These so-called expendable items must be periodically replaced by the owner or user and must be readily available to keep the product in an operational condition. The business organization that is interested in maintaining growth or expansion through satisfied customers must include the logistics system to provide these spare parts in the cost proposal. To a similar degree nonexpendable parts must also be stocked to replace those damaged because of wear, malfunction, or misuse. The best procedure for developing and pricing an initial spare parts list is to review the complete list of parts, subassemblies, and assemblies, and to make a decision concerning the spares level for each based on an anticipated or assumed failure or use rate. This completed spares list can be used in developing the total quantities needed, the manufacturing rate, and the resulting resource estimate. The method of estimating spare parts costs as well as the reasons and supporting data for this method should be included in the cost proposal.

Maintenance and Repair Manuals. Virtually every work activity or work output is accompanied by some sort of assembly, maintenance, repair, or instruction manual. Assembly instructions, maintenance manuals, and repair manuals usually take a significant amount of time and effort to develop because their content must be extracted, condensed, simplified, and clearly described based on more complex engineering and manufacturing documentation. In establishing initial cost-proposal ground rules, then, specific assumptions must be made as to the quantity, quality, number of pages, types of illustrations and artwork, and distribution of these instructions and manuals. These ground rules should be included in the proposal.

Optional Equipment and Services. In industries where products are standardized to reduce production costs and to increase the benefits of mass production, there are few options provided, but the trend toward customization, individuality, and adaptability of a work output to a specific individual's or company's taste or circumstances has resulted in numerous products that have optional extra equipment or services. In some products the list of options is even longer than the list of parts in the original item. For this reason it is important to consider carefully what optional equipment and services are to be included in the inventory, to develop a resources policy and resource estimate for these items and services, and to state these resources and their rationale in the cost proposal.

Estimating Materials Costs

When a product is first designed, a key part of the design documentation is a bill of material. A bill of material or parts list is usually included in or with the initial detailed design drawings of an item, and it is updated as the design changes. This bill of material is a valuable source of material quantity information on which proposed material costs can be based.

The pounds, cubic or square yards, board feet, square feet, gallons, or linear feet of the required materials are usually obtained by determining or computing quantities directly from the "bill of materials" or parts list or from detailed drawings and specifications of the completed item with added sufficient allowance for waste or scrap. The next step is to apply the appropriate material unit price or cost to this quantity to develop the final material costs. The costs of procuring, handling, storing, and maintaining materials stocks and supplies can be included in the material costs or may be included in material overhead costs. Usually, materials are classed as those items purchased rather than made by an organization.

Drawing Takeoff. The most precise means of determining the actual quantity of materials required to do a job is the extraction or calculation of material quantities from drawings of the item or specifications for the process or service, if these detailed drawings already exist. Calculation of material quantities involves such considerations as the anticipated method of manufacturing the item, conducting the process, or delivering the service; the size or quantity of uniform purchase lots; and the anticipated scrap, waste, boiloff, or leakage. Since the term "materials" covers a wide range of substances varying from raw materials to completed parts, there is often a delicate balance between the type, shape, and kind of material purchased and the labor to be performed on it after it is purchased.

Careful analysis of any existing drawings and specifications of the work output is required to determine (1) the best state or condition in which to purchase the material; (2) the optimum size or quantity of material to be bought; (8) the method of fabrication that will best use the full quantity of purchased material; and (4) the expected quantity of scrap or waste resulting from successful manufacture of the product. A certain amount of waste can be expected because the material sizes do not usually conform to the shape of the completed part. Good manufacturing design, however, will make maximum advantage of available material sizes and shapes to minimize waste. It should be pointed out in the proposal how the careful analysis and estimating of these factors have saved resources in the overall work output or work activity. If detailed drawings and specifications for the work do not exist, then the estimator must forecast the labor hours and materials required to produce these drawings and specifications.

Material Handbooks and Supplier Catalogs. Once material quantities have been developed from drawings, specifications, and parts lists, costs can be derived from material handbooks, supplier catalogs, or supplier quotes. The most highly refined catalogs and handbooks are available for architectural building construction, but ample catalogs are available for the manufacturing and process industries, and catalogs and handbooks are continually becoming available for high-technology industries such as the aerospace industry. It is always desirable to use the latest available catalog prices.

A good estimator will obtain and keep a complete file of catalogs and handbooks containing descriptions and prices of materials, parts, supplies, and subsystems and components of the major product or products being estimated and will reference the handbooks used in deriving cost estimates in the cost proposal.

Quantity to Buy and Inventory Considerations. The quantity or number of supplies or parts purchased or stocked strongly affects the materials costs. The study of materials, supplies, and goods, and their handling, transportation, packaging, shipping, and storage is called *logistics*. To stock just the right quantity of materials or supplies, a logistics study must be done of the economics of inventory systems.

The basic decision that faces a company in developing materials costs for a proposal is whether to take advantage of the lower costs of materials made possible by buying materials in large quantities and saving these materials until they are needed, or to wait until the materials are needed before purchasing them. The benefits of buying materials in large quantities are fourfold: suppliers and producers can offer these materials at lower prices when larger quantities are purchased because of the production economies of scale; the buyer can avoid future cost or price increases for the purchased material that result from inflation or escalation; the ready availability of vital materials can often serve as a cost avoidance in producing a work output where there is a fluctuating demand; and the purchase of larger quantities of needed materials at one time rather than in separate lots reduces the procurement and transportation costs per unit of material.

These four benefits of buying and stockpiling quantities of materials larger than those needed immediately for the work activity are counteracted by other costs associated with the storage, maintenance, handling, and use of the larger quantities of materials; and by the "opportunity cost" of having capital tied up in inventory. The real costs of carrying an inventory include insurance and taxes on the stockpiled material and the land or building it is stored in; breakage, deterioration, and pilferage of the material; and heating, light, and security for the warehouse or storage area. Reduction in costs of operation to obtain a competitive cost estimate can be achieved through "Just-in-Time" methods originally developed by the Japanese and now widely used in the United States. See "JIT Factory Revolution," Hiroyuki Hirano, Productivity Press, 1989.

Scrap and Waste Considerations. The most important thing that can be said about scrap and waste in materials estimating is that these factors must not be forgotten or omitted in the estimator's analysis of the job to be done. A normal amount of waste or scrap is encountered in almost every manufacturing or production process because materials must be changed in form, shape, or volume in some manner to arrive at a final product. In this process of converting the shape or form of a material, scrap material is produced by the machining process, by-products are formed, and the inspection process will reveal and cause rejection of a certain portion of the work. Standard sizes or quantities of materials should be proposed wherever possible.

The estimation of scrap and waste factors can be done most effectively by reviewing the actual manufacturing or production process, observing actual scrap and waste factors on previous projects, by judiciously applying these factors to the activity being proposed, and by referencing the factors used and the method of application of these factors in the proposal itself.

Treatment of Other Resource Components

Since the preparation of a cost estimate often brings up questions about the work that have not been asked or discussed, the cost proposal team must often define or force the definition of other general ground rules that have not been developed by other members of the proposal team or by the company's management. These ground rules and assumptions can include areas such as the policy for providing spares, spare parts, warranties, maintenance manuals, repair manuals, optional equipment and services, and customer services. The ever-present areas of scrap, waste, and human error allowances in themselves can cause large errors in estimating and must not be overlooked when formulating estimates of each work element.

ESTIMATING CONSTRUCTION ACTIVITIES

One of the most structured, documented, and well-thought-out activities that proposers deal with is construction. In a construction proposal, it is important to recognize and to include adequate labor hours and materials for each step in the construction process, and to meticulously lay out and identify each construction activity.

The need for onsite physical access to the construction element being performed at a given time dictates a general flow of activities for the construction process. The unique feature of an integrated onsite manufacturing and assembly process performed by diverse skills makes it particularly important to lay out the construction flow sequence as a prerequisite to and an integral part of the estimating of labor hours and materials. In the fields of residential and industrial construction alone, there are nearly 3,000 categories of material and labor. Many construction labor standards are tied to the item being installed or material being used. Construction-estimating manuals usually include average wage rates for various construction skills and trades, wage modification factors that adapt these wage rates to various geographical areas, and material and labor costs associated with each construction activity. Any construction manuals used in the proposal rationale should be listed or referenced.

IN-PROCESS INSPECTION

The amount of in-process inspection performed on any process, product, project, or service will depend on the cost of possible defects in the item or scrappage of the item as well as the degree of reliability required for the final work output. In high-rate production of relatively inexpensive items, it is often economically desirable to forego in-process inspection entirely in favor of scrapping any parts that fail a simple "go, no-go" inspection at the end of the production line. On the other hand, expensive and sophisticated precision-manufactured parts may require nearly 100% inspection. A good rule of thumb is to add 10% of the manufacturing and assembly hours for in-process inspection. This in-process inspection does not include the in-process testing covered in the following paragraph.

TESTING

Testing usually falls into three categories: (1) development testing, (2) qualification testing, and (3) production acceptance testing. Rules of thumb are difficult to come by for estimating the resources required for development testing because testing varies with the complexity, uncertainty, and technological content of the work activity. The best way to estimate the cost of development testing is to produce a detailed test plan for the specific project and to cost each element of this test plan separately, being careful to consider all skills, facilities, equipment, and materials needed in the development test program. Again, rationale should be included in the cost proposal.

Qualification testing is required in most commercial products and on all military or space projects to demonstrate adequately that the article will operate or serve its intended purpose in environments far more severe than those intended for its actual use. Resources required for this qualification testing and the estimating rationale should be included in the cost proposal.

Receiving inspection, production testing, and acceptance testing can be estimated using experience factors and ratios available from previous similar work activities. Receiving tests are tests performed on purchased components, parts, and/or subassemblies prior to acceptance by the receiving department. Production tests are tests of subassemblies, units, subsystems, and systems during and after assembly. Experience has shown, generally, that test labor varies directly with the amount of fabrication and assembly labor. Rationale for estimating receiving and production testing resources should be included in the cost volume along with the rationale for the resources required for other forms of testing.

Special Tooling and Test Equipment. Special purpose tooling and special purpose test equipment are important items of cost because they are used only for a particular job; therefore, that job must bear the full cost of the tool or test fixture. In contrast to the special items, general-purpose tooling or test equipment is purchased as capital equipment and its costs are spread over many jobs. The type, quantity, rationale, and resources required for special-purpose tooling and equipment should be included in the proposal cost volume.

COMPUTER SOFTWARE COSTS

Because of the increasing number and types of computers and computer languages and the advent of new computer software-development tools, it is difficult to generate overall ground rules or rules of thumb for computer software cost estimating. Productivity in computer programming is greatly affected by the skill and competence of the computer analyst or programmer and by the types of productivity tools used. Usually, a company will have an in-depth experience base for estimating the systems analysis and computer program coding time required for various types of machine languages and high-order languages. The advent of Computer-Aided Software Engineering (CASE) tools has dramatically changed the potential resources required to produce software.[1] It is important to include advanced software techniques and rationale in the cost proposal, since software costs are becoming an ever-larger part of proposed costs for many high-technology projects.

ESTIMATING SUPERVISION, DIRECT MANAGEMENT, AND OTHER DIRECT CHARGES

Two cost elements of "other direct costs" that are becoming increasingly prominent are travel and transportation costs. A frequent check on public and private conveyance rates and costs is mandatory. Most companies provide a private-vehicle mileage allowance for employees who use their own vehicles in the conduct of company business. Rates differ and depend on whether the private conveyance is being used principally for the benefit of the company or principally for the convenience of the traveler. Regardless of which rate is used, the mileage allowance must be periodically updated to keep pace with actual costs. Many companies purchase or lease vehicles to be used by their employees on official business and sometimes on personal travel. Per-diem travel allowances or reimbursement for lodging, meals, and miscellaneous expenses must also be included in overall travel estimates. These reimbursable

[1] See R. D. Stewart, *Cost Estimating*, 2nd ed., New York, John Wiley & Sons, 1991.

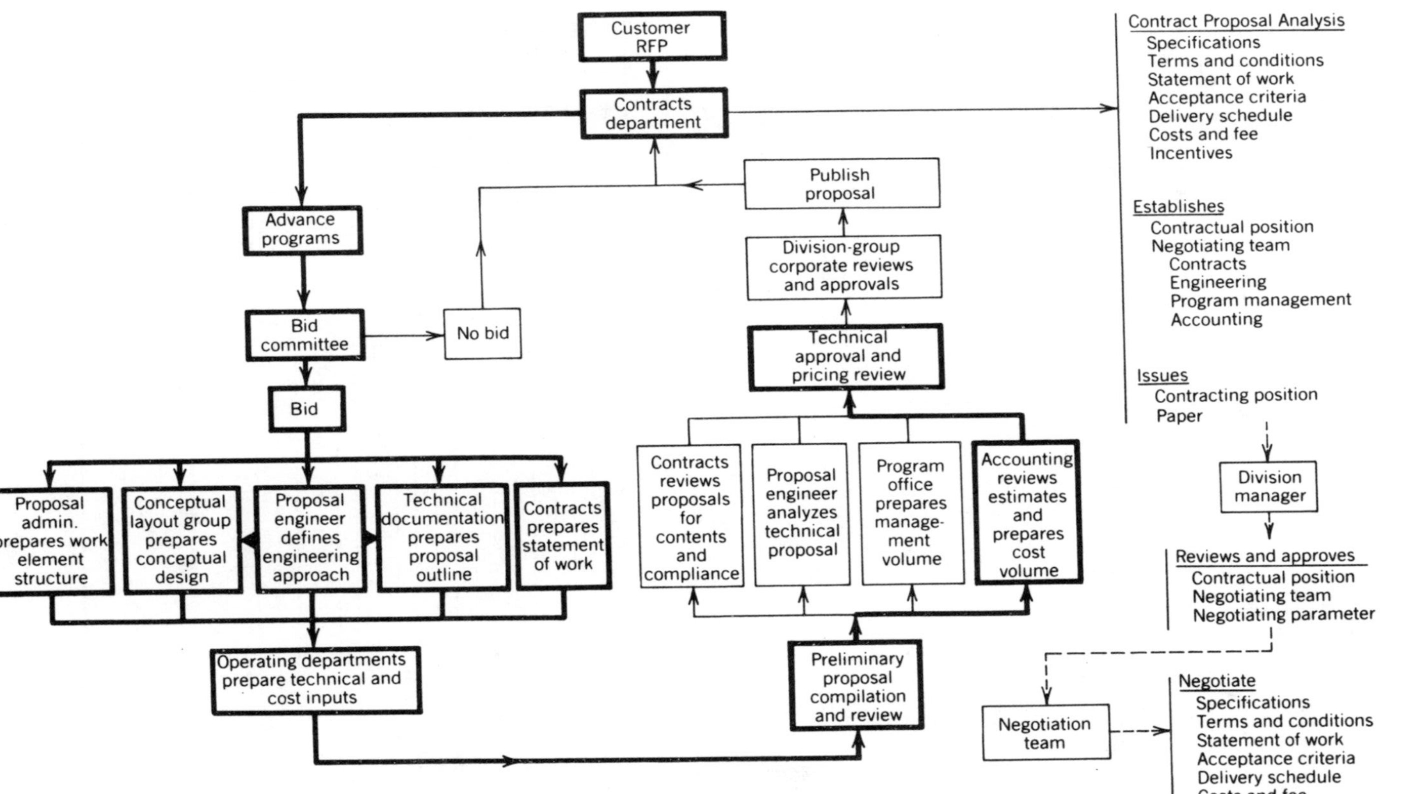

FIGURE 9.3 Proposal flow chart showing cost volume preparation.

DEPARTMENT OF DEFENSE **CONTRACT PRICING PROPOSAL** *(RESEARCH AND DEVELOPMENT)*		*Form Approved* *Budget Bureau No. 22-R0100*
This form is for use when *(i)* submission of cost or pricing data *(see ASPR 3-807.3)* is required and *(ii)* substitution for the DD Form 633 is authorized by the contracting officer.	PAGE NO.	NO. OF PAGES
NAME OF OFFEROR	SUPPLIES AND/OR SERVICES TO BE FURNISHED	
HOME OFFICE ADDRESS *(Include ZIP Code)*		
DIVISION(S) AND LOCATION(S) WHERE WORK IS TO BE PERFORMED	TOTAL AMOUNT OF PROPOSAL $	GOVT SOLICITATION NO.

DETAIL DESCRIPTION OF COST ELEMENTS

1. DIRECT MATERIAL *(Itemize on Exhibit A)*			EST COST ($)	TOTAL EST COST[1]	REFER-ENCE[2]
a. PURCHASED PARTS					
b. SUBCONTRACTED ITEMS					
c. OTHER - (1) RAW MATERIAL					
(2) YOUR STANDARD COMMERCIAL ITEMS					
(3) INTERDIVISIONAL TRANSFERS *(At other than cost)*					
TOTAL DIRECT MATERIAL					
2. MATERIAL OVERHEAD 3 *(Rate % X $ base =)*					
3. DIRECT LABOR *(Specify)*	ESTIMATED HOURS	RATE/ HOUR	EST COST ($)		
TOTAL DIRECT LABOR					
4. LABOR OVERHEAD *(Specify department or cost center)* 3	O.H. RATE	X BASE =	EST COST ($)		
TOTAL LABOR OVERHEAD					
5. SPECIAL TESTING *(Including field work at Government installations)*			EST COST ($)		
TOTAL SPECIAL TESTING					
6. SPECIAL EQUIPMENT *(If direct charge) (Itemize on Exhibit A)*					
7. TRAVEL *(If direct charge) (Give details on attached Schedule)*			EST COST ($)		
a. TRANSPORTATION					
b. PER DIEM OR SUBSISTENCE					
TOTAL TRAVEL					
8. CONSULTANTS *(Identity - purpose - rate)*			EST COST ($)		
TOTAL CONSULTANTS					
9. OTHER DIRECT COSTS *(Itemize on Exhibit A)*					
10. *TOTAL DIRECT COST AND OVERHEAD*					
11. GENERAL AND ADMINISTRATIVE EXPENSE *(Rate % of cost element Nos.)* 3					
12. ROYALTIES 4					
13. *TOTAL ESTIMATED COST*					
14. FEE OR PROFIT					
15. *TOTAL ESTIMATED COST AND FEE OR PROFIT*					

This proposal is submitted for use in connection with and in response to *(Describe RFP, etc.)*

and reflects our best estimates as of this date, in accordance with the instructions to offerors and the footnotes which follow.

TYPED NAME AND TITLE	SIGNATURE
NAME OF FIRM	DATE OF SUBMISSION

DD FORM 1 APR 68 633-4 REPLACES EDITION OF 1 JAN 67, WHICH IS OBSOLETE.

FIGURE 9.4 Government contract pricing proposal.

EXHIBIT A - SUPPORTING SCHEDULE *(Specify. If more space is needed, use blank sheets)*		
COST EL NO.	ITEM DESCRIPTION *(See footnote 5)*	EST COST ($)

I. HAVE THE DEPARTMENT OF DEFENSE, NATIONAL AERONAUTICS AND SPACE ADMINISTRATION, OR THE ATOMIC ENERGY COMMISSION PERFORMED ANY REVIEW OF YOUR ACCOUNTS OR RECORDS IN CONNECTION WITH ANY OTHER GOVERNMENT PRIME CONTRACT OR SUBCONTRACT WITHIN THE PAST TWELVE MONTHS?

☐ YES ☐ NO *If yes, identify below.*

NAME AND ADDRESS OF REVIEWING OFFICE *(Include ZIP Code)*	TELEPHONE NUMBER/EXTENSION

II. WILL YOU REQUIRE THE USE OF ANY GOVERNMENT PROPERTY IN THE PERFORMANCE OF THIS PROPOSED CONTRACT?

☐ YES ☐ NO *If yes, identify on a separate page.*

III. DO YOU REQUIRE GOVERNMENT CONTRACT FINANCING TO PERFORM THIS PROPOSED CONTRACT?

☐ YES ☐ NO *If yes, identify:* ☐ ADVANCE PAYMENTS ☐ PROGRESS PAYMENTS *OR* ☐ GUARANTEED LOANS

IV. DO YOU NOW HOLD ANY CONTRACT *(or, do you have any independently financed (IR & D) projects)* FOR THE SAME OR SIMILAR WORK CALLED FOR BY THIS PROPOSED CONTRACT? ☐ YES ☐ NO *If yes, identify*

V. DOES THIS COST SUMMARY CONFORM WITH THE COST PRINCIPLES SET FORTH IN ASPR, SECTION XV *(See 3-807.2 (c) (2))*?

☐ YES ☐ NO *If no, explain on a separate page.*

INSTRUCTIONS TO OFFERORS

1. The purpose of this form is to provide a standard format by which the offeror submits to the Government a summary of incurred and estimated cost *(and attached supporting information)* suitable for detailed review and analysis. Prior to the award of a contract resulting from this proposal the offeror shall, under the conditions stated in ASPR 3-807.3, be required to submit a Certificate of Current Cost or Pricing Data (see ASPR 3-807.3(e) and 3-807.4).

2. As part of the specific information required by this form, the offeror must submit with this form, and clearly identify as such, cost or pricing data (that is, data which is verifiable and factual and otherwise as defined in ASPR 3-807.3(e)). In addition, he must submit with this form any information reasonably required to explain the offeror's estimating process, including:
   a. the judgmental factors applied and the mathematical or other methods used in the estimate including those used in projecting from known data, and
   b. the contingencies used by offeror in his proposed price.

3. When attachment of supporting cost or pricing data to this form is impracticable, the data will be specifically identified and described *(with schedules as appropriate)*, and made available to the contracting officer or his representative upon request.

4. The format for the "Cost Elements" is not intended as rigid requirements. These may be presented in different format with the prior approval of the contracting officer if required for more effective and efficient presentation. In all other respects this form will be completed and submitted without change.

5. By submission of this proposal, offeror, if selected for negotiation, grants to the contracting officer, or his authorized representative, the right to examine, for the purpose of verifying the cost or pricing data submitted, those books, records, documents and other supporting data which will permit adequate evaluation of such cost or pricing data, along with the computations and projections used therein. This right may be exercised in connection with any negotiations prior to contract award.

FOOTNOTES

1 Enter in this column those necessary and reasonable costs which in the judgment of the offeror will properly be incurred in the efficient performance of the contract. When any of the costs in this column have already been incurred (e.g., on a letter contract or change order), describe them on an attached supporting schedule. Identify all sales and transfers between your plants, divisions, or organizations under a common control, which are included at other than the lower of cost to the original transferor or current market price.

2 When space in addition to that available in Exhibit A is required, attach separate pages as necessary and identify in this "Reference" column the attachment in which information supporting the specific cost element may be found. No standard format is prescribed; however, the cost or pricing data must be accurate, complete and current, and the judgment factors used in projecting from the data to the estimates must be stated in sufficient detail to enable the contracting officer to evaluate the proposal. For example, provide the basis used for pricing materials such as by vendor quotations, shop estimates, or invoice prices; the reason for use of overhead rates which depart significantly from experienced rates (reduced volume, a planned major rearrangement, etc.); or justification for an increase in labor rates (anticipated wage and salary increases, etc.). Identify and explain any contingencies which are included in the proposed price, such as anticipated costs of rejects and defective work, or anticipated technical difficulties.

3 Indicate the rates used and provide an appropriate explanation. Where agreement has been reached with Government representatives on the use of forward pricing rates, describe the nature of the agreement. Provide the method of computation and application of your overhead expense, including cost breakdown and showing trends and budgetary data as necessary to provide a basis for evaluation of the reasonableness of proposed rates.

4 If the total royalty cost entered here is in excess of $250 provide on a separate page (or on DD Form 783, Royalty Report) the following information on each separate item of royalty or license fee: name and address of licensor; date of license agreement; patent numbers, patent application serial numbers, or other basis on which the royalty is payable; brief description, including any part or model numbers of each contract item or component on which the royalty is payable; percentage or dollar rate of royalty per unit; unit price of contract item; number of units; and total dollar amount of royalties. In addition, if specifically requested by the contracting officer, a copy of the current license agreement and identification of applicable claims of specific patents shall be provided.

5 Provide a list of principal items within each category indicating known or anticipated source, quantity, unit price, competition obtained, and basis of establishing source and reasonableness of cost.

☆ U.S. GOVERNMENT PRINTING OFFICE: 1977—260-810/8097

FIGURE 9.4 *(Continued)*

RFP. No. ____________

Offeror ____________

Yearly Cost Summary

	Precontract Costs	First Year	Second Year	Other Years	Total
Labor hours "Productive" (hands-on labor) "Nonproductive" (other direct labor) Total straight time					
Overtime					
Total labor hours					
Costs Direct labor—straight time Direct labor—overtime Direct labor—shift premium					
Total labor cost Payroll additives Fringe benefits Overhead Other direct costs					
Subtotal					
G&A expense Total estimated cost Fee/profit					
Cost of facilities capital (cost of money) Total estimated cost & fee/profit					

FIGURE 9.5 Yearly cost summary. G&A: general and administrative costs.

RFP No. ______________

Offeror ______________

ST = Straight time
OT = Overtime

Labor Costs

☐ Prime
☐ Subcontract

☐ Exempt
☐ Nonexempt, Nonunion
Precontract ☐
Year ☐ #1 ☐ #2

Labor Classification	Avg. Headcount	Shift	Labor Hours		Labor Rates		Labor Costs			
			ST	OT	ST	OT	ST	OT	Shift Premium	Total
Totals										

FIGURE 9.6 Labor costs worksheet.

RFP No. ____________

Offeror ____________

Labor Costs Summary

☐ Prime
☐ Subcontract

Labor Classification	Avg. Head-count	Labor Hours		Labor Costs			
		Straight Time	Overtime	Straight Time	Overtime	Shift Premium	Total
Precontract Exempt Nonexempt, Nonunion Nonexempt, union							
Total							
First year Exempt Nonexempt, nonunion Nonexempt, union							
Total							
Second year Exempt Nonexempt, nonunion Nonexempt, union							
Total							
Remaining years							
Total							

FIGURE 9.7 Labor costs summary.

RFP. No. ____________ Payroll Additives ☐ Prime

Offeror ____________ ☐ Subcontract

Element	Base	Rate	Cost	Labor Hours	Cost per Hour
Precontract & First year Federal insurance compensation Act Federal unemployent insurance State unemployment insurance Workmen's compensation Other (specify)					
Total					
Second year Federal insurance compensation act Federal unemployment insurance State unemployment insurance Workmen's compensation Other (specify)					
Total					
Other Years					
Grand total					

FIGURE 9.8 Worksheet for payroll additives.

RFP No. ____________

Offeror ____________

Fringe Benefits

☐ Prime
☐ Subcontract

☐ Exempt
☐ Nonexempt, nonunion
☐ Nonexempt, union
Year ☐ #1 ☐ #2

Element	Base	Rate	Cost	Labor Hours	Cost per Hour
Priced fringe benefits Group insurance Retirement Savings plan Education assistance Other					
Total priced fringe benefits					
Fringe benefits included in labor costs Sick leave Civic and personal leave					
Fringe benefits not priced Severance entitlement Other (identify)					
Total fringe benefits					

*Exclude vacation and holidays

FIGURE 9.9 Worksheet for calculating fringe benefits.

RFP No. ________ Overhead and G&A
☐ Prime
☐ Subcontract
Offeror ________ Expense Summary

Offeror's fiscal year begins ________, and ends ________

Description of burden distribution bases
Overhead
G & A

Proposed Rates (Percentages)		First Year	Second Year	Third Year	Fourth Year	Fifth Year	Total
Overhead	Pricing						
	Ceiling						
G & A	Pricing						
	Ceiling						
Costs (Dollars)							
Overhead	Amount proposed						
	Amount at ceiling rate						
G & A	Amount proposed						
	Amount at ceiling rate						
Allocation base (amount)	Overhead						
	G & A						

Historical information (Percentages)		Recorded rate	DCAA* audited rate	Final negotiated
Overhead	Most recent year 2nd most recent 3rd most recent 4th most recent 5th most recent			
G & A	Most recent year 2nd most recent 3rd most recent 4th most recent 5th most recent			

*Defense contract audit agency

FIGURE 9.10 Overhead and general and administrative (G&A) expense summary.

RFP No. ______________________

Offeror ______________________

Overhead Expense Forecast

First contract year	☐	Prime	☐
Second contract year	☐	Subcontract	☐

Account	Amount

Total overhead expense pool ______________________

Distribution base (identification and amount)

Overhead rate

For indirect personnel included in the expense pool, specify functions included and the staffing and rates within each function.

FIGURE 9.11 Overhead expense forecast form.

RFP No. ______________________

Offeror ______________________

G & A Expense Forecast

First contract year	☐	Prime	☐
Second contract year	☐	Subcontract	☐

Account	Amount

Total G & A expense pool ______________________

Distribution base (identification and amount)

G & A rate

If local indirect personnel are included in the expense pool, specify functions included and the staffing and rates with each function.

FIGURE 9.12 General and administrative (G&A) expense forecast form.

RFP No. ______________ Other Direct Costs ☐ Prime

Offeror ______________ ☐ Subcontractor

Element	Precontract	First Year	Second Year	Other Years	Total
Travel Relocation Recruitment Training—contract related Other (list)					
Total					

FIGURE 9.13 Worksheet for calculating other direct costs.

expenses include costs of a motel or hotel room; food, tips, and taxis; local transportation and communication; and other costs such as laundry, mailing costs, and onsite clerical services. Transportation costs include the transport of equipment, supplies, and products, as well as personnel, and can include packaging, handling, shipping, postage, and insurance charges. Rationale for travel and transportation costs should be included along with their corresponding resource estimates.

Cost Growth Allowances. Occasionally a cost proposal will warrant the addition of allowances for cost growth in both labor and material costs. Cost growth allowances are best added at the lowest level of a cost estimate (in the labor and materials estimates) rather than at the top levels. These allowances include design growth allowances; reserves for possible misfortunes, natural disasters, and strikes; and other unforeseen circumstances. Reserves should be used to account for design growth, since cost growth with an incomplete design is a certainty. The cost proposal should include a discussion of the above-mentioned labor allowances and cost growth allowances, along with the computation method and rationale for their application.

COSTING IN THE PROPOSAL FLOW

As shown in Figure 9.3 when technical and cost inputs are completed by the operating departments, the accounting or finance department assembles the cost proposal or cost volume. The accounting department uses predeveloped labor rates for the various skills that will be used in performing the job; applies labor burden, material burden, and overhead costs; calculates cost-of-money; and proposes a fee structure. The forms shown in Figures 9.4 through 9.13 are typical forms used by the accounting, finance, or pricing organization in accumulating costs for the cost volume. Figure 9.4 is the contract pricing proposal that accompanies proposals to the government. (Recently, a more simplified version has been developed, but it requires backup information of the same type as shown on this earlier version of the form.) The other figures shown can be used for yearly cost summaries, labor costing backup, labor cost summary, description of payroll additives and fringe benefits (labor burden), overhead, general and administrative costs; and other direct costs such as travel, relocation, recruitment, and training. The priced proposal, after management review and approval, is forwarded to the proposal publishing activity along with backup rationale, appendices, and supporting data for incorporation into the published proposal.

10 PROPOSAL WRITING

Write the things which you have seen, and the things which are, and the things which shall be hereafter.

—Revelation 1:19

Great was the company of those that published it.

—Palms 68:11

The culmination of the proposal preparation activity is the writing and publication of the data, information, and backup material that have been collected, organized, and iterated during the proposal preparation process. Since written and published words are being used in the proposal as the means of communication to the reader or evaluator, the choice, organization emphasis, and interrelationship of these words are of vital importance for straight-line control. Proper selection of the right amount of text, illustrations, tables, and graphs and the presentation of this material in an attractive, readable, easily referenced format are the keys to stimulating the evaluator's positive reaction to a proposal. All of the material presented in the proposal must be useful to the reader or evaluator, and this material should work together in a synergistic way to depict clearly and concisely that: (1) the proposer has a thorough understanding of the job requirements, (2) the proposer has the capability and experience required to perform the work in a successful manner, and (3) the job will be carried out on schedule and within cost constraints in a high- quality manner.

As discussed in Chapter 4, the Graphic Representation of Work (GROW) method of initiating the proposal-writing process is exceedingly helpful in accomplishing all of the objectives stated in the previous paragraph. If storyboards have been generated for the proposal in question, and if the preparation, review process, updating and approval have proceeded using the GROW approach, the actual writing process will be relatively easy. The use of storyboards encourages an integrated team approach and storyboards represent a formal means for early feedback, and early change and updating of thesis, illustrations, and points to be emphasized. Since they provide a means for early team review and high-level management review and approval, storyboards should be well-coordinated, compatible, and should be made consistent before the actual writing process commences.

Since the major points to be made as well as the supporting illustrations have already been established and approved by the proposal team and by management, proposal writers have ready-made outlines and roadmaps to follow in the writing process. The time spent staring at a blank sheet of paper, wondering how to start and which points to emphasize will diminish dramatically, and the whole proposal-writing task will flow together smoothly and comfortably. Team reviews of the storyboards should have already ironed out incompatibilities and inconsistencies in the basic points to be covered and in the approach to be advanced in the various volumes or portions of the proposal. Writers and editors must now articulate and expand upon these major points to produce cohesive, compatible, integrated proposal text and graphics.

As mentioned earlier, "The quality of the proposal itself is the single thing, in or about a proposal that will most convincingly show the reader or evaluator that the company will provide a high-quality work activity or work output." The quality of writing, organization, illustrations, printing, and even the quality of the paper and cover are important in conveying the overall impression to the customer that a company exerts meticulous care in doing its work. Elaborate or fancy proposals without high-quality content or technical excellence, of course, are of little value. In-depth quality can be easily observed and readily recognized by the experienced proposal evaluator.

THE WRITER'S ROLE IN PROPOSAL PREPARATION

Synergism in Proposal-Writing

Unfortunately, many proposals fail to present a cohesive, integrated flow of material from start to finish. This is usually caused by the fact that several or many individuals, organizations, and departments contribute to the proposal, often making inputs just prior to the publication date, and because insufficient effort and time is applied by the proposal writer and proposal manager in tying together the entire package into a consistent and integrated story of how the company can and will do the work.

Writing a proposal is much like baking a cake. Each ingredient in and of itself would be rather tasteless and uninteresting. But when the ingredients are combined in a precise manner and allowed to interact and unite into one entity, the whole becomes attractive, palatable, and highly desirable. The combination of ingredients under the right interactive conditions has formed a whole new product. This synergistic approach must be used in proposal writing if a high potential of capture is expected. The proposal writer must avoid having the proposal appear as merely a group of ingredients that have been collected from several sources and have not been correlated in writing style, mathematical consistency, or technical content. When inputs are derived from several sources, the information must be thoroughly digested and reconstituted into a total cohesive package through the proposal consolidation, review and reconciliation process. This reconstituted information must then be put into words by a skilled writer or team of writers who have an overall appreciation for the importance of the work to the company and a positive and almost aggressive optimism related to acquisition of the new work. All of this must also be presented in a manner that meets good writing practice on the areas of style, grammar, punctuation, spelling, and word usage.

Words Stimulate

The proposal writer must remember that throughout the proposal the objective is to stimulate the reader or evaluator to make a positive decision relative to the company's expertise and ability to do the work successfully and in an outstanding manner. This ability is usually reflected in the writer's words, by an underlying confidence and assurance that the company will win and by the documentation of authentic and convincing experience. When writing about specific activities or outputs to be provided to the customer, it is usually most effective to use the first person future plural tense in describing work to be done. For example, "we will perform the following packaging and handling activities," is a positive, future-tense statement that presumes that the proposing company will be awarded the contract. Avoid statements like "we would" or "if the contract is awarded to us we would . . . ," which imply doubt as to confidence in acquiring the work. The best practice to follow is to write the work statements and plans *as if the contract has already been received*. This writing style reinforces the reader's assurance that the proposing company has both the desire and ability to do the work and that the company has taken a winner's attitude and approach to acquiring and performing the work successfully.

Another example of a positive writing style that will assist in stimulating the reader or evaluator to make a favorable selection recommendation is to use past-tense statements when it comes to reflecting commitments to the project. For example, "We have assigned Mr. Roger Scott as project manager for this work"; or "We have set aside 2,600 square feet of our K–10 building shop area for the tooling and equipment described below." Both indicate a degree of company precommitment to the work. The third-person singular past tense is best used when describing actual past accomplishments of personnel, equipment, or company. As an example; "Mr. Scott successfully managed the REVAMP and CONTROL projects during 1990. He completed these projects within cost targets and on schedule."

Documenting and Explaining a Record of Success

A proposal is no place to be modest. A proposal is one document in which it is not only desirable but absolutely necessary to document and explain successes in areas directly or indirectly related to the proposal task. Flowery superlatives relative to a company's ability should be avoided in the proposal itself. However, they can be used sparingly in the letter of transmittal. A well-documented historical record of successes, along with positive testimonials of third parties, will give a firm the competitive edge over others who have equal technical excellence but who have failed to describe and document their successes.

What about past failures? Abilities and capabilities should be emphasized, rather than limitations and weaknesses. Being honest does not mean pointing out one's faults. It does mean being completely truthful about capabilities and not misleading the customer into thinking that facilities, personnel, finances, or resources are available to do the job when they do not exist. If a past major program failure does exist, it is desirable to point out the corrective measures that have been taken to preclude recurrence of the malady. But this description must not be overly complex or cumbersome, as it will detract from the positive aspects of a company's capabilities.

The Proposal Writer as a Storyteller

The proposal writer must be sufficiently familiar with the request for proposal or request for quotation and the overall project plans to assess and bring into focus the relative importance of each phase of the work and to generate an overall consistency and flow of information in the proposal to make it both interesting and persuasive. The proposal writer is usually responsible for finalizing and coordinating a detailed outline of the proposal's contents, a glossary of terms, and a bibliography of reference books, periodicals, reports, and other documents. The proposal writer is not only the grammarian but the integrator of words and ideas, imparting just the right emphasis and explanation to each part of the written proposal. The writer is the storyteller who makes the proposal stimulating, even exciting, to the reader or evaluator.

The Customer Should Experience the Proposal

Whether the proposal is one of just a few pages, or one of the mammoth proportions, a skillful job of organizing, writing, editing, and publication will allow the customer to experience the proposal rather than to just reading it or scanning it. In writing and publishing a high-quality proposal, the evaluator's attention is not just attracted but demanded by the careful, meticulous, and accurate presentation of the facts—and the proposal has been presented in an interesting and compelling way that is designed to stimulate interest, evoke a decision, and to culminate in action: the selection of the firm to do the work. As shown in Figure 10.1, publication of the proposal closes the loop in the proposal flow cycle and triggers prenegotiation activities that precede actual contract award. The published proposal is sent to the customer and is distributed to those in-house organizations that have to prepare for action when the firm is selected and the contract is awarded to do the work.

IN-HOUSE PUBLICATION OF THE PROPOSAL

If a company is likely to submit proposals on a periodic basis and not on a one-time-only basis, it is most practical, desirable, and efficient to do proposal publication

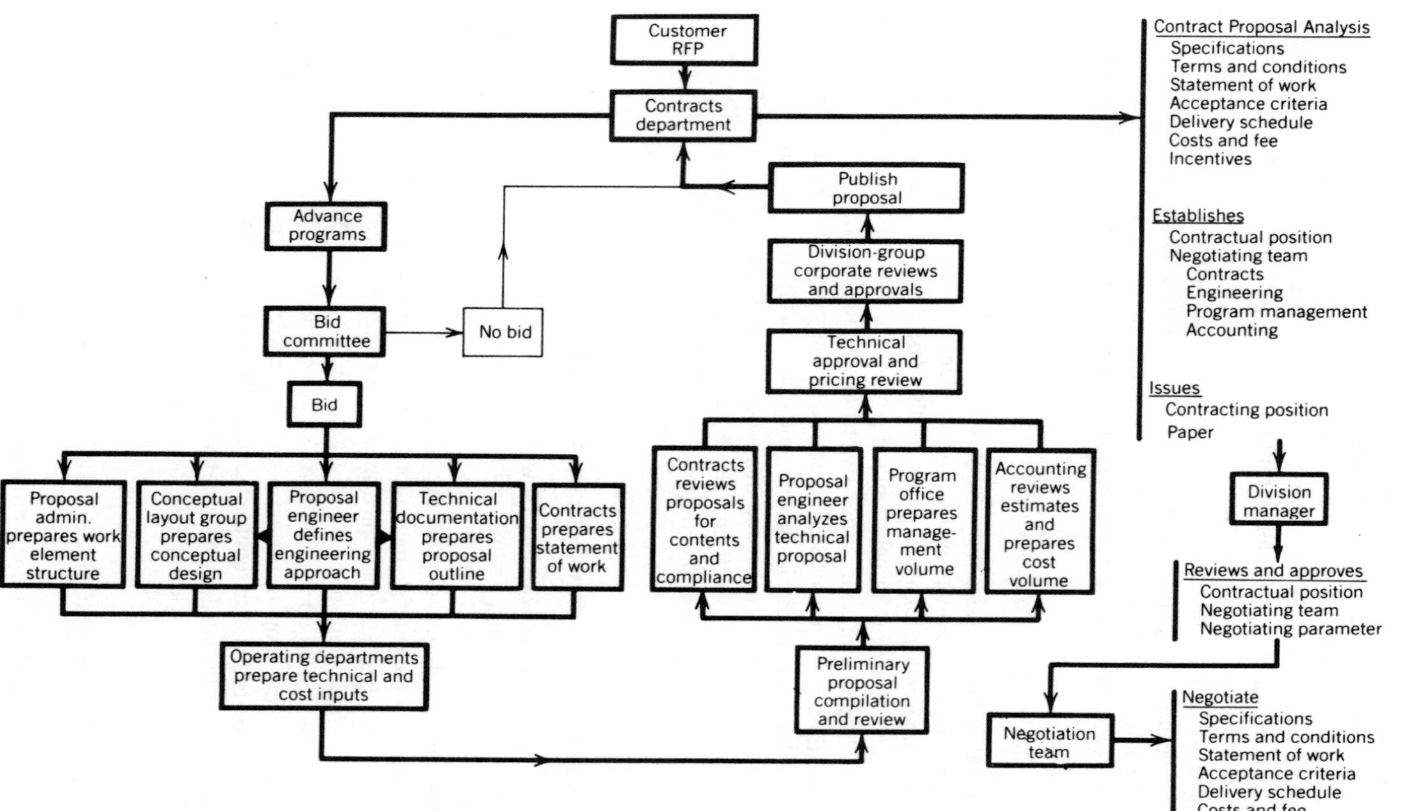

FIGURE 10.1 Proposal flow chart showing completion of proposal cycle.

work within the company's organization. In-house publication of proposals, using company personnel and equipment, is desirable because (1) the potential performers of the work should be involved as closely and as integrally as possible with the publication process; (2) the confidentiality of proposal information should be preserved; and (3) the timeliness of publication should be ensured. Other benefits of in-house publication of proposals are immediate access to the publication staff and to the proposal material to incorporate changes and maximum responsiveness of the publication staff to company objectives. In our opinion, neither proposal writing nor proposal publication are appropriate activities to subcontract to another company.

To publish a proposal in house, personnel, skills, and equipment must be surveyed to determine the methods of composing and reproducing printed text and illustrative matter that are available within the company. The next chapter (Chapter 11), on computer-aided proposal preparation, provides a view of the present as well as a view into the future of some of the advanced tools that are and will be available to proposal writers.

WORD PROCESSING: A BOON TO PROPOSAL PUBLISHERS

By far the predominant method of composition of proposals is the use of the word processor or desktop publishing system. The greatest advantage of computerized text management systems is that the basic text need only be entered once. Many proposal writers enter the text themselves. There is no need of typing and retyping preliminary drafts, rough drafts, final drafts, and final copy. A highly legible hard copy draft can be produced rapidly on a high-speed laser printer; edited or modified by the originator, writer, and/or editor; and modified by adding, deleting, moving, or changing only those words or sections affected. Desktop printers can be given instructions to produce double or multispaced drafts for ease of editing, or single-spaced, formatted final copy incorporating graphics for photocopy reproduction.

Proposals input on the proposer's word processor can be output by phototypesetting equipment if correctly coordinated with the typesetter to provide compatible control codes. Phototypesetting equipment can provide a wide variety of text designs, fonts, and type styles. Since most proposals also employ computing equipment in deriving the final formats of the cost estimates and resulting price estimates, it is relatively easy to integrate tabular and graphic pricing and resource-estimate information with related text.

THE PUBLICATION DEPARTMENT'S ROLE IN PROPOSAL PREPARATION

The company's publication department usually has responsibility for organizing, publishing, and sometimes even writing the proposal document or documents. Of immediate and prime concern to the company publication manager is the planning and scheduling of publication activities and the availability and allocation of personnel. The proposal publication manager must be a person who is fully acquainted with all phases of publications work; who is available to devote full time to the effort, if required, during the publication phase; and who can work under pressure. Above all, this person must be one who is aware of the magnitude of the work being proposed and who appreciates the efforts of, and pressures on, the other members of the proposal preparation team. The publication leader, knowing the situation, must be able to obtain cooperation and meet schedules. The publication manager should schedule other department work so that additional people will be available when needed. This usually means that the entire publication department of a small company will be tied up for the last several days of the proposal effort.

Once a publication manager has been selected, he or she should devote full attention to the proposal. The first task of the publication manager is to see that standard formats, forms, software, and desktop publishing equipment are provided to proposal team members. Then he or she must attend proposal conferences and briefings and apprise members of the team of their specific responsibilities in the overall effort. During the initial stage of development, contributions will be received from individuals preparing the technical volume and organization and management volume. In the meantime, the publication manager will prepare corporate background material and organize outlines, progress charts, and other standard material.

The publication manager, as well as the writers, should become thoroughly familiar with the request for proposal or request for quotation. The RFP or RFQ may specify a particular format or other publication requirements. A review of this document will familiarize the publication manager with the technical aspects of the proposal and will put this person in a position to allocate attention to the various phases of the work being proposed.

The proposal publication manager must work very closely with the proposal publication team to plan the editing and the physical word processing or desktop publishing, as well as the printing, binding, and mailing. Proposal production includes integration of text material, illustrations, tables, graphs, photographs, indices, appendices, bibliographies, tables of contents, summaries, inserts and foldouts, and special covers. To avoid excessive work during the last few days prior to submitting the proposal, lengthy sections can sometimes be reproduced and held for collation.

A word-processing or desktop publishing editing format sheet should be prepared, since final text preparation will probably require several keyboard operators, who may be unfamiliar with the company's or project's format requirements. The format should be based on two precepts: readability and ease of input. Abbreviations and often-used special words must be standardized as soon as possible in the proposal cycle, as should reference symbols, mathematical terms, and format for the preparation of formulas. This guide matter should be distributed to all parties concerned before preparation of final material is started.

While preparation of the technical volume or volumes is in its early stages, the publication department can prepare such sections as company background and history, experience in similar projects, résumés of key personnel, and facilities descriptions. In a well-organized publications department, this material can be lifted from a hard copy or word-processor library and tailored to suit the particular proposal. If it is not available as such, it must be gathered from previous proposals or obtained from company management and from other departments. If these sections are voluminous, they can be printed in advance, as can covers, separators, appendices, and indices. By the time these tasks are completed, information will be coming in from the technical volume and organization and management volume teams and proposal writers. If it is not being received, gentle (or not so gentle) prodding may be in order. The originator of the written material should not spend time polishing up the grammar and syntax, as this can be done in final copy editing of the entire document. Photographs or other material requiring special artwork must be prescheduled and received on time, and text inputs must be received in time to prepare a final draft for management review prior to final text and graphics integration, polishing, and reproduction.

PHYSICAL CONSIDERATIONS

Tables and Illustrations

Schematic and block diagrams, flow diagrams, charts, and graphs are absolutely essential to understanding how a system or piece of equipment functions. Nothing is more fatiguing to a proposal reader or proposal evaluator than to have to wade through numerous pages of technical data that require the reader to form mental pictures when real pictures would have been so much better. The reader can be helped to understand the work activity or work output being proposed by being given visual aids frequently. Text should be converted to tabular form wherever possible. Tables are more easily seen at a glance and allow the reader to find what is being looked for quickly. If the storyboard approach has been used, integrated, compatible, and approved graphics will already be available.

If available, hardware photographs can add a touch of realism that cannot be duplicated with mere words. If photographs are to be used, however, they must be reproduced in such a way as to have the greatest impact. Extraneous material that detracts from the important features of the work activity or work output should be either cropped out or retouched.

Sometimes there is an opportunity to achieve great impact and drama with three-dimensional shaded graphics. For example, for a piece of equipment or an operation that is not yet built or is in process, a detailed conceptual illustration can never be construed to be unnecessarily elaborate. A detailed rendering can be very effective since it imparts a feeling of reality that could not be achieved with a line drawing. Some computer-aided design programs can produce highly realistic shaded three-dimensional representations that appear almost photographic. This type of illustration can be used if it serves the purpose of stimulating reader interest, understanding, or knowledge of the work being proposed. Resolution in dots per inch of the reproduction of these artforms should be such that it will preserve the quality of the original as much as possible.

Illustrations and tables should be located as closely as possible to the point of text citation. Placing the smaller illustrations in the text provides the ultimate in readability. Full-page illustrations should be avoided if one-sided printing is used, since it is desirable to have the referencing text and the figure on the same page. If pages are printed on both sides, however, a full-page (or smaller) illustration can face its text citation. In instances where extremely large flow diagrams, tables, charts, or graphs are required, a foldout page is the only solution. These large pages can be more easily referenced by the reader if an 8½ × 11-inch blank apron is provided as the left panel of the foldout and the figure is inserted as an appendix at the rear of the volume. These sheets may then be folded out and referenced beside the text in any portion of the proposal volume.

Proposal Covers and Bindings

The cover is most likely the only part of the proposal that will resemble a brochure. A good cover may win a significant psychological advantage by its impact and provide a chance to gain points even before the proposal is opened. The most objective evaluator cannot fail to be motivated to read further when confronted with an interesting and well-executed cover. The cover will gain attention, while the material inside the volume must hold attention. The cover, however, need not be overly expensive to be effective. With the imaginative use of paper stock and two colors of ink, an impressive cover can be produced. A photograph, combined with good color balance and creative typography, can often be eye catching as well as meaningful to the evaluator. The creative nature of the company preparing the proposal will generally be reflected in the proposal cover design. Full-color photographs and multicolor covers are now practical with the advent of new high-speed, low-cost color copiers. Covers should be of adequate weight to sustain repeated handling. Generally, a 100-lb. cover or a double weight (135-lb.) cover is used.

Except for proposals of less than 50 pages, the best methods of binding are plastic combs or multiring notebooks. For proposals where a large number of volumes and/or copies are required, the former is a more practical binding method. Proposals with few copies required can be bound in multiring notebooks. Many interesting styles of notebooks are available, some of which contain clear plastic covers on the front, rear, and binding edge for insertion of a specially printed inserts. Some notebooks come equipped with D-rings that allow pages to lie with even edges; some have pockets on the inside covers for insertion of supplemental information. Although these notebooks are very attractive and handy for evaluation, they have the disadvantages of high price and a greater possibility of lost, removed, or miscollated pages than more permanently bound volumes. Comb-bound proposal volumes may be more than one inch thick, yet their facing pages lie flat when the proposal is opened, just as is the case with the more expensive multi-ringed notebooks. Pages may be removed and replaced if necessary in comb-bound documents, but are less likely to be inadvertently removed. Pages may be printed separately, collated, and then bound. Plastic combs come in an assortment of colors, sizes, and lengths to adapt to almost any binding situation. This binding method can be accomplished in-house with relatively inexpensive equipment. Proposals of less than 50 pages may be bound like a brochure. A one-piece cover is used and the pages are wire or thread-stitched or stapled to the cover. This method, however, is not geared to the rapid, in-house processing demanded by most proposals.

Index Tabs and Figure Positioning

The practice of including index tabs or index tab sheets at major subdivisions within each proposal volume will make the evaluator's job easier and will provide an ability to more quickly and readily flip to a given portion of the proposal for a quick look or in-depth study. Index tabs on major headings and even subheadings should be clearly marked, durable, and color coded if possible to provide a ready visual reference. Tabs should be printed right-side up when the proposal volume is rotated one-fourth turn clockwise. Figures or tables, if not placed on the page vertically, should also be readable right-side up when the volume is rotated clockwise one-fourth turn. Two major irritants to proposal evaluators are: (1) difficulty in finding and cross-referencing information and (2) having to turn the book with the binding toward them to read illustrations and tables. The above positioning of index tabs and figures will reduce or eliminate these irritants.

STRUCTURE OF THE PROPOSAL

The Proposal Outline

An outline of a typical multivolume proposal format is shown in Figure 10.2. Whether the proposal is a multivolume proposal or a single-volume proposal, it will contain most of the subjects listed on this outline. The first step of overall proposal writing and publication that is taken jointly by the publications leader, writer, and proposal manager is to prepare and approve a detailed outline. If the proposal itself is voluminous, this outline could extend to as much as 10 pages. If the storyboard method has been used, outlines should be compatible with the storyboard outlines. The proposal preparation, writing, assembly, and publication processes will be much easier if this detailed outline is prepared at the beginning of proposal development and kept up to date as the proposal takes shape. This outline is used as a detailed guide for writing, assembly and integration of all parts of the proposal, but can also be used to make a rough estimate of the number of words, pages, illustrations, foldouts, or tables needed in each section and subsection. This information will allow the publication leader to decide how many volumes and covers are needed and the approximate size of each volume. The detailed outline is the framework of the total proposal and the structure, which permits the writer to present proposal information with a common thread of continuity. A prearranged outline numbering and format system should be developed in advance of storyboard preparation to assure outline numbering or lettering consistency. The detailed outline of the proposal should be accompanied by a cross-reference matrix that indicates the RFP statement of work being addressed and the work breakdown structure element that is applicable.

Ia Letter of Transmittal (sometimes put in each volume)

A summary of what is being proposed
Company's confidence and support in accomplishing job
References to RFPs, letter requests, phone calls, etc.
When the notice to proceed must be given
Time duration of proposal validity
Payment provisions
Special stipulations and conditions

Ib Executive Summary

Ic Front Matter

Title page, foreword, table of contents (table of contents for all volumes should be in *each* volume), list of illustrations, list of tables

II Technical Proposal

Scope of work
Description of work
Schedule for the work

III Organization and Management Proposal

Company organization
Project organization & key personnel duties
Résumés
Experience and past performance

IV Cost Proposal

Resources estimates
Rationale and backup information
Cost summary and pricing information

V Appendices

Draft contract
Company financial report
Facilities descriptions
Supplemental material

FIGURE 10.2 Proposal outline for multivolume proposal.

The Letter of Transmittal

Although the letter of transmittal of a proposal is often written last, it is probably next in importance to the proposal cover in stimulating the evaluator to read further. The letter of transmittal should, in a one-paragraph opening, describe the work activity or work output that is being proposed. This summary can be further supported by an "Executive Summary" that is included in a separate volume or at the very beginning or very end of the proposal material itself. The letter of transmittal should convey the fact that the company has an unshakable confidence in its ability to perform and complete the work on schedule, within cost, and in a high-quality manner and that the project has the total support and backing of corporate management. If any superlatives are used in the total proposal package, the letter of transmittal should include them. Statements such as "In view of this company's outstanding past successes in performing similar and identical tasks, as well as our competitive price estimate, it is appropriate and prudent that our firm be employed to perform the work," are not unheard of and often are acceptable in a letter of transmittal, where as they would be overbearing or presumptuous when included in the proposal text material.

The letter of transmittal should also include references to the request for proposal or request for quotation, telephone conversations, or previous correspondence if appropriate and other important information to be highlighted such as time duration of proposal validity, payment provisions, and any other special stipulations or conditions. For example, if the entire proposal is based on the customer's supplying drawings, technical support, parts or materials, facilities, equipment, goods, or personnel on specified dates, this should be clearly stated and the appropriate proposal volume and page should be referenced in the letter of transmittal.

Front Matter

The front matter of a proposal volume generally consists of a title page, a proprietary information statement page, a foreword, a table of contents, a list of illustrations, a list of tables, and an executive summary (if not included in a separate volume). The title page includes the name of the work activity or work output; the organization to whom the proposal is submitted; the name, address, and telephone number of the organization submitting the proposal; the date submitted; the security classification (if any); and a document or library number. The proprietary information statement page states that the proposal information cannot and should not be transmitted to other companies. The foreword is a one-paragraph statement serving to identify whether the proposal is internal, external, unsolicited, or responsive to a request for proposal or request for quotation. If the proposal is in response to a request, the requesting document title and number should be referenced.

Table of Contents, Index, List of Abbreviations. The table of contents is a roadmap to help the reader determine the major and minor elements of the proposal and their interrelationship with each other. It should include a cross-reference list to pages or paragraphs of the request for proposal or request for quotation. This cross-reference list will help the evaluator quickly identify which part or parts of the proposal is responsive to which RFP/RFQ requirement and will assist in determining responsiveness to each requirement. It is sometimes desirable to include the table of contents and cross-reference list in all volumes of a multivolume proposal, to allow the reader to find material in volumes other than the one currently being evaluated. As will be shown in Chapter 12, large proposals are usually evaluated by source-evaluation boards consisting of several teams. Each team may evaluate its respective volume (technical, organization and management, and cost) and not have an opportunity to view the overall proposal content except in the form of a table of contents or index. Also, terms and abbreviations that are consistent between all volumes often appear in more than one volume, necessitating the repetition of lists of terms and abbreviations in each volume. A list of illustrations and a list of tables are included to permit a rapid scanning of pictorial or graphic material in the proposal. If there are relatively few tables in the proposal, the table list may be combined with the list of illustrations.

The Proposal Executive Summary

The executive summary, which can and often does appear at the front of every proposal volume, is a brief presentation containing the desirable results the customer will obtain, how these results will be ensured, and why this particular company should be selected to do the work. It strongly emphasizes the key sales points of the proposal.

It is a tried and true principle of good salesmanship that first impressions are both crucial and persistent. The initial impression gained by the customer from a proposal is derived from the proposal executive summary. The effects of this impression can influence the evaluation of the entire proposal. This is the evaluator's first encounter with the text. This portion of the proposal should be the ultimate in clear, persuasive prose.

The executive summary will be read by individuals at several decision levels in the customer's company. A few of these decision-makers will be technical experts. The majority of them are more concerned with basic relationships than with technical detail. Major new-work acquisitions are of interest to other divisions within the customer's organization. The executive summary can be used by the customer to inform these interested parties about the new work. The proposal executive summary is an ideal vehicle to inform, persuade, and educate this group of people. The summary is simply a capsule version of the new-work activity.

Objectives of the Executive Summary. The summary sets the scene. It should convince the customer that it is worthwhile to read, study, and constructively evaluate the proposal. To accomplish this, the client's own self-interest should be appealed to by highlighting those key benefits (three or four at most) that will be gained by selecting this firm to perform the program. If these benefits are among the two or three that the customer has indicated are the most desirable, they will be doubly effective.

Preparation of Executive Summary. The executive summary can best be prepared in the final stages of the proposal preparation process. At this point in the preparation cycle, the solution is fully developed and all of the relevant points about the program have been defined. These points form the basis for writing the summary. The proposal writer has been thoroughly indoctrinated during the preparation of the proposal and at this point in the cycle should be free to prepare the executive summary.

The executive summary, being of such crucial importance to the proposal, demands the best writing talent available. The best talent can produce a better job if an organized approach is used. These are some steps to use in writing an executive summary:

1. The entire document should be read carefully and brief notes made of the important points.
2. These notes should be put on 3″ × 5″ index cards or entered into a computer database or desktop manager program, one item per entry. Other important items can be added as required. The entries can then be arranged in a logical sequence.
3. With these notes as a guide, the writer can construct sentences that express the ideas put forth in the proposal.
4. Order and sequence should be evaluated and changes made as necessary. Then the sentences can be expanded into paragraphs, along with connectives and transitions as appropriate.
5. The entire executive summary should be reread to ensure that it does not contain any statements that are not substantiated by the source. A reader unacquainted with the original should be able to glean the essential facts from the executive summary and form valid conclusions.

Technical, Organization and Management, and Cost Volumes

The contents of these volumes is discussed in detail in Chapters 7, 8, and 9. The writers, editor, and publisher must be sufficiently familiar with these three volumes to assure that there are no overlaps, duplications, inconsistencies, or omissions in the three volumes and that all three sections of the proposal are consistent in format, content, grammar, abbreviations and names used, references and cross-references, page, figure, and table numbering, and printing style. The writing and publishing team must also assure that the three proposal sections are compatible with other proposal material, such as the letter of transmittal, appendices, and any supplemental material requested by and supplied to the potential customer.

Appendices

Appendices are normally used to include a draft or final contract document, company financial reports, facilities descriptions, and supplemental proposal backup material. Appendices and referenced documents are often used to avoid exceeding proposal page limitations and can be provided, if desired, only on request, in order to reduce the volume of material initially transmitted to the customer. Including a draft or final contract document as an appendix, enclosure, or reference has the tremendous psychological advantage of giving the customer the opportunity to immediately respond positively with a contract award or request for a negotiation session if the customer likes the contract document, however. The disadvantages of including a proposed contract document are: (1) not enough is usually known about final contract provisions at the end of the proposal cycle, and (2) without sufficient explanatory words, the contract document, although a draft, may be construed as a final non-negotiable position, whereas the proposer actually intends to leave the door open to reasonable contract changes.

PUBLISHING THE GIANT PROPOSAL

The size of a proposal can vary from just a few pages to many volumes of pages. Even the best-managed publications department can be completely thrown into disorganization by the sudden demand for thousands of pages in a limited number of days. It is not uncommon for proposals for large, complex, high-technology projects to require up to 50 volumes of proposal material for each copy of the proposal. Usually the beleaguered manager of such a proposal cannot hope to reduce the deluge of paper or extend the time. Time can be bought, however, by using every clock hour and by adding manpower: creating a corps of deputies and supporting organizational elements. A large ad hoc organization is usually required to meet the demands of a giant proposal. Not every company will produce one of those classic piles of three-ring binders, but whatever the company produces, it should be braced for the day when it is asked to turn out an immense document in a very limited time. Then pages will be calculated and people will be counted and there will not be enough clock hours available for the regular staff in the usual facilities to just pick up and lay down the sheets, to say nothing of reading them, sorting them, or making marks on them.

This company will need help—a giant ad hoc organization to cope with a giant assignment. An effective ad hoc publication organization can often be created by merely expanding and reinforcing existing publication units. But it should be remembered that all the internal help that is to be concentrated on a special job must be obtained at the expense of the other activities of the company. It may be necessary to resort to a bit of temporary distortion of job descriptions and functions to assure that maximum benefit is extracted from the talent available.

Although proposal preparation, writing, and publication are inherently and necessarily in-house tasks because of time and company security limitations, the publication of an immense proposal is an occasion to deviate from an in-house proposal-development policy and to use outside help in selected areas and for selected tasks. The most appropriate tasks to subcontract are the physical aspects of proposal publication, such as final typesetting, printing, binding, packaging, and preparation for mailing. All writing, editing, and final mailing must be done by company personnel to assure technical and business accuracy and timeliness of mailing. If a vendor is used, security and work quality should be closely controlled by onsite monitoring of the vendor's activities to ensure that publication support is not shared with a business competitor.

Irrespective of the chain of command or the titles bestowed on groups and individuals, activity throughout the ad hoc proposal organization will fall into three categories: developing content, processing, and production. The special demands of the giant job on each of these functions can be described without attempting to define any rigid organizational structure.

Developing Content

Required: Strong Editorial Leadership. Heightened interest and involvement on the part of the proposal technical staff will bring new and able people into the act as authors and reviewers. The storyboard method is essential to the exercise of strong editorial leadership on a giant proposal effort. The storyboards of the original authors will be reviewed by their superiors and by the review teams at an early stage in the proposal preparation process. Writing will also be reviewed by supervisors and review teams when the proposal is in the preliminary draft, final draft, and final form. The publication staff must devote itself principally to the mammoth task of editing, sorting out, and checking all the input information. For a giant proposal, it is important that early in the planning stages a top level proposal manager and technical editor work closely with the proposal authors to provide leadership on such things as making the storyboard outlines, establishing relative emphasis of the various proposal elements, apportioning content among volumes, determining format, and setting ground rules for the graphic material. Poorly conceived instructions to the originating authors can complicate processing operations far downstream. Within the restrictions imposed by the request for proposal or request for quotation, a company should get things off on the right foot mechanically. New and special formats, nonstandard page sizes, special type faces, or anything else that will work a hardship on the group or on vendors should be avoided.

The Very Detailed Outline. Developing a very detailed outline down to a low level and "freezing" it early in the proposal preparation process is particularly important for the giant proposal. Starting with a general format outline such as that shown on Figure 10.2, a more detailed outline can be constructed. An overall outline is sometimes indicated by the requesting organization in the request for proposal or request for quotation. Given some freedom in the detailed structure of the outline, most technical people can construct an array of topics and subtopics and apply the prearranged outline numbering structure to show interrelationship and subordination.

The challenge for the helpful proposal leader and editor is to supply lively, active-voice titles that in turn challenge the technical author or inputting organization, as well as constrain the subject matter to its specific contribution to the overall proposal. The outline for a giant proposal must give many far-flung authors the proper perspective on their individual contributions. The outline of storyboards is also the basis of checklists and status charts of authors' contributions, which will be needed in the processing operations. The very first outline draft must contain detailed page quotas; in general 2 or 3 pages will result from each storyboard.

Previous proposals on similar subjects and detailed reports on constituent subjects of the giant proposal will provide first-cut estimates of page quotas. If this outline is well-done and not allowed to fluctuate, at least in its major subdivisions, the publication department can prepare, in advance, the assortment of covers and tabbed dividers that will be required for the multitude of volumes and sections of the proposal.

The Logistics of Handling Large Proposal Inputs. Although it should be one's intent to avoid including large volumes of support material if not needed for complete understanding of a proposal, often the customer requires this data because entities within the customer's organization have requested it. The practice of consigning everything that can be interpreted as support material to appendices not only helps the reviewer by reducing the amount of material to be waded through to obtain continuity, but facilitates production by allowing transfer of existing materials from previously published documents or word-processing files. If extracts or complete issues of reports by a company or by a subcontractor are to serve as appendix material, every effort should be made to use them in their original form and to run them from the original stored information. It is no problem or violation to insert an opening page in the appendix stating that "Appendix C is identical to Section 5 of the June 1982 report on . . . , and the format of the original report is retained."

When providing large quantities of data such as computerized cost printouts, reduced copies of these machine-produced plots and tabulations can be included to provide the customer with authentic original information. Considerable time can be saved in producing these sections if patching or mounting of such tables is resisted. A single lead page can point out what the column heads are and describe the content and sequence of the sheets that follow. Bulk should be reduced and duplication of excessive or repetitive material should be avoided. Sometimes a selected computer run can be followed with text or a table that summarizes its results.

Simplifying Illustrations. Creators of content for the giant proposal must be made aware early in the proposal development process of the graphics requirements. General ground rules such as size, clarity, position on page (horizontally or vertically), labeling, etc., should be provided to all members of the proposal-writing team. As the proposal process continues, the graphics department will begin to assess the types, sizes, and complexity of figures, photographs, tables, charts, and graphs that are to be included in the proposal. Standardization of format can be used to advantage when including huge numbers of illustrations. Hard copy or computer-based blank forms can be prepared, for example, if many different calendar bar chart schedules or spreadsheet tables are to be produced. If a number of graphs are to be included, standard grids can be established. Mass production of figures can employ many artistic tricks of the trade to reduce illustration generation labor hours for duplicating routine formats. Photocopy techniques can produce neat, clean, legible reduced or enlarged figures and diagrams without compromising overall proposal quality. Computer graphics outputs in standard formats speed the production of figures and charts by at least an order of magnitude and can reduce updating and revision times to near zero. Individuals who are preparing inputs to proposals should be cautioned to avoid the practice of placing too much text in a graphic illustrations.

Processing the Giant Proposal

The activities that are contained in the conversion of raw content to final copy ready for printing are commonly grouped together under the term *processing*. If the groundwork mentioned earlier has been thoroughly done, processing of the information will be much easier. The huge proposal still imposes some heavy loads on editors, graphics personnel, and word processor operators. The processing phase not only includes the assembly and integration of the total package and preparation for printing, but also includes the review, correct, and approve cycle. The approval cycle itself can cause changes and iterations in the proposal material, particularly if last-minute management or technical decisions are made to adjust the company's competitive or technical posture.

Coordinating the Processing Operations. Normal publication operations use editorial groups to guide the technical authors, compile and sort out the pieces, and steer things through the production process. The volume of work in a giant proposal both permits and necessitates the separation of these functions. First, there is true editing, which is the skill of making the words flow together and conform to requirements for accuracy, grammar, and format. Then there is an enormous amount of checking and proofing required, followed by actual formatting and styling of the final document. This last step requires steady and patient workers adept at screening out the confusion without relaxing the pressure to meet the publication schedule. Mobilizing processing operations, then, will recognize the different qualifications of the existing staff as well as those who have been recruited from other organizations within the company. There may be some editors who can bring order to any pile of paper and some coordinators who can do a good job of copy editing, but it is usually better to have a sharp break between these functions, even to underscore it with a slight separation in physical work location. People who excel at reading and writing can often be found in the library, publications, training, and personnel departments of the company.

The Special Style Manual. Every company has its own editorial guide book or style manual; for the enormous proposal it is worthwhile to reissue it in parts to cover specific editorial problems and procedures. There will be sections of interest only to the editors and originating authors. Others, like the section on format, abbreviations, and special symbols, for example, must go to all word-processor operators and graphics personnel. This special style manual or group of manuals can contain other detailed editorial guidelines and ground rules that are unique to the proposed work or that have been specified or requested in the request for proposal or request for quotation.

Checklists and Tracking the Processing Schedule. The separation of processing operations and the possibility of several sequential operators working on any one parcel of proposal manuscript make it necessary to use a covering checklist to assure that everything gets done in the proper sequence. This is true particularly when several shifts are being used to perform the work, a situation that is not uncommon in giant proposal preparation. There needs to be a checklist for the first run of sort–edit operations and perhaps separate ones for the review–approve cycle and for the final checks before production. The coordinator of processing activities needs every possible relief from explaining things to each worker: work must be parceled out quickly to the next processing operation. Professional scheduling personnel may be available within the company. If so, these people can help in scheduling and tracking the proposal-processing activities for a giant proposal. The giant proposal itself is a work activity that must be planned, scheduled, estimated, and carried out with split-second timing and precision. The processing, as well as the reproduction and binding functions, can use conventional planning and scheduling tools to accomplish this objective. Status and load charts can be used to keep a real-time check on the progress of each proposed element. These charts not only provide visibility of the overall publication effort, but provide a basis for real-time last-minute changes and updates to the publication schedule. The overall proposal outline is the natural base for status charts and checklists. The outline can be listed on the left margin of the checklist or time chart and the responsible individual or organization or calendar time frame can be placed along the top. Computer-generated printouts of all text drafts, graphics, and proposal status reports should have date and time information in the heading or footer of each page to facilitate identification of the most recent version.

These checklists and charts can then be used in assuring that all processing operations are done by the appropriate individual or organization and at the required time.

Production Operations

Final desktop publishing and reproduction work for the giant proposal, as mentioned earlier, may require the help of an outside vendor or subcontractor. Although the work can and sometimes must be done outside the company, the existing company personnel responsible for these functions should act as a channel for the work. The company's purchasing department, the regular production supervisors, and those who will process the huge document must select and contract with these vendors well in advance, brief them thoroughly on what is coming, and keep them advised every step of the way. Final publishing can be done directly from the same basic computer magnetic disks or tapes that were used for work from the original draft through final copy editing. Word-processing conversion to final printed copy, however, must be tested thoroughly beforehand to be sure that the vendor's computer equipment and software are compatible with that of the company's and that the digital information can be transferred with great speed and with no difficulty. It is a good idea to go through a dry run of this final computerized publishing process with the publisher using sample proposal information to assure that there will be no bugs in this final phase of proposal production. After output copy is received, a final proofing is required to check the consistency and quality of output.

Distribution and Delivery

Assuming that sufficient time in the overall publication schedule has been allowed to take care of the sheer physical job of assembling, packaging, wrapping, and shipping the giant proposal to its destination, this phase should pose no difficulty. Some companies feel at this point that duplicate shipments are justified, with separate messengers and separate transport routings to assure that the proposal will get to the customer and that it will get there on time.

11
COMPUTER-AIDED PREPARATION OF REQUESTS FOR PROPOSALS (RFPs) AND PROPOSALS

Who says, "Let Him make speed, let Him hasten His work, that we may see it?"
—Isaiah 5:19

Except for the very smallest of proposals, automated proposal preparation has become a necessity in order to maintain a competitive posture in the current market, which has an increasingly higher technology content. Automated management information systems and computer-aided techniques are already commonplace throughout the entire procurement process starting with pre-RFP business and data searches and ending with contract award and performance. Many government agencies and private corporations are planning a completely electronic procurement process. Figure 11.1 is a simplified representation of a typical procurement process which, combined with this description, shows how electronic data processing is now being used effectively in virtually every step of the procurement cycle and how seamless connections can be made between the steps. Systems are being planned and developed at this very writing that will integrate many of the required functions. Fully integrated computer-aided procurement processes are beginning to come on line. The last section of this chapter describes some of these fully integrated computer-assisted procurement systems.

NEW COMMUNICATIONS AND MICROPROCESSING ENVIRONMENTS

What if we could use the same electronically generated and stored data (text, graphics, and tabular material) from the beginning of the procurement process through final project closeout, only making changes or additions where necessary to match our capabilities to customer requirements? And what if the customer received uniformly formatted proposals from multiple customers that could be compared, adjusted, and scored in a computer-aided proposal evaluation system to aid in the selection process? Although this scenario seemed farfetched only several years ago, elements and links have been developed and software is presently available that will perform many of the vital solicitation and proposal functions in a smooth flow from market survey and opportunity announcement to project completion. Several agencies and large commercial customers request detailed cost-proposal information in electronic spreadsheet or database format on magnetic disk. Some solicit entire proposals on magnetic media or through telecommunications lines. Several government agencies (the Department of Health and Human Resources is one) are planning a completely paperless system for transmittal of proposal requests, receipt and evaluation of proposals, and award and monitoring of contracts and grants.

For many years, NASA has had access to online electronic pricing factors and labor rates from its prime contractors for use in checking allowability of proposal and contract costs. Sophisticated and powerful document-management tools are being used at the very beginning of projects when requirements are being formulated. Computer networks and conference room video displays are used to speed up the urgent and complex personal and organizational interactions that must occur to provide credible proposals in the short time allocated. New software with inter-applications communications; expanded voice-notation capabilities, applications and file linking; and professional graphics and typesetting capabilities is available on single and multiuser systems. The word processor of the 1980s and the desktop publishing systems of the 1990s have evolved into complete document management systems for the year 2000 and beyond. These sophisticated systems, available on powerful desktop, laptop, and notebook-sized computers, do more than mere formatting of text, checking of spelling, and grammar analysis. They permit the rapid assembly of documents from portions of many other source documents; they assist in the writing of requirements through the use of elaborate cross-referencing of text and numerical data; and they permit the rapid construction of consistent, complete, and error-free technical, cost, and organization and management volumes. These features are not just highly desirable, but are essential in today's proposal preparation process.

ONLINE AND IN-HOUSE CLIENT DATABASES

Computer-assisted business development and marketing aids are already available, and more are coming on line in increasing frequency. Online databases and voice-mail messages are available for prospective suppliers of goods and services to city, county, state, and federal agencies. Several large agencies and companies collect vendor product descriptions and prices in electronic format to aid in making purchasing decisions and advertise their present and future needs through online electronic systems. Other companies make it their business to synopsize the procurement and

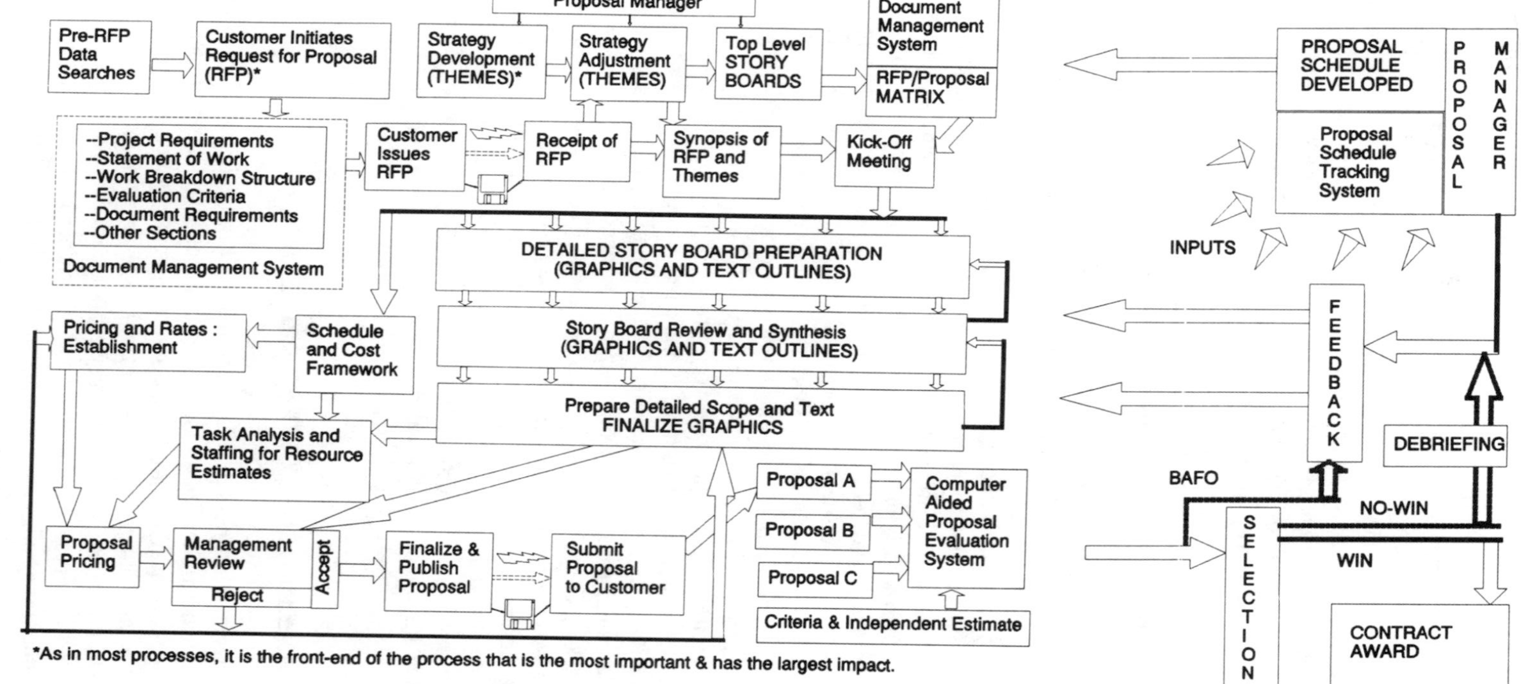

FIGURE 11.1 Diagram of the computer-aided procurement cycle. BAFO: best and final offer.

purchasing plans of large organizations and firms and to provide subscription services for key-word searches that zero in on specific product or service lines. Mercury Electronics Publishing Company (717-367-4200), Softshare, Inc. (805-683-3841), and Sales Opportunities/Services, Inc. (202-575-0185), for example, prepare excerpts of items in the *Commerce Business Daily* every day. They are available for batch and online key-word searches, which allow businesses to quickly locate opportunities in agencies of the federal government. The *Commerce Business Daily*, a 40- to 50-page small-print document issued every workday of each year, contains 24 categories of goods and services to be purchased by the government consisting of over 300 subcategories ranging from high-technology scientific research to utilities and housekeeping services. Online key-word searches of this document produce customized reports to each business that contain: (1) the procuring agency; (2) the type of procurement; (3) a description of the work; and (4) names and telephone numbers of persons to contact for more information.

The Government Printing Office provides a key-word search service of the *Commerce Business Daily* on a quarterly or yearly basis and maintains an online database that includes the contents of each day's issue. Individual users or networks can subscribe to this online database and perform their own searches (call 703-450-1882 for more information). Online *Commerce Business Daily* services can also be obtained from the United Communications Group (301-961-8700), and P.M. Services (215-289-9400). J. Thomas, Inc. (517-750-1271) provides software that will search these databases to identify applicable prospects.

In-house client databases can be easily constructed using user-friendly development tools such as "Application Express" for RBase.™ These client databases can perform the important functions of tracking and updating the status of potential contracts from each client. Typical fields in such a database (see Figure 11.2) are the customer's name, agency or organization; task description; customer contact's name; customer contact's telephone number; date of announcement; date RFP is requested; RFP number; proposal due date; award date. Other information can also be included to help sort out business potential, such as new business category, product/service category, name of the internal company contact, new-business priority, acquisition strategy, remarks, estimated funding, win/lose designator, name of incumbent, expected award date, name of winning contractor, and recompetition date(s). A "T number" is assigned to each prospect as it is entered to keep track of the number of prospects entered to date.

In building the database, the business development analyst enters prospects identified through searches of procurement and purchasing announcements, requests for proposals received, and personal contacts (see Figure 11.3). Procurement announcements and RFPs received or scanned in electronic format can be excerpted and entered directly through dynamic application integration rather than through keyboard entry. The database is updated as new developments occur. Rescheduled solicitations, canceled procurements, or modified customer addresses and telephone numbers can be entered at this update. Feedback from regularly scheduled business development meetings provide database updates relative to the business acquisition priority, potential competitors, and acquisition strategy.

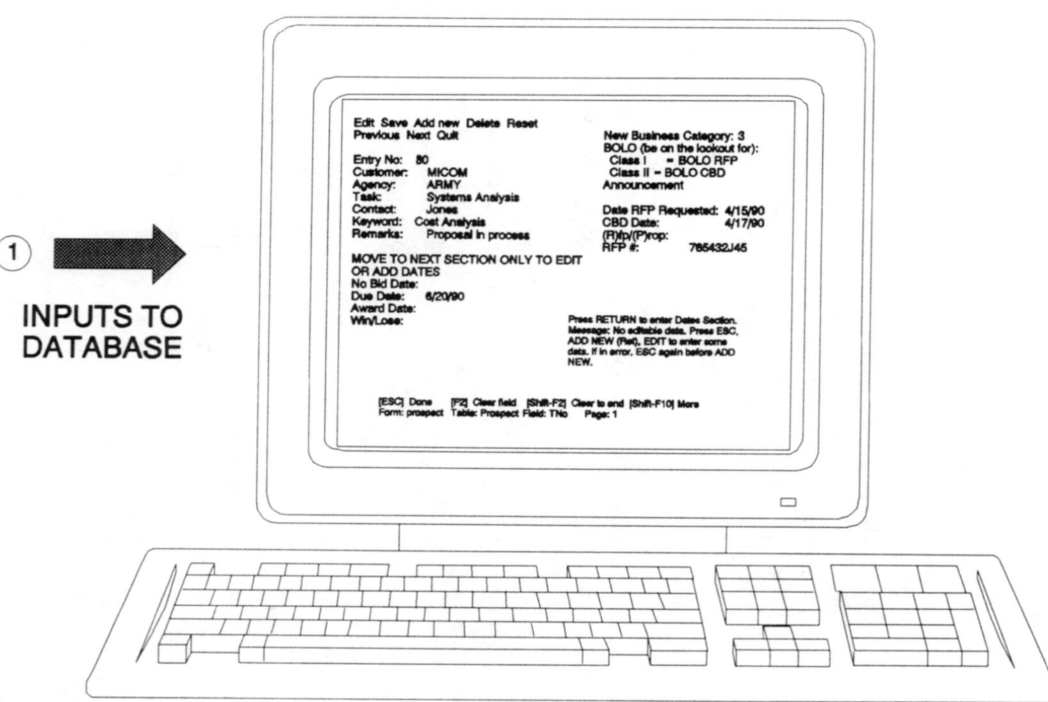

FIGURE 11.2 Typical screen and selected contents of business development database.

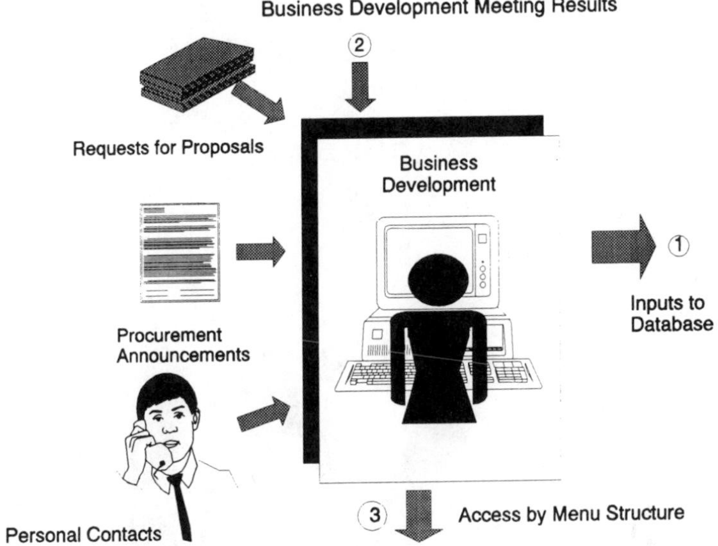

FIGURE 11.3 Inputs to business development database.

The analyst has access to a menu structure that permits the editing or viewing of prospect reports, or the printing of reports (see Figure 11.4). Prospects can be selected on the screen based on the name of the internal company contact, the task description, or the T number. Preconfigured reports of several categories can be viewed or printed. Prospects can be listed by internal company contact, by procuring agency or organization, by company priority, by corporate growth category, or by prospects added in the past week. These reports are useful in weekly or monthly business development meetings to determine how each manager is progressing in acquiring new prospects and turning these prospects into sales, to determine the customer and business category mix of prospects, to flag prospects of the highest priority, and to determine the number of prospects added in total for each customer and each corporate growth category. Proposal reports that can be viewed on the screen or printed for use in the business development meeting provide a listing of proposals in process, RFPs being evaluated for potential bid, proposals submitted and awaiting award, number and type of no-bid decisions made in total and in the past reporting period, and the no-bids from any specified date. Using these data, the marketing department, the sales department, and company management can keep close tabs on the exact status of all business-acquisition efforts.

Contract reports include the contracts awarded since a specified date (wins), and the contracts awarded to competitors (no-wins) since a specified date. The success of marketing and sales activities, then, can be evaluated on a continuing basis in terms of contracts won versus business development effort expended. Figure 11.5

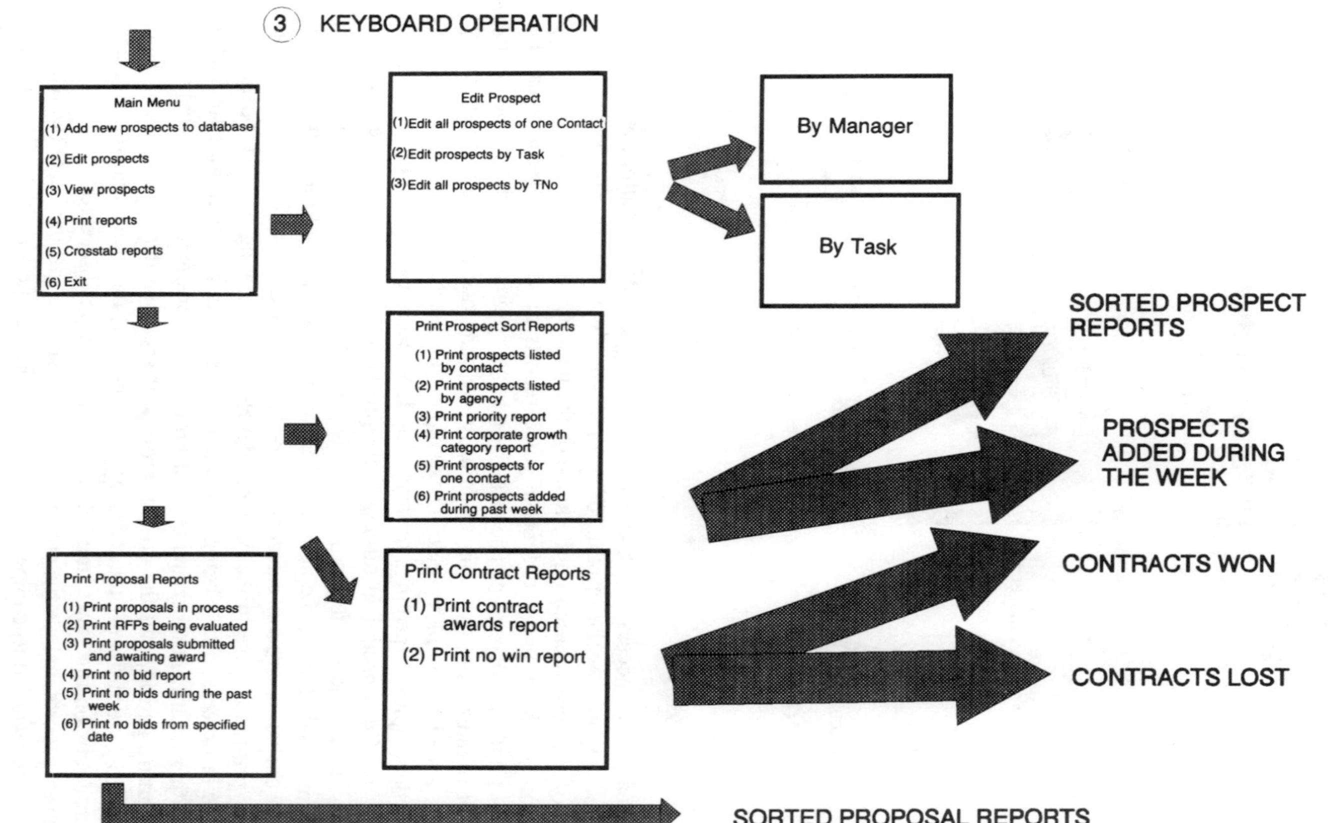

FIGURE 11.4 Menu structure for business development database. TNo: "T Number."

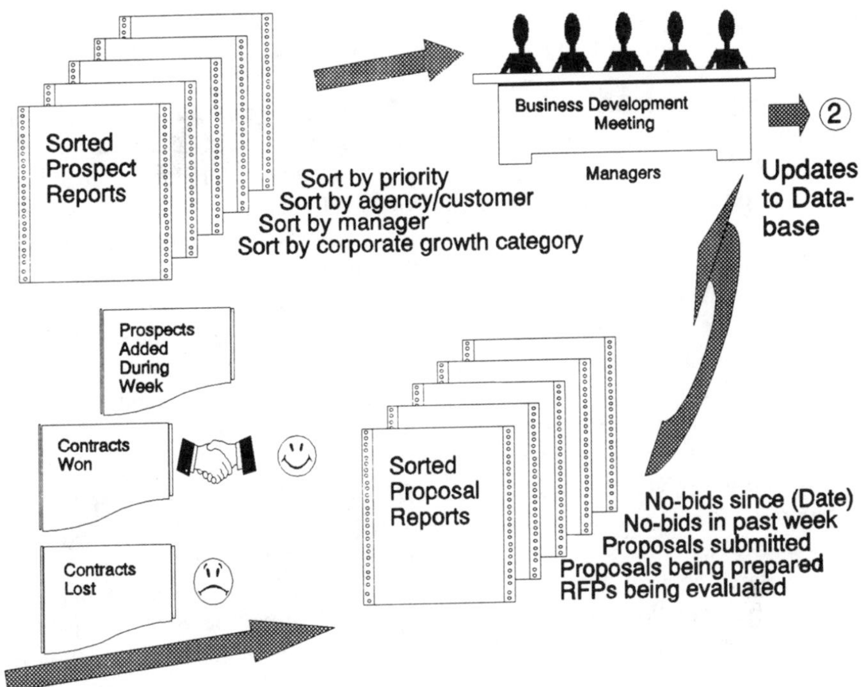

FIGURE 11.5 Report outputs of business development database.

shows typical reports that can be generated by such a system and their flow to the managers before or in the business-development meeting. An organized, systematic, methodical computer-based system, when combined with an aggressive and proactive marketing plan that is consistent with overall corporate objectives and goals, can provide management with indications of the trends and projections for new business and can assist in promoting sales in keeping with company objectives.

COMPUTER-AIDED RFP PREPARATION

To receive requests for proposals, your firm must be on a "bidders' list" of the procuring agency or company. Some bidders' lists are available on electronic bulletin boards accessed by computer modem, and others allow real-time entry of your firm's name, address, telephone number, fax number, and product lines. Organizations that require written mail requests for entry on bidders' lists usually enter these lists on computer files for easy updating, mail-label production, or phone/fax transmittal of bid-request information to potential sellers of services or goods.

Organizations that issue requests for proposals or invitations to bidders on a regular basis generally rely on existing files of "boilerplate,"or standard, information that can be selectively and electronically transferred to each solicitation document and formatted to become an integral part of the purchase order or proposal request. Standard forms are provided in electronic format on disk or through online bulletin boards. Electronic blank forms and spreadsheets provided by the requestor of cost and text information offer two significant enhancements to the procurement process: ease of entry and ease of evaluation and comparison, as discussed below.

Ease of Entry and Reduced Workload for Preparing Organizations. Receipt of a standard format for cost and pricing information on disk, with all spreadsheet or database rows, columns, or fields named, formatted, and sized, reduces preparation time and effort of the firm that is preparing the bid. Equations for multiplying rates and factors by labor hours and material quantities can be built into the electronically formatted and transmitted media, as well as summing and roll-up equations. If the procuring organization performs this "front-end" part of the job, the jobs of each and every bidder are made easier with an existing formatted electronic spreadsheet on hand. All that remains is for the bidder to enter the resource estimates for labor and materials (pricing formula variations among proposers, of course, would necessitate unique form changes by the proposers).

Blank forms in electronic format are also much easier to fill out by computer than typeset and printed forms are to fill out by either typewriter or computer. In computerized forms, updates and changes to field entries can be made easily, and the completed form ends up with a much more attractive appearance, as entries can be centered between lines and type sizes can be selected to enter the required information in the allocated space. (In prepublished forms, the allocated space is often too small or inappropriately shaped for entry of the required information in standard typewriter fonts.) Figures 11.6 and 11.7 are facsimiles of computer-generated forms produced in electronic media on WordPerfect 5.1. Figure 11.7 is virtually identical to the original printed form supplied by the Department of Health and Human Services. Computer-based forms are easily filled in because the entry fields are automatically tabbed in sequence and highlighted on the computer screen as each entry space is encountered.

Ease of Evaluation and Comparison. Submission of cost, technical, schedule, and organizational information on prepared electronic forms furnished by the proposal requestor allows the requesting organization to more easily and quickly compare and evaluate multiple proposals against independently estimated costs and against preestablished scoring criteria. (A potential computer-aided proposal evaluation system based on the use of electronically transmitted proposal data is discussed later in this chapter.)

Computer-aided Document Management

In addition to rapid boilerplate (standard clause) retrieval from prestored libraries and electronic forms, another major software-based methodology is becoming available that could significantly aid in the RFP preparation process as well as the proposal preparation process to follow: computer-aided document management (CADM). Until recently, there have been few affordable software tools available for both small and large proposers that would ensure common threads of thought, approach, and ideas throughout requirements documents and procurement documents designed to implement projects.

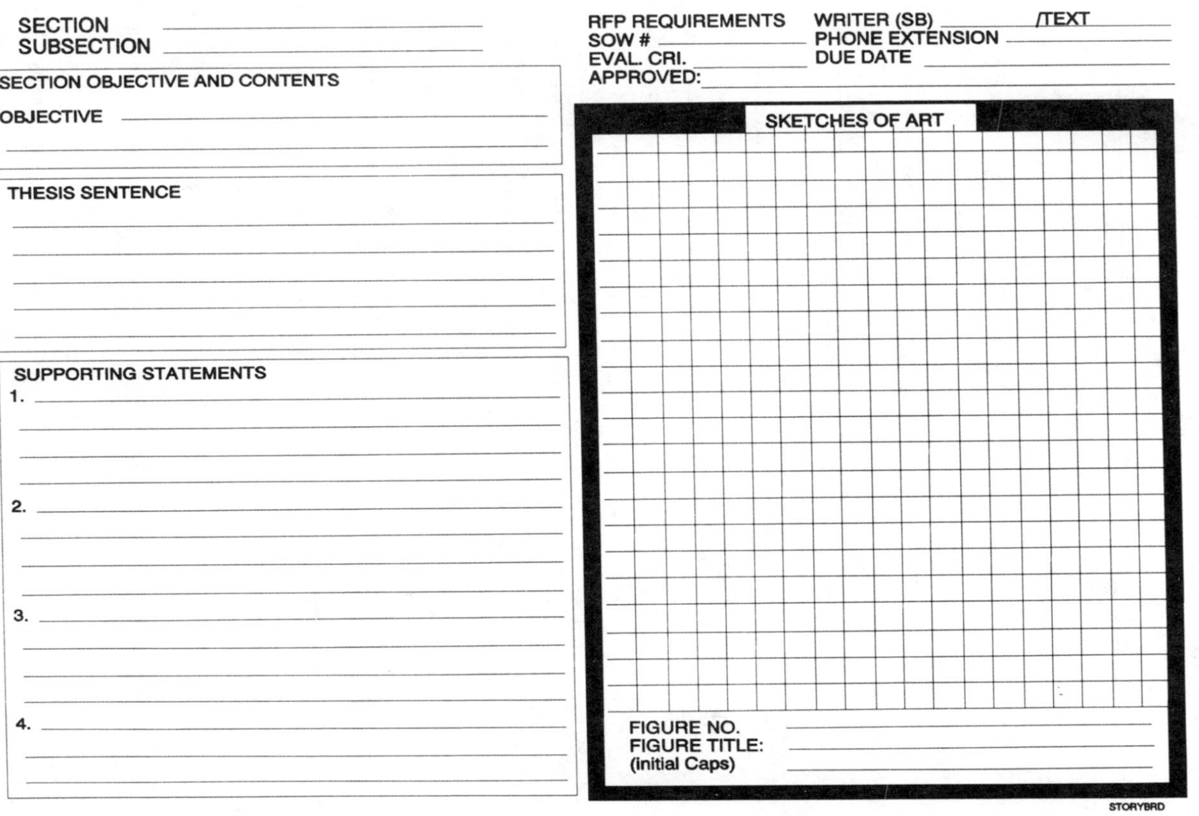
MOBILE DATA SERVICES

PROPOSAL NAME: ____________

SECTION ____________
SUBSECTION ____________

RFP REQUIREMENTS
SOW # ____________
EVAL. CRI. ____________
APPROVED: ____________

WRITER (SB) ________ /TEXT ________
PHONE EXTENSION ____________
DUE DATE ____________

SECTION OBJECTIVE AND CONTENTS

OBJECTIVE ____________

THESIS SENTENCE

SUPPORTING STATEMENTS

1. ____________

2. ____________

3. ____________

4. ____________

FIGURE 11.6 Electronically generated storyboard form.

OMB No. 0925-0195
Expiration Date 1/31/92

DEPARTMENT OF HEALTH AND HUMAN RESOURCES PUBLIC HEALTH SERVICE **SMALL BUSINESS INNOVATION RESEARCH PROGRAM PHASE 1 GRANT APPLICATION** FOLLOW INSTRUCTIONS CAREFULLY	LEAVE BLANK		
	TYPE	ACTIVITY	NUMBER
	REVIEW GROUP		FORMERLY
	COUNCIL/BOARD (month/year)		DATE RECEIVED

1. TITLE OF APPLICATION (Do not exceed 56 typewriter spaces)

2. SBIR SOLICITATION NO. **SBIR 91-2**

3. PRINCIPAL INVESTIGATOR ☐ New Investigator

3a. NAME (Last, first, middle)	3b. SOCIAL SECURITY NO.
3c. POSITION TITLE	3d. MAILING ADDRESS (Street, city, state, zip code)
3e. TELEPHONE (Area code, number and extension)	
4. HUMAN SUBJECTS ☐ NO ☐ YES { ☐ Exemption # ______ OR ☐ Form HHS 596 enclosed	5. VERTEBRATE ANIMALS ☐ NO ☐ YES
6. DATES OF PROJECT PERIOD From: Through:	7. COSTS REQUESTED 7a. Direct Costs $ 7b. Total Costs $
8. PERFORMANCE SITES (Organizations and addresses)	9. APPLICANT ORGANIZATION (Name, address, and congressional district)
	10. ENTITY IDENTIFICATION NUMBER
	11. SMALL BUSINESS CERTIFICATION ☐ Small business ☐ Minority and disadvantaged ☐ Woman-owned

12. NOTICE OF PROPRIETARY INFORMATION
The information identified by asterisks (*) on pages ______________ of this application constitutes trade secrets or information that is commercial or financial and confidential or privileged. It is furnished to the Government in confidence with the understanding that such information shall be used or disclosed only for evaluation of this application; provided that, it a grant is awarded as a result of or in connection with the submission of this application, the Government shall have the right to use or disclose the information herein to the extent provided by law. This restriction does not limit the Government's right to use the information if it is obtained without restriction from another source.

13. DISCLOSURE PERMISSION STATEMENT If this application does not result in an award, is the Government permitted to disclose the title only of your proposed project, and the name, address and telephone number of the corporate official of your firm, to organizations that may be interested in contacting you for further information? ☐ YES ☐ NO	14. CORPORATE OFFICIAL (Name, title, address and telephone number)	
15. PRINCIPAL INVESTIGATOR ASSURANCE: I agree to accept responsibility for the scientific conduct of the project and to provide the required progress reports if a grant is awarded as a result of this application. Willful provision of false information is a criminal offense (U.S.Code, Title 18,Section 1001).	SIGNATURE OF PERSON NAMED IN 3a (In ink. "Per" signature not acceptable)	DATE
16. CERTIFICATION AND ACCEPTANCE: I certify that the statements herein are true and complete to the best of my knowledge, and accept the obligation to comply with Public Health Service terms and conditions if a grant is awarded as the result of this application. A willfully false certification is a criminal offense (U.S. Code, Title 18, Section 1001).	SIGNATURE OF PERSON NAMED IN 14. (In ink. "Per" signature not acceptable)	DATE

PHS 6246-1 (Rev. 12/87)

FIGURE 11.7 Department of Health and Human Services grant application form (sample of computer-generated form).

Recently, a small firm of aerospace professionals in Houston, Texas, Bruce Jackson Associates (713-486-7817) developed and is rapidly upgrading and expanding the sales of a unique software package named "Document Director." Document Director™ is a microcomputer-based computer-aided document management tool that combines many of the features of a word processor and a database. It is able to link ideas, thoughts, and requirements from one part of a document to another part of the same document or from one part of a document to one or many parts of another document. These linkages would be particularly valuable for the development of a request for proposal, particularly for one that is large and full of complex requirements, because of the potential interaction of these requirements and because of the need to maintain consistency and compatibility among various parts of the request for proposal.

For example, in a request for proposal it is necessary to ensure compatibility, traceability, and completeness among the project requirements, statement of work, work breakdown structure, evaluation criteria, project document requirements, and other sections such as the instructions to proposers. A computer-aided document management system will help the authors of the RFP link all related information among these RFP sections, verify that linked information is compatible, ensure that all essential information is covered, and ensure that nonessential information is omitted. New interapplications computer architectures will permit the dynamic linking of the RFP to other customer source documents to ensure that each RFP contains the latest updates of all related specifications, policies, and contracting requirements of the procuring entity.

Preparation and maintenance of requests for proposals in digital electronic format in a linked computer-based environment becomes particularly beneficial when requests for proposal are amended, modified, or updated. Not only will this permit rapid impact assessment on other parts of the RFP and timely issuance of changes, but also the impact on other computer-based documents, policies, and schedules can be rapidly identified and assessed as well. The existence of built-in pointers and remarks that assist in identifying other parallel actions that need to be taken is very useful to the procuring organization.

The final step in the RFP preparation is the simultaneous electronic issuance of requests for proposals to those on the bidders' list. Several long-distance telephone companies offer this service. Alternatives to voice phone-network transmittal are either faxing or mailing of computer diskettes. We do not know of an occasion where it has been done, but a reasonable step in the transmittal of a request for proposal in electronic format would be to simultaneously transmit the electronic linkage files of the computer-aided document management system to each proposer to permit easy linking of proposal responses to RFP requirements as the proposals are being prepared.

Pre-RFP Data-processing Activities

As emphasized in earlier chapters, straight-line control must start long before the request for proposal is received. Straight-line control of the automation and data-processing activities leading to the submission of a winning proposal starts long before the RFP is received—even long before the company's name is entered on a bidders' list for a specific customer or product/service line. Straight-line control in preparing for proposal data-processing activities includes:

- Acquisition of computer software for computer-aided document management, graphics, word processing, scheduling, and cost estimating of the proposed activities; training of personnel in using this software; and entering key preliminary data for use in proposals. The computer type and software type must be compatible or interconnectable among the various organizational entities preparing the proposal.

- Collection, organization, verification, expansion, writing, and editing of a comprehensive company background and experience database or choosing word-processing files for quick computer-based accessing and publication in proposals.
- Interviewing and entering data on key personnel, technical personnel, administrative personnel, and support personnel to obtain comprehensive information on the education, background, experience, and capabilities of personnel at all skill levels and skill categories; this information is for use in automated storage, retrieval, and selective composition of *targeted résumés* for each procurement.
- Preparation of computerized spreadsheet forms, computerized formats, and computerized style sheets, if they are not included by the customer with the request for proposal.
- Pre-RFP entry of generic scheduling, cost, and pricing information in the automated cost-estimating and proposal-pricing system. This information includes: (1) the company's fiscal and holiday calendars; (2) cost breakdown structure (elements of cost, cost categories, skill categories, skill levels, resource types); (3) overheads and general and administrative (G&A) rates; (4) capital cost of money factors; and (5) project organizational structure (if available).
- Preparation of computerized storyboard forms, résumé formats, scheduling formats, and text outline formats.
- As the date of release of the RFP nears, and as specific strategies and themes for the upcoming proposal emerge, these strategies and themes can be converted to computerized format through the production of top-level electronic storyboards that will be used to "set the tone" of the detailed proposal activities that are to follow receipt of the RFP. These top-level storyboards can be reviewed, refined, and enhanced for final update when the RFP is received.
- For large proposals, a preliminary electronic scheduling network is developed for the proposal schedule itself.

Receipt of RFP

Receipt of the RFP touches off a whole chain of activities in a number of areas for both the large and the small proposer:

- *Update of Proposal Schedule/Network (for Large Proposals).* The proposal schedule and accompanying critical path network are updated, based on actual date of RFP release and required date and time of proposal submission; the schedule is distributed to all members of the proposal team.
- *Confirmation of Strategy and Themes.* Proposal and corporate managers study the RFP in detail to determine if preconceived strategies and themes are consistent with the letter of the RFP requirements. The process of linking RFP requirements

to proposal contents starts, using computer-aided document management techniques and other computerized cross-referencing methods.

The Kickoff Meeting. If a computer-driven video display conference room facility is to be used for the kickoff meeting presentation, and if the RFP has been issued in electronic format, sections of the RFP, key specifications, evaluation criteria, and other vital information can be displayed directly to the proposal team and discussed in a group forum. The computerized schedule and milestones for proposal preparation, as well as predeveloped and adjusted themes and strategies, can be displayed on the computer projection screen and updated in real time if required. Magnetic diskettes containing outlines, computer forms, and preconfigured spreadsheets are issued at this time or preissued through a network to all those who are involved in computer-based storyboard preparation and cost-estimate preparation. If the RFP is not received on computer disk or over modem lines, excerpts of key information can be electronically scanned and provided on disk or through a network along with copies of the appropriate sections of the RFP and accompanying instructions for internal proposal preparation and evaluation. From this point, individual "book bosses" and people charged with writing various sections of the proposal start the electronic storyboarding activities.

COMPUTER-AIDED PROPOSAL PREPARATION

Electronic Storyboard Preparation

After previously prepared electronic storyboard forms have been made available, the graphic representation of work (GROW) process is started by the proposal writers, working on their own computers or workstations.

As in manual storyboard preparation, the graphics are prepared first, then the thesis sentence and supporting statements are written to support the graphics. Writing (authors' data entry) does not begin until all storyboards are completed and reviewed in a "review and synthesis" step (refer to the central portion of Figure 11.1).

Full use of up-to-date electronic proposal preparation techniques and software for this storyboarding and storyboard review/synthesis step will depend upon the degree of automation of the organization preparing the proposal and the mode selected for internal communications and review. A network of computers or workstations, which permits team members to review and comment on the material prepared by other team members, is ideal for this application. In addition, a video display conference, where each member of the proposal team presents his or her storyboard and supporting statements before the key proposal-team members is important to achieve team interaction and rapid integration of information to assure compatibility, cohesiveness, and supportability to all sections of the proposal. An electronically generated storyboard form will permit quick and legible entries and updates. An example is shown on Figure 11.6.

Preparation of Detailed Supporting Text

Most organizations have a mix of persons who regularly use the computer to directly input text and those who prepare hand-written or dictated information to be entered by a data entry person. Those who have been associated with proposal preparation activities for very long know how important it is to know how to type. Technicians, engineers, managers, administrators, and even company executives involved in proposal preparation are increasingly keyboard literate. Knowledge of at least one word-processing software program is becoming important if not essential when one becomes involved in the proposal preparation process.

The modern, powerful word processors have many more features than mere text entry and manipulation. Spell-checking, grammar analysis, and thesaurus features can be very helpful to the proposal writer who is proficient in word processing or desktop-publishing literate. Special characters are available on word processors that never have been available on typewriters. These include foreign letters, mathematical symbols, copyright and trademark characters, bullets, diamonds, arrows, and many other symbols. Characters known as "dingbats" (character-sized symbols) are available to liven up documents. Many word processors have built-in equation-forming methods that will produce complex mathematical, chemical, and statistical formulas and equations in a professional-appearing format. Built-in decimal alignment for columns of numbers, mathematical equations that can create mini-spreadsheets and accurately added columns within the word processor, sorting techniques, and standard tabular and column entry procedures make it easy to build tables and alphabetically or numerically sorted columns of information. Pages can be formatted to contain page numbers, headers, footers, or footnotes. Indexes, tables of contents, and bibliographical listings can be automatically generated. The automatic outlining feature available in most word processors is exceedingly valuable in the construction of dynamically changing documents like a proposal because insertion of a new section in the outline causes an automatic roll-down of paragraph numbering. Similar features are available for figures and tables. Figure, table, footnote, and bibliographical reference numbers can also be linked to their text location. The appropriate numbers will change throughout the document when a new figure, table, footnote, or reference is inserted.

Another feature of current high-end word processors that is useful to proposal writers is the COMPARE function. This feature compares two documents and automatically redlines the words added and strikes out the words deleted in the original document. The resulting display will show the writer exactly where changes have been made to a section of the proposal from an original version to an updated version. The SEARCH function, which locates certain words, phrases, combinations of words or numbers, acronyms, and abbreviations is also useful if changes must be made in the surrounding text. The REPLACE function automatically replaces any prespecified set of adjacent characters throughout the document, a handy feature when last-minute changes are made in nomenclature, titles, or acronyms. Split-screen capabilities permit the simultaneous viewing of two documents or pages at once for comparison, editing, transfer, copying, or deletion of information.

Passwords can be assigned by the author of a section of the proposal, which will prevent the alternation of a portion of the document without specific knowledge of that password. This feature is particularly important when other writers have access to the document on the same computer or on a network.

The word-processing features mentioned in the foregoing are those which would be most important to a proposal document writer. Many other features such as type sizes, fonts, file manipulation, page formatting, and insertion/manipulation of graphics are useful for the proposal publication and printing process, which, for a large proposal, would be performed by data analysts, copy editors, and word-processing professionals.

AUTOMATION IN PROPOSAL PRICING

The automation of proposal pricing has reached a high state of perfection through the current market availability of a number of sophisticated proposal scheduling and pricing systems. Micro-Frame Technologies of Ontario, California; Primavera Systems, Inc., of Bala Cynwyd, Pennsylvania; Boeing Aerospace of Seattle, Washington; and Timberline Software of Beaverton, Oregon, are examples of companies that supply integrated scheduling, cost-estimating, and proposal-pricing systems. These and other software vendors and suppliers have produced sophisticated computer software packages that perform all aspects of cost-proposal preparation, including work-breakdown-structure-based estimating and pricing, scheduling, and performance measurement systems to be used after the contract is awarded. To simplify this discussion we will use the AWARD™ product of Micro-Frame Technologies (714-983-2711) as an example of what can be accomplished in computer-based estimating and pricing. Micro-Frame's AWARD™ estimating/pricing program has the following features:

- Top-down estimating: enter total cost and let the system calculate hours and other values.
- Bottom-up estimating: automatically spreads hours, skills, cost, and calculates all other values.
- Rolls up all values through the WBS, two organization structures, and a cost-breakdown structure.
- Integrates task descriptions (text), resource estimates (data), and estimate rationale (text and data).
- "What if" capabilities include repricing, start date shift, and increase or decrease by cost element.
- Up to 1000 automatic cost-spread functions.
- Over 1000 standard and user-defined reports.
- Unlimited user-defined resources, cost elements, bid codes, equipment, personnel, and overheads.
- Percentage adjustment of any WBS element or cost element.
- Flexible, user-defined rates, overheads, and cost escalation.

- Enter any pricing unit such as quantity, equivalent persons, labor months, hours, direct costs, overhead costs, total costs, or priced costs and immediately calculate, roll up, and store all other values.
- Immediately display calculation and automatic time spread results.
- Operates on a local area network with up to 100 simultaneous users.

The computer program will assist the user in building a detailed work breakdown structure (up to 99 levels), preparing and adjusting estimates at the WBS element level, and in preparing preconfigured and user-defined (customized) cost and pricing reports that can be entered directly into the cost proposal. The user defines working calendars, direct resource rates, burden rates, and fee structures. Information can be imported to the program from company accounting or engineering estimating files, and the user defines up to two organizational breakdown structures and one cost breakdown structure. Fee structure can be established for each work element if required.

Estimates can be made in hours, dollars or number of equivalent persons, or input by the month and automatically spread over the WBS element time period. The statement of work for each work element is immediately available to the estimator on screen as he or she performs the estimate, and text rationale that supports the estimated labor hours or material quantities can be entered into an on-screen text file. "What-if" analyses include the capability of repricing of the cost proposal with different direct rates or indirect rates, shifting the period of performance of the entire proposal or of individual work elements forward or backward in time, and expanding or reducing the proposal resources by adjusting the element of cost, WBS element, contract line item, statement of work item, department function, recurring costs, and nonrecurring costs by a given quantity or by a percentage.

The program allows users to establish global files that can be used by all projects. These include the accounting calendar and the resource library. In addition, the organizational breakdown structure can be created and copied into each project. These files are established before proposal setup. Once created, they are simply maintained as changes are made.

Fiscal calendars are included that specify the number of labor hours contained in each fiscal month, as shown in Table 11.1.

Labor class codes are assigned in keeping with the organization's skill category and skill level breakdown such as engineering labor, quality labor, tooling labor, administrative labor, purchased parts, subcontract items, catalog items, and manufacturing labor.

Burden templates (tables showing burden percentages for each cost element) are established for burden elements such as department burden, engineering burden, in-plant labor, out-of-plant labor, material, material burden, and manufacturing burden.

Escalation rates, steps, duration, and precision can be established for each resource category. Escalation can be applied monthly, quarterly, semiannually, or annually. Department numbers, department titles, department manager's names, and burden templates are added for each department (see Table 11.2).

TABLE 11.1. Computerized Fiscal Calendar Specifying Monthly Labor Hours.

FISCAL CALENDAR		
Fiscal Start	Month Label	Equiv. Labor Hours
31-DEC-94	JAN 95	160
28-JAN-95	FEB95	160
25-FEB-95	MAR95	158
01-APR-95	APR95	160
29-APR-95	MAY95	155
27-MAY-95	JUN95	163
01-JUL-95	JUL95	160
29-JUL-95	AUG95	145
26-AUG-95	SEP95	152
30-SEP-95	OCT95	161
28-OCT-95	NOV95	156
25-NOV-95	DEC95	160

The program enables users to define a WBS of up to 99 levels using their own coding structure. This WBS forms the primary data organization for all resource, work statement, and rationale information loaded into the system. In addition to retrieving information by WBS, the computer software provides multiple alternative data selections and sorts.

Element *type* (cost account, work package, planning package, and any others needed) can be specified for later selective retrieval and reporting of specific cost or pricing information.

The computerized proposal-estimating and pricing system contains an "Estimate Manager" module in which the user may:

- Schedule resource estimates manually on a day-to-day or month-to-month basis.
- Schedule resource estimates automatically in terms of standard hours, work hours, prime dollars, total burdened dollars, total cost dollars, labor months, equivalent persons, and units.

TABLE 11.2. Department Burden Template Table.

Department	Description	Manager	Burden Template
1000	Program Management	G. Patton	013
2000	Engineering	T.A. Edison	214
3000	Material	B. Franklin	112
4000	Purchasing	A. Alphonse	010
5000	Tooling	K. Zeringue	060
6000	Manufacturing	T. Brown	134

- Normalize estimates to the accounting calendar.
- Instantly review estimate results in the estimate summary window.
- Formulate resource estimates automatically with a variety of distribution curves
- Build comprehensive online descriptions for each task
- Record a narrative justification of the method used to arrive at time and dollar estimates for each work element (this narrative is often called the *estimate basis* or *estimate rationale*).

Graphic displays are provided for the following standard resource spread curves: (1) linear; (2) bell curve; (3) front load #1; (4) front load #2; (5) back load #1; (6) back load #2; (7) double peak; (8) early peak; (9) late peak; and (10) trapezoid. Curve spread factors for these standard ten curves can be edited on screen, or the user can build any number of his or her own spread curves for storage and later use. These spread curves can then be applied to any resource over its active duration.

For proposal preparation in a dynamic environment where internal negotiations of each resource allocation proceeds right up to the final publication date, the "what-if" analyses feature is extremely valuable. Using this feature, the estimator/analyst can (1) globally adjust start and finish dates for all tasks; and (2) reprice or adjust an entire project or portions of a project; indicating with a positive or negative percent factor against the selected element of cost, the program enables the user to reduce estimate values not only at their summary level, but at the detailed grass-roots level at which they were estimated. This is a powerful tool for responding to management challenges, best and final offers, or "what-if" exercises.

Reporting and graphing specifications and templates can be developed for a wide variety of reports using a data-conditioning system. In this system, any of the sort fields such as recurring costs, nonrecurring costs, element type or contract line item can be selectively included in the report. Report conditions such as report start date, number of years to be reported, incremental (monthly, quarterly, semiannual) values, cumulative values, and/or grand totals only can be identified and selected. Report values may be varied to show hours, dollars, and/or any level of burdened or unburdened costs.

A number of administrative resource library reports are available including resource library, element of cost class breakdowns, and burden template methodologies. These reports can be used to satisfy requirements for providing up-to-date forward pricing rates in the proposal; they are shown on Table 11.3.

Standard baseline report formats provide the basis for the majority of reports required for today's proposals. Resource information is available in detailed form by WBS element or summarized across all WBS elements by resource itself. The user can modify the standard reports or produce customized reports if the proposal needs are not met by the standardized reports.

A typical specialized report is a *compliance report*, which shows the linkages between the contract line item, the statement of work, and the work breakdown structure. All elements are listed by contract line item with corresponding statement of work and work element, as well as hours and dollars. Simple scanning of the

TABLE 11.3. Available Administrative Resource Library Reports.

Accounting calendar
Burden templates
Contract line items
Elements of cost and class
Estimate rationale
Management reserves
Organizational breakdown structure text: indented
Organizational breakdown structure graphics
Program baseline
Resource library
Responsibility assignments
Spread curves
Task descriptions
Undistributed budget
Work breakdown structure: indented
Work breakdown structure graphics

report will ensure that no line items or statements of work references were omitted; this will avoid costly omissions.

Other special reports, designed to meet the requirements of specific customers, include information such as the type of contract, customer-furnished property, payment schedules, and guaranteed loan requirements.

As can be seen from the preceding description of capabilities, use of a program such as Micro-Frame's AWARD or one of many other scheduling, cost-estimating, and pricing systems can virtually automate the production of the cost volume of a large proposal. Reports can be exported to word-processing and desktop-publishing software for final formatting and publication in bound volumes.

VERIFYING PROPOSAL'S COMPLIANCE TO RFP REQUIREMENTS

For medium-sized to very large proposals, a *compliance matrix* should be built, which relates proposal contents with request for proposal requirements. Some requests for proposals require that this matrix be submitted with the proposal. Even if it is not required as part of the proposal, a compliance matrix is a handy tool that will discriminate between (and compare) requirements that were generated from the RFP and those that were assumed in the proposal. This procedure will bring to the surface any RFP requirements that have not been adequately addressed and will highlight proposal elements or tasks that inappropriately exceed RFP requirements. A computer-aided document management (CADM) system is ideal for this purpose. In the CADM system, one-to-one, one-to-many, many-to-one, and many-to-many relationships can be built between the RFP and the proposal. Reports can then be generated that show exactly where in the proposal each RFP requirement has been addressed. Document Director,™ mentioned earlier, is a typical microcomputer-based program that will display both the RFP requirement and the proposal response (or responses) on the same screen in a two-window environment. The proposal writer can then compare the requirement to its associated response or responses to determine compatibility and compliance.

MANAGEMENT REVIEW, FEEDBACK, AND FINAL PUBLICATION

Data-processing techniques can play a significant role in the various management reviews and team reviews of proposals, as this is the point where numerous changes and adjustments invariably enter into the proposal preparation process. Computers have the ability to quickly change documents and tabular resource information and to convert this information into attractive printed formats. Completely automated proposal integration systems are on the way and are beginning to receive wider acceptance. These systems permit real-time interaction of proposal participants through menu-driven, computer screen graphics and text displays, and promise to speed up the proposal review and updating process.

PROPOSAL SCHEDULE TRACKING SYSTEM

A computer-network-based proposal preparation procedure would lend itself easily to a rigorous computer-based proposal schedule tracking system. The system would receive inputs from all activities in the proposal process, as shown in the upper right hand corner of Figure 11.1. Actual proposal progress would be tracked against a previously prepared proposal schedule developed for the initial kickoff meeting. The proposal manager can use this system to detect delays as they occur and to feed schedule correction information into the ongoing proposal preparation process.

COMPUTER-AIDED PROPOSAL EVALUATION

Provided that in-depth planning, formatting, and programming has been accomplished prior to the evaluation phase, the customer can, on receipt of electronically transmitted proposal data, begin the evaluation process by comparing all proposals with pre-configured criteria-scoring techniques and by making comparisons of the proposals regarding the independently prepared (by the customer) resource, schedule, and cost information.

Possession of rapidly displayed and easily compared data permits evaluators to distinguish discrepancies in resource allotments to each proposal element and, hence, aids in the identification of weaknesses in understanding of the RFP requirements. From identified resource allocation strengths and weaknesses, adjustments can be made to the proposal resources and costs to derive a "most probable cost of doing business" with each proposer. If this process can be accomplished in a computer-aided proposal evaluation system, the process will be faster, more accurate, and will permit rapid printout and display of proposal assessment results. The remaining portion of the procurement cycle, shown on the lower right-hand corner of Figure 11.1, can also take advantage of text and data in electronic format. If a best and final offer (BAFO) is requested, rapid manipulation of scope and resources can be accomplished through effective use of the already existing graphics and text on computer.

In the event of a no-win (bid rejection) and subsequent debriefing, feedback can be provided for the proposal team to permit a rerun of the proposal in an exercise that will provide information to reduce the probability of losses in future similar procurements. A win will automatically trigger the process of negotiation and subsequent performance measurement and monitoring. These functions can be aided by computer manipulation of proposal data and by connection to parallel data that tracks actual schedules, expenditures, and work accomplishment.

THE FUTURE OF COMPUTER-AIDED PURCHASING AND PROCUREMENT

Several government agencies and many private firms are developing methods of using computers and high-speed electronic communications to speed up the purchasing and procurement process. The main barrier to the adoption of automated systems in procurement is the habit of dealing with paperwork rather than electronic media. Somehow it just doesn't seem like a contract, purchase order, or payment if you can't touch it or feel it. Seeing it on the terminal's video screen, as Jack Bartley of the Department of Defense told the authors, is not enough for most people. They can't seem to resist the urge to print it out . . . and then they have another piece of paper to mark on, file, copy, and send to others. With the advent of computerized paychecks sent directly to the bank and other modern electronic marvels, however, people are realizing that the spending power of electronically transmitted money is the same as that of real money. The electronic culture is becoming more widely accepted, and fully electronic purchasing and procurement are coming on the scene, at least for small purchases. Larger purchases will inevitably follow.

In a telephone interview with the authors, John Romer of the General Services Administration (GSA) described an electronic catalog and ordering system for stock items called MUFFIN. Vendors are allowed to enter their inventory lists and prices into an electronic bulletin board, which can then be accessed by potential buyers. This procedure not only saves the buyers' time and effort but also saves the vendor money in catalog printing, updating, and distribution. Buyers can shop the electronic catalog and order items interactively from their computer terminal. For special items, electronic proposals can soon be requested from selected suppliers. With purchase order transmittal and electronic payment, the purchasing loop is closed. The only nonelectronic action is the physical shipment of the goods to the customer. Some firms provide computer disks with their products and prices listed and accessible through a menu-driven ordering system, but this procedure is rapidly being replaced by continuously updated online systems with easy search methods and invoice totaling. Some products also include precalculated life cycle cost and value engineering information, which permit the buyer to evaluate different alternatives in real time before making the final purchase decision. Also, special bulletin board announcements are available on new product lines, features, etc., to aid in purchasing decisions. These systems are available on a 24-hour basis, permitting round-the-clock access to the information and smoothing the time-zone interface to permit flexible work hours and a greater span of time for inputs and access. According to our personal contacts in the Department of Defense, the Department of Defense Office of Automation anticipates that an end-to-end digital information system will soon be available to firms doing business with the government. Pilot tests are being made for small purchases but will be soon expanded to larger procurements. Their first pilot site is Wright Patterson AFB in Dayton, Ohio. Other sites that have been put on line are the Army's Fort Belvoir in Virginia and the Sacramento Air Logistics Center in California. The U.S. Air Force is working on a Corporate Information Management (CIM) system that will not only automate purchasing but many other administrative functions as well. Legal aspects of such systems are being worked out; and, according to Jack Bartley, we are moving into a "trusted" mail environment in which digital encryption and signature authentication are secure. On large procurements, all technical data may not yet be feasibly transmitted in one package. In these instances, fax or microfiche transmittals are feasible in concert with digital data transmittal. More information on automated purchasing can be obtained from the Coalition for Government Procurement at 1990 M Street, Suite 400, Washington, DC 20036.

The use of a computer-aided proposal evaluation process requires an understanding of the overall proposal evaluation process. Also, understanding the proposal evaluation process as it is carried out by most customers is a valuable asset in preparing the proposal itself. The next chapter delves into the process of source evaluation and source selection as it is carried out by most customers for high technology, multidisciplinary projects.

12

HOW THE PROPOSAL IS EVALUATED

. . . but glory, honor, and peace to every one that worketh good.

—Romans 2:10

Now that the proposal has been prepared, it would be useful to have the knowledge of how it most likely will be evaluated. This knowledge will help a proposer to exercise wisdom in the preparation of the proposal and will provide an understanding of the entire acquisition process. If the proposal, bid, quote, or presentation to management is more than merely a bottom-line cost or price, chances are that it will encounter a structured evaluation process. Certain basic factors are considered in both structured and unstructured evaluations. The knowledge of these evaluation factors, how they are treated and scored, what adjustments will be made to the proposal, and what comparisons are likely to be made between one proposal and those of competitors will be invaluable to the preparer of proposals.

SOURCE EVALUATION VS. SOURCE SELECTION

Before describing the details of the source evaluation process, it is necessary to define the difference between source evaluation criteria and source selection criteria. The source evaluation board, committee, or team chairperson has the responsibility for evaluation of the proposal and for presenting the results of this evaluation to a source selection official. In order to be selected as the chosen contractor or firm to perform a given job, the proposal must rate high in the evaluation process. But a high rating in the source evaluation process does not necessarily guarantee a contract award. The reason for this is that the source selection official may not elect to follow the source evaluation board chairperson's recommendations. The source selection official may elect to give a higher weight to unscored criteria or criteria not contained in the request for proposal in making a decision. In any event, the source selection official has final authority and is authorized to select whichever firm he or she chooses. *Source* selection could include many factors (such as geopolitical factors) beyond those reviewed in source evaluation. It should be kept in mind, however, that the official is unlikely to choose a firm that develops and presents an inferior proposal. The purpose of the following discussion is to provide the reader with the maximum amount of information about source evaluation and source selection, which if carefully studied, will help in preparing a winning proposal.

SOURCE EVALUATION BOARDS: SOURCES OF INFORMATION

The most structured and organized proposal evaluation procedure comes from the establishment and operation of a formal source evaluation board. Formal source evaluation boards are created to make recommendations to source selection officials for government procurements, but the same formalized process can be used in commercial selections, such as choosing of a geographical location for a new production plant, expansion of a product line, or introduction of a new product or service into the marketplace. The evaluation process itself, if not properly organized, adequately staffed, and efficiently carried out, can take an inordinately long time and can result in inconclusive results. A good rule of thumb is to provide at least as much time for proposal evaluation as was expended in proposal preparation. Often, excessive proposal information is requested that cannot feasibly be digested in a reasonable time period or with a reasonable amount of resources. The procuring organization, however, will learn a lot about the work or product that it can expect to receive from a supplier if a thorough evaluation is accomplished using a structured evaluation process and evaluation organization, whether it is called a source evaluation board, committee, or team. The principal function of source evaluation is to collect the knowledge necessary to allow or permit a wise and objective selection recommendation to be made.

To obtain the wisdom required to evaluate a proposal and to recommend selection and contract award, it is necessary for the source evaluation board, committee, or team to obtain knowledge about the potential ability of a firm to perform the work in a manner that will be more advantageous to the procuring organization than that of any competitors. The source evaluation organization collects this knowledge from several sources, described below:

1. *The Proposal.* As mentioned in previous chapters, the proposal is the principal and sometimes the only source of knowledge provided to the proposal evaluation and selection individual or organization. As a supplement to the information in the proposal, however, evaluators often obtain additional information and knowledge from the other sources listed below.

2. *Plant or Onsite Visits.* Proposal or source evaluators frequently make visits and inspections to the site or sites where the work will be performed. Hence, it is necessary to be accurate in matching the claims in a proposal to the actual onsite situation. Credibility can be seriously damaged if the onsite visit reveals a situation that is not acceptable or that is not accurately depicted in the written proposal.

3. *References and Experience Verification.* Proposal and source evaluators will usually call, write, or visit previous customers to determine the quality, timeliness, accuracy, and overall responsiveness of previous work. This type of investigation will sometimes lead to a different evaluation than that which results from the written submission. The best policy is to pursue markets where a reputation for outstanding or at least acceptable performance has already been established; this will assure good recommendations from these references.

4. *Oral Presentations.* Occasionally, evaluation organizations will request that a proposer provide an oral briefing to accompany or supplement the proposal. This oral briefing gives the evaluator or evaluators an opportunity to personally observe the confidence, demeanor, technical expertise, and credibility of the proposing organization's proposed technical and management team. Since a selection to perform the work will result in a close working relationship between supplier and customer, these oral presentations can become key elements in imparting and maintaining a personal, living credibility to what is otherwise merely a written document.

5. *Written Questions and Answers.* After proposal evaluators have had an opportunity to initially review a proposal, questions may arise in their minds relative to the proposed technical approach, the organization proposed to accomplish the job, or the proposer's understanding of the resources required. Frequently these questions (and their answers) are of such importance to the evaluation and selection process that the procuring company or organization will submit one or more questions in writing to one or more of the proposers. Each proposer will be allowed sufficient time to prepare a clarifying response and will submit the answers as a supplement to the proposal.

6. *Best and Final Offers.* In negotiated procurements it is a common practice to allow the proposers to submit best and final offers. These are final price quotes based on updated information developed by the proposer prior to completion of the evaluation process. Best and final offers must be made based on a company-developed cost estimate and should not be overly influenced by perceptions of competitor's prices.

ORGANIZATION OF A SOURCE EVALUATION ACTIVITY

To collect and digest the information needed to make a knowledgeable recommendation, the source evaluation activity usually has source evaluation ground rules (a plan); source evaluation personnel (a team); source evaluation board functions; a physical location for the evaluation; and a source evaluation schedule. These are discussed below.

Source Evaluation Ground Rules. Source evaluation ground rules usually take the form of a source evaluation plan, which spells out the activities of the source evaluation board, committee, or team. In many instances this plan calls for the evaluation board, committee, or team to prepare requests for proposal or requests for quotation that are issued to call for the proposals. Structuring the request for proposal or request for quotation in a knowledgeable way prior to release can save considerable effort in the evaluation process by eliminating unnecessary requirements and assuring that certain vital information and data needed for a wise and objective evaluation are provided in the proposals. These ground rules will contain the scoring factors, criteria, and the importance of each criterion in advance of the evaluation process. Numerical scoring of factors and subfactors is often used to reduce the subjectivity of recommendations that might be made by the source evaluation board or team.

Source Evaluation Personnel. Qualified personnel are chosen by the procuring or acquiring organization to evaluate the technical, organizational, and business aspects of the proposal. These individuals are usually among the best available in a given discipline and are given considerable latitude and authority to draw on the resources of the entire evaluating organization to thoroughly evaluate the proposal. Although procuring organizations seldom have the personnel or expertise to match the proposal preparation team on a one-for-one basis, they do have technical experts and business managers with considerable expertise, and their access to multiple proposals for the same or similar jobs allows them to gain a rapid familiarity with a specialized subject. They can cross-check and compare a proposal with others submitted by other companies and/or organizations. A source evaluation board, team, or committee usually has sufficient expertise to locate weaknesses and/or strengths in a proposal and to identify "discriminators" between two competitors. A typical proposal evaluation organization is shown on Figure 12.1.

Functions of the Source Evaluation Board. Many of the functions of the source evaluation board are administrative in nature and are similar to the functions that were performed in the proposal evaluation process. In general, the three basic volumes or sections of a proposal will be evaluated by the three committees the organization and management, technical, and cost committees, shown in Figure 12.1. There are some important interactions between these committees, however, that may affect the style or approach to proposal development.

First, the cost committee does more than just check the credibility of a cost proposal. The cost committee is usually charged with the important and sometimes controversial task of making cost adjustments to a proposal to develop the "most probable cost of doing business" with a company. Some of the cost committee's inputs to make these adjustments will come from the correction of pricing errors or from labor rate or factor adjustments; but the major adjustments will come from resource adjustments (labor hours and materials) derived by the organization and management and technical committees.

The organization and management committee will make adjustments, if appropriate, to management labor hours based on major or minor weaknesses they find in staffing level, skills, or percentage of management attention to the project. They will be asked by the cost committee to estimate the impact of correcting these weaknesses in terms of labor hours, materials, and/or subcontracts and will be required to submit time-phased estimates of the recommended changes, along with detailed rationale

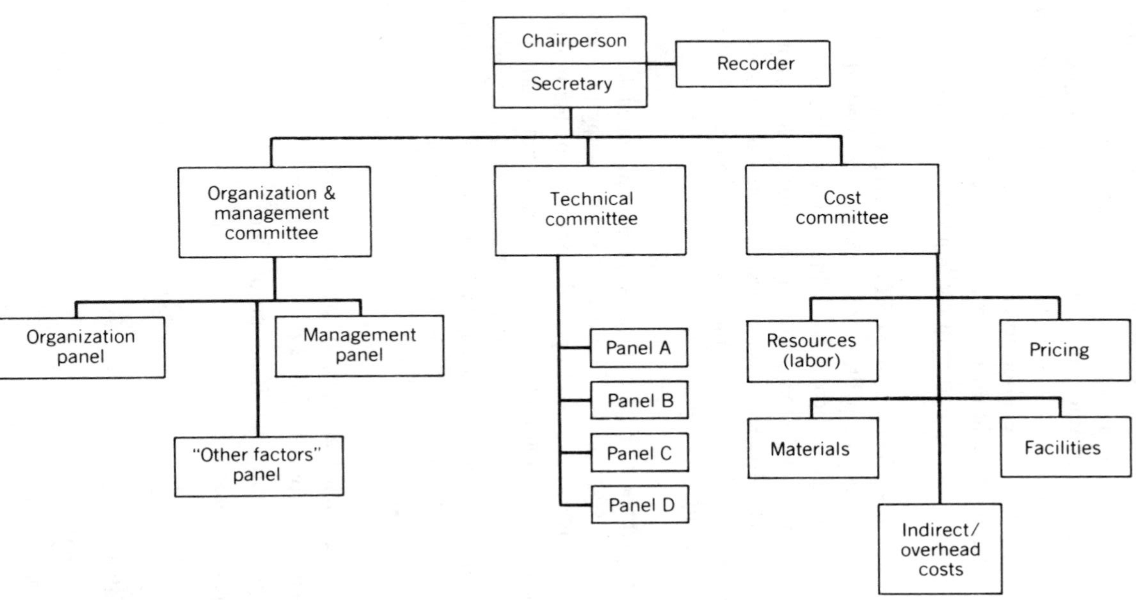

FIGURE 12.1 Diagram of proposal evaluation organization.

for the changes, to the cost committee. The organization and management committee will also identify any needed reductions in estimated effort due to perceived overestimates on the part of the proposer. These adjustments are then priced by the cost committee and used to develop the "most probable cost." Any adjustments, upward or downward, are reviewed and approved by the source evaluation board chairperson as well as the cost committee chairperson, prior to incorporation in adjusted costs.

Likewise, the technical committee, in identifying major or minor strengths and weaknesses in a technical proposal, will be asked to estimate the labor hours, materials, or subcontract costs required to correct weaknesses or to reduce overestimates. Technical committee adjustments are also approved by the source evaluation board chairperson prior to incorporation in the adjusted cost estimate.

Physical Location for the Evaluation Process. Since proposal evaluations are usually conducted in a highly competitive atmosphere, a secure and isolated office space is chosen for the evaluation board, committee, or team. When a large amount of money is at stake in a given competition, industrial espionage is not uncommon, and protection of sensitive information is a necessity. Planned or inadvertent release or disclosure of sensitive information could result in a company having an unfair advantage and in the filing of a legal protest.

Source Evaluation Schedule. Figure 12.2 is an overall schedule showing how proposal-related activities are phased into the activities leading to a new contract. The schedule shown covers a 10-month period from the time the project advocate within a customer's organization requests management's approval to proceed with a project until the time contract is awarded to the selected company. This time period will vary significantly and depends upon the size, complexity, and importance of the work output or work activity being evaluated. In general, the sequence of events will be in the order shown in Figure 12.2:

1. Request management's approval of project
2. Management's approval to proceed
3. Preparerequest for proposal (if not unsolicited)
4. Issue request for proposal
5. Proposal preparation
6. Proposal submission
7. Source evaluation
8. Source selection
9. Contract negotiation
10. Contract award.

The source evaluation board is sometimes organized during the time that the request for proposal is being prepared so that key members of the evaluation team can participate in and/or approve the format, content, and requirements in the proposal

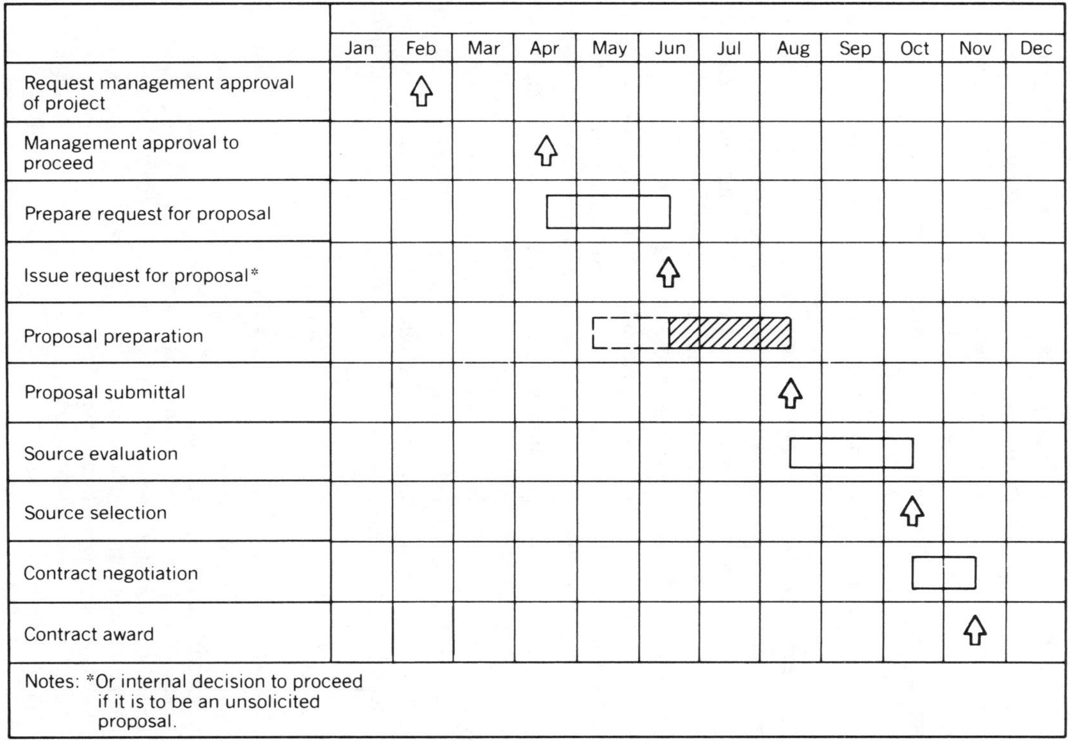

	Jan	Feb	Mar	Apr	May	Jun	Jul	Aug	Sep	Oct	Nov	Dec
Request management approval of project												
Management approval to proceed												
Prepare request for proposal												
Issue request for proposal*												
Proposal preparation												
Proposal submittal												
Source evaluation												
Source selection												
Contract negotiation												
Contract award												

Notes: *Or internal decision to proceed if it is to be an unsolicited proposal.

FIGURE 12.2 Overall schedule showing proposal-related activities.

request. Figure 12.3 is a typical sequence of proposal evaluation for an 8-week source evaluation activity.

EVALUATION FACTORS AND CRITERIA

Most proposal and source evaluation organizations establish evaluation factors and criteria before they solicit proposals and then continue to use these preestablished evaluation factors and criteria throughout the evaluation process. The purpose of establishing these factors and criteria is to designate what areas of schedule, cost, or performance are considered important in the planned work output or work activity and to provide a structure for the assignment of weights and scores to be applied to each factor and criterion. A scoring scheme or plan is developed for the purpose of reducing or at least leveling the effects of subjectivity in the evaluation process. By employing a numerical scoring process, each evaluation team member is required to rank the expected work accomplishment against a given standard. Although this method does not completely eliminate personal bias, opinions, and preferences, it has a tendency to produce a more equitable rating, particularly if a number of persons are doing the rating in adjacent or overlapping areas of performance.

There are three major categories of evaluation factors: (1) performance suitability factors, (2) cost factors, and (3) other factors. In general, the only factors that are weighted and scored are the performance suitability factors, although this policy has been found to vary between organizations. Performance suitability factors consist of those factors that reflect the capability of an organization to perform to a specified set of requirements; they include both technical and management factors. Cost factors, usually not weighted and scored, include a detailed cost evaluation using an independent cost estimate as a baseline and a "most probable cost of doing business," which is an adjusted cost estimate developed by the cost evaluators.

Preliminary review of proposals	1st week
Questions submitted to proposers	2nd week
Identification of major strengths and weaknesses	3rd week
Preliminary cost adjustments identified	3rd week
Preliminary oral team reports to board	3rd week
Preliminary board report to selection official	3rd week
Answers received from proposers	4th week
Oral discussions with proposers	4th week
Best and final offers received	4th week
Identification of major and minor strengths and weaknesses	5th week
Resource assessment of strengths and weaknesses	6th week
Final pricing of adjustments—reports to board chairperson	7th week
Final board report to selection official	8th week

FIGURE 12.3 Typical sequence of proposal evaluation on 8-week source-evaluation activity.

"Other factors" include a company's experience and past performance; the financial condition of the company; labor relations, the proposer's anticipated usage of small businesses and the proposer's minority business record; and the geographical location and distribution of the work. Socioeconomic and geopolitical objectives are given particularly important emphasis if the proposed work is to be performed for federal, state, or local governments. Each of these factors and the criteria under which they are evaluated is discussed individually below.

Performance Suitability Factors

Performance suitability factors, sometimes called *mission suitability factors*, are selected and weighted to provide an indication of how well the offeror or proposer understands the requirements for the work output or work activity. The evaluation of performance suitability factors and criteria requires the evaluators to technically penetrate the proposal and to discern if and how much the proposer understands the intent, purpose, and method of successful completion of the work. Using these factors and criteria, the evaluator can and should determine if the proposer is merely repeating the request for proposal requirements or if the requirements are truly understood in a way that will permit successful performance of the project's objectives. The criteria set forth by the evaluation team are applied to determine the offeror's comparative rating in performance suitability potential, including the offeror's understanding of the requirements, approach to the work, and the competence of the personnel to be directly involved.

There may be a great many varied and complex areas that bear on how well an organization can be expected to produce a work output or perform a work activity. The evaluating organization must be able to identify, analyze, and score those discrete criteria that determine how well the product or service can be expected to meet their performance demands. If individual criteria and weights have been prudently determined, the summation or integration of all of the scores of the various criteria will give a representative picture of the relative merit of each offeror from the standpoint of performance suitability.

In developing performance suitability criteria, emphasis has been on the identification of significant discriminators rather than on a multitude of criteria that tend to average out when integrated. Too many criteria have proven as detrimental to the effective evaluation of a proposal as too few. To the proposer, this means that the significant discriminators must be identified. Some performance suitability factors that have been found by experience to be relevant to virtually all procurements and that are generally included in the evaluation process by most evaluation teams are described below.

Understanding the Requirements. A proposer's understanding of the requirements depends on the comprehension of what the work is; how it should be most effectively performed; and what hardware, software, or data should be generated, controlled, or submitted in the performance of the work. A proposal will be closely scrutinized to determine if the proposer truly comprehends both the content and magnitude of the task. Although costs are usually analyzed separately by the evaluation board, committee, or team, they are of significant value to the evaluator in indicating understanding of the resources, both human and material, required for performance of the work. The assessment and estimate of the resources required is also an indication of an understanding of the work itself. Too high an estimate or too low an estimate will give an indication to the evaluators that a proposer may not have the proper conception of the magnitude, complexity, or content of one or more segments of the work. Because of this, proposal and source evaluation organizations usually make full utilization of the cost proposal to help them determine the understanding of the requirements spelled out in the request for proposal, drawings, or specifications for the job, hence the importance of cross-referencing and assuring the compatibility of the technical and cost volumes, as emphasized in Chapters 7 and 9.

The Management Plan. The management plan in a proposal is also a frequently scored performance suitability factor. The evaluators will score the proposed organization, the recognition of essential management functions, and the effective overall integration of these functions. Evaluators will be assessing the proposer's arrangement of internal operations and lines of authority, as well as the proposer's external interfaces with the customer and other organizations. Clearly defined lines of authority, a closely knit and autonomous organization, and identification of all essential management functions will be important in providing a high score in this area. Particularly important will be the relationships with subcontractors and associate contractors. The authority of the project manager and this person's position in the company hierarchy, the project manager's relationship with the next higher echelon of management, and the project manager's command of company resources also will be the subject of key criteria in the evaluation. Since the management plan usually also includes various schedules for the logical and timely pursuit of the work, accompanied by a description of the offeror's work plan, the organization, logic, sequence, and flow of this plan are also likely to be the subjects of a scored evaluation.

Excellence of the Proposed Design. If the project or product is hardware-oriented, the excellence of design and the detail presented in the design and specification portions of the technical proposal volume are major factors in the competition. In order to arrive at an informed judgment, the source selection official requests the evaluators' views on the merits of competing designs—both against the stated requirement and against each other. Generally, the best design will be the one that promises to provide the required performance at a reasonable cost. Evaluation of proposed designs varies from a top-level evaluation to a detailed evaluation of each subsystem and component, depending on the expertise of the evaluation team, the time provided for evaluation, and the expected unknowns in product/project design.

A large and important indicator of the excellence of a proposed design will be the amount of laboratory or prototype testing that has been done to demonstrate that the design concept is feasible and workable. A high score in this area means that the source evaluation activity is convinced not only that a proposer has a unique and innovative approach, but that sufficient work has been done to demonstrate to a reasonable extent that the design will perform its function in an operational environment. A primary indicator of the maturity of the design is the degree of actual testing in scale model, mockup, or prototype form and the maturity of the subsystem, parts, or components of the design. If the design includes mature, commercially available, well-proven components, evidence to this effect will be a persuasive influence on the technical committee of the source evaluation.

Key Personnel. The qualifications, education, experience, and past performance of the proposed key personnel usually represent a large percentage of the scored points in a major competition. Evaluators look for a proven record of high-quality performance by the top six or eight managers, engineers, scientists, technicians, administrators, or analysts who will manage and perform the work. Written résumés play their part in the evaluation of the quality of key personnel, but they are usually only considered as baselines from which the real evaluation of these individuals begins. Personal reference checks are made with people who are knowledgeable of a given individual's training, experience, and performance. These checks are made at levels commensurate with the proposed status or proposed role of the individual being checked. First-hand observations of the personality, performance, knowledge, and competence of key personnel can be, and usually is, derived from oral discussions with the proposer's team.

The principal place where qualifications, experience, and quality of key personnel will be displayed and evaluated is in he résumés that accompany the management plan. There are many good publications on the effective preparation of résumés, some of which are referenced in the bibliography of this book. Most résumé writers emphasize the importance of adapting the résumé to the proposal to properly emphasize the individual's applicable experience. In most cases this means more than just a change in writing style or content. It usually means a direct face-to-face interview with the proposed performer to draw out verifiable, applicable information from past experiences and activities. Many individuals have had applicable experience in a wide variety of areas that they do not even remember unless some memory jogging is done by an inquisitive and positive discussion. The evaluators of résumés can determine rapidly if background and experience are falsely expanded or "stretched" to cover a job. But factual information is easily verified and always appreciated by the evaluator.

Skill Categories and Skill Levels

A proposal will be evaluated to determine if the company has convincingly demonstrated that the appropriate skill categories are available to the project and will be assigned to the project. Skill categories will also be scrutinized to determine if a reasonable mix of skill levels within that category are available and will be used in performance of the work. Utilization and application of advanced technology equipment in the project will also be highly regarded by evaluators.

Corporate or Company Resources. An evaluation board, committee, or team comparatively assesses the resources proposed by each offeror in the general areas of manpower and facilities. Sample questions posed by the evaluation group in this area are: Are the proper numbers and types of skills available in the company and will they be assigned in a timely fashion to perform the work? Are the general type and capacity of facilities and (where required) special test equipment that are offered suitable and adequate to assure timely performance of the work? If the proposer does not possess adequate resources, has the ability to acquire them been demonstrated through subcontracts or otherwise? Succinct and demonstrable answers to these questions are found in excellent proposals.

Cost Factors

Cost factors indicate what the offeror's proposal will most probably cost the customer or procurement organization. Unless a proposal is a fixed-price bid and a company has a known reputation for being able to perform and deliver acceptable performance under fixed-price conditions, the evaluating organization will analyze the cost proposal in depth to: (1) determine the credibility of the cost estimates, (2) determine if necessary cost elements have been omitted, and (3) establish a "most probable cost of doing business" for the given work activity. The most probable cost is determined by an in-depth analysis and evaluation of the proposal to identify strengths and weaknesses and by adding additional expected costs, if necessary, to the bid price to correct the weaknesses and take care of obvious omissions. In some fixed-price bid situations, procuring companies or government agencies have found it necessary to perform an independent cost analysis of the work to determine if a company is underbidding or overbidding. Acquisition of a fixed-price contract at low cost is not beneficial to the customer if the job is not completed because of the proposer's lack of funds or if the supplier goes out of business. (More will be said about the evaluation of cost factors later.) Even though cost factors themselves are not scored, they can have a significant impact on a company's standing in the competition.

Experience, Past Performance, and Other Factors

Many proposers overlook the impact of the "experience and past performance" portion of their proposal. As mentioned in Chapter 3, a company should not even bid unless its experience and past performance indicate that it can accomplish the job. Hence, it is necessary not only to possess this experience and record of outstanding performance, but to convey this information to the evaluator in a way that it can be analyzed, scored, and verified. Since this section of a proposal is similar to the résumés of key personnel in that it is a résumé of a company, the evaluation team will amend and supplement their knowledge of the capabilities through their own investigations and outside contacts. If the purchasing organization has had experience with the proposer's company on other projects, the evaluators will collect and organize information related to quality of production, timeliness of completion, and responsiveness to direction. If there is no experience record between the proposing and evaluating organization, the evaluator(s) will be polling previous and current customers to determine the quality, timeliness, and efficiency of performance.

Experience is the accomplishment of work by an organization submitting a proposal that is comparable or related to the work or effort required under the proposed work. Activities of a comparable magnitude that include technical, cost, schedule, and management elements or constraints similar to those expected to be encountered in the proposed work are clearly relevant. Other experience or past performance (favorable or unfavorable) is indicative of the proposer's overall motivations, capabilities, and abilities. The evaluation of experience and past performance is usually not limited to scrutinizing the experience (or inexperience) of the design, management, or production team specifically committed in the proposal to performance of the resulting contract. The evaluation is of the overall corporate or company experience.

Past performance is especially pertinent: how well a company has performed previously on similar jobs is considered to be a strong indication of how well it can perform the task at hand. It is perceived by most evaluators of proposals that many organizations exhibit characteristics that persist over time. For example, resiliency in the face of trouble, resourcefulness, determination to live up to certain commitments or standards, skill in the development of key people, and acquisition of new high-technology equipment and machines are indicative of desirable traits in a performing organization. The evaluation team will be collecting and developing information to establish and verify performance in these areas.

Other Factors

The evaluation team and the selection official will consider factors other than performance suitability factors, cost factors, and experience and past performance when evaluating proposals. These other factors and their treatment by evaluators are each described briefly below.

Other factors that are evaluated but not scored are the so-called socioeconomic, geographic, and geopolitical factors. They include considerations such as labor relations, geographic location and distribution of the work, financial condition of the company, and small business or minority performance—the latter being most important in government procurements.

Financial Condition and Capability. Although evaluators can usually obtain the financial background of a company from a number of financial institutions and/or publications (Dunn and Bradstreet, Standard and Poor's, *The Wall Street Journal*), the insertion of the latest financial statement or report to stockholders in a proposal usually satisfies this requirement. This information is all the more important to an evaluation team if a company is not well-known.

Importance of the Work to the Company. The priority or importance that the corporate level of the proposing company places on the work being proposed is an important "other factor." Indications of this priority, if not clearly stated in the proposal or its transmittal letter, can be gleaned from various parts of the proposal. The evaluators must often read between the lines to gain a knowledge of the emphasis and importance that the parent company, home office, or corporate management attaches to the proposed work. This can be seen in the quality of personnel assigned, the level of the proposed project organization in the company, the corporate resources dedicated to the job, and the overall quality and content of the proposal itself. Evaluation of this other factor is highly subjective and, therefore, great care is taken to avoid giving it too great a place in the overall evaluation—but it is a factor that is evaluated.

Stability of Labor–Management Relations. This factor usually pertains to the historical and present quality of relations between labor and management in a company, particularly as it relates to the potential for labor unrest; poor worker morale; or an extended, prolonged, and costly strike. If there is a recent history of poor performance in this area, special assurance from management and labor alike that performance will proceed on a highly efficient, uninterrupted basis will probably be necessary to retain a high likelihood of selection to do the work.

Extent of Minority and Small-Business Participation. Minority and small-business participation is particularly important for government prime contractors and subcontractors, but this factor is often considered in commercial activities as well. The effectiveness of the Equal Employment Opportunity (EEO) program is usually considered as an "other factor" in the evaluation.

Geographic Location and Distribution of Subcontracts. Geographic location of a plant or office and location of subcontractors can often have a bearing on the capability of a company to carry out a task, because of the possibility of labor shortages, natural disasters, and transportation disruptions. Also, certain geopolitical influences in government contracts address the importance of providing more employment in areas of high unemployment. Although this is not usually a deciding factor, two equally qualified proposers may find that this factor will come into play in the evaluation and selection process.

WEIGHTING AND SCORING PROCEDURES

When a formalized proposal evaluation or source evaluation procedure is used, the evaluation board, committee, or team usually assigns a weight to each factor, then grades the proposal based on observed strengths and weaknesses to determine how much of that weight is to be added to obtain the total score. In a typical example, shown in Figure 12.4, the evaluation team has established a maximum of 1,000 points and has distributed these 1,000 points among five performance suitability factors. Other point totals and factors can be used depending on the type, complexity, and importance of the activity being evaluated. In the example shown in Figure

Performance Suitability Factors	No. of Points Assigned	PROPOSAL SCORING Grade	Grade Percent	Grade Points
Understanding of the requirement	125	10	100%	125
Management plan	100	5	50%	50
Excellence of proposed design	500	9	90%	450
Key personnel	200	9	90%	180
Corporate or company resources	75	6	60%	45
Total score	1000	—	—	850
(Minimum acceptable grade)	(600)			

Major strengths: Well developed, detailed design completed, with prototype hardware demonstrated.

Company is well versed in the product line, having already produced other similar items.

Minor strengths: Project manager is experienced in one other identical program.

Major weaknesses: None.

Minor weaknesses: Management team lacks complete cohesiveness: must be corrected prior to negotiations.

Company is small with fewer assets than competitors. Present growth rate makes this a minor weakness.

FIGURE 12.4 Typical example of evaluation team scoring.

12.4, however, the evaluation team considered *Excellence of the Proposed Design* as a predominantly important factor, with *Key Personnel*, *Understanding of the Requirement*, *Management Plan*, and *Corporate or Company Resources* of lesser importance, in that order. The proposal of Company A was then graded on a scale of 1 to 10 under each factor, and points were computed from a possible 10% to 100% for the proposer for each performance suitability factor. The grades received for each factor are a result of analysis of major and minor strengths and weaknesses found through an in-depth study of the proposal and a comparison of the proposed work activity with the stated work requirements. Some evaluation boards employ even more complex and sophisticated weighting and scoring techniques, which include subfactors and subcriteria, assignment of points to major and minor strengths and weaknesses, and curvilinear averaging of grade points to achieve an overall score. These high degrees of sophistication of weighting, grading, and scoring are of marginal value, since the evaluation process is itself based on subjective judgments at the lower level. Some form of structured weighting, grading, and scoring process is helpful, however, to reduce personal biases of evaluators and to provide a convenient numerical method of determining the approximate relative standing of several proposals.

Figure 12.5 is a comparison of the grading and scoring of four hypothetical companies for a sample project. Although the example is fictitious, it is not uncommon

	Company A	Company B	Company C	Company D
Understanding of the requirement	125	115	50	75
Management plan	50	75	100	75
Excellence of proposed design	450	500	250	400
Key personnel	180	160	160	140
Corporate or company resources	45	75	55	65
Total score	850	925	615	755
Average score	85%	92.5%	61.5%	75.5%
Bid price ranking (percent of independent estimate)	110%	150%	85%	105%
Most probable cost ranking (percent of adjusted independent estimate)	95%	130%	105%	100%

FIGURE 12.5 Example of comparison of evaluation team scoring of four proposals.

to see wide variations in grading and scoring of companies bidding for the same task. Also shown on this figure are bid price ranking and most probable costs as a percentage of an adjusted independent estimate.

COST PROPOSAL ADJUSTMENT AND MOST PROBABLE COST

One of the functions of the technical evaluation panels on a proposal or source evaluation board, committee, or team is to identify resource adjustments that must be made to correct any weaknesses found in the proposed plan of action. Another function is to recommend adjustments that must be made in the customer's independent cost estimate based on strengths, weaknesses, and comparisons found in the proposals. These adjustments are used to develop a most probable cost of doing business with each proposer and to update the customer's independent estimate after receipt of proposals. Experience has shown that bids will be lower than an independent estimate made by the procuring organization, primarily because the bidders are acutely aware of price competition. The procuring organization itself does not encounter competitive pressures to reduce resource projections from an estimated cost to a competitive bid price, so it can obtain a more accurate estimate.

Adjustments to proposal cost estimates resulting in most probable costs can be large enough to change the relative cost ranking of bidders, because proposers sometimes omit vital cost elements (or add unnecessary items) in their cost structure. On several occasions, firms that bid on a price-negotiable contract found that the lowest bidder did not get the job because the computed most probable cost of doing business with that bidder was actually more than that for a proposer with the higher bid price. It is obviously very important for the evaluation team to document the rationale and support for any adjustments to bid price, because selection of other than a low bidder could cause a protest and ensuing investigation of the evaluation procedures. Evaluation teams are very aware of this precaution and do not make adjustments to proposers' cost estimates unless these adjustments can be fully substantiated and backed up by factual information and the best expert opinion and advice available.

The two lines at the bottom of Figure 12.5 labeled "Bid Price Ranking" and "Most Probable Cost Ranking," illustrate how the ranking can change after adjustments are made. There have been frequent instances where cost adjustments by evaluators have caused a change in cost ranking, and, in many of these instances, this change in ranking has resulted in a different contractor selection than would occur if the selection decision were made on bid price alone. In the example shown, Company C's bid was the lowest, being 85% of the government's estimate. After the evaluation process had been completed, however, and cost adjustments made, the most probable cost of Company A's proposal was the lowest when compared to the adjusted independent estimate. Use of the adjustment process and the development of a most probable cost are prudent actions and can be important factors in avoiding under-estimates and overruns. The knowledge that this practice may be used will provide an incentive to a proposal team to leave no stone unturned in identifying all work elements, schedule elements, and cost elements. As mentioned in Chapter 1, the acquisition of new work must be based on providing cost-effective work rather than on oversight or excessive competitive optimism. Since cost is an important nonscored factor in many procurements, it is essential to supply as much credible and substantiated backup data and rationale for the cost estimate as possible, within the confines of the proposal page limitations. If a company is bidding for a fixed-price contract, it is still important to record the rationale and backup data for all cost estimates for use by the company management, for use as a takeoff point for estimates in future proposals, and for use in tracking actual occurrences against the estimate.

In source evaluation activities, the cost is a highly important but not always an overriding criterion. For example, a proposal rated last technically, or one with an unacceptable score on the management volume, or even one found unacceptable and nonresponsive will seldom win just because the dollar bid or the adjusted most probable cost is the lowest. For most procurements complex enough to require a proposal, that proposal must first be technically acceptable and then deemed satisfactory regarding management aspects. Only those proposals which have survived the technical and management hurdles will be evaluated from a cost or most probable cost point of view. Among those proposals that are still in contention after the technical and management hurdles are successfully passed, the most probable cost to the customer becomes the all-important criterion.

The amount of cost-sharing (percentage of costs to be supported out of company funds) proposed by the bidder can also be an important but not necessarily overriding, factor in the cost evaluation. For example, say Company A and Company B are both acceptable technically and from a management point of view. The cost-evaluation comparison is as shown in Table 12.1.

TABLE 12.1. Cost-Evaluation Process re: Cost-Sharing

Item	Company A's Proposal (millions)	Company B's Proposal (millions)
Total proposed price	$20	$16
Cost share dollar amount	$ 6	$ 4
Price to customer	$14	$12
Estimated customer in-house costs	$15	$ 1
Most probable cost to customer	$15	$13

Company A has proposed the greatest cost-sharing dollar amount both in absolute dollars ($6 million compared to $4 million) and percentagewise (30% over 25%). But Company B still has the lowest most probable cost to the customer.

LEARNING FROM PAST MISTAKES

Organizations that prepare and submit proposals may become overly discouraged if and when their firm is not selected for the job. It is in this "nonselection" area that the structured evaluation process can provide a greater service to the proposer's organization than is usually realized. Most evaluation proceedings provide for a debriefing if it is requested. It is through a debriefing that the proposer can determine a great deal of what went wrong and how to make improvements in the next proposal. In the instance that a company does not win the contract, the entire process can be looked upon as a learning experience. The "we play until we win" approach will pay off as a company continues to learn from its experiences. This does not negate the advice given in Chapter 1 that a company go into the proposal as if it is going to win the contract. Further, the debriefing will not only point out the negative aspects of the proposal, but also the positive aspects—strong points—which can be maintained and can further strengthen a firm as it competes for new work. Assuming that a good work output or activity design was presented, several factors should be thoroughly reviewed, such as: (1) the original decision to bid; what were the influences and motivations? (2) what skill mix and skill level matrix were presented? (3) were there any overriding geographical or political influences on the selection decision? and (4) did the organization and its individuals have the appropriate kind and depth of knowledge and experience about the work output or activity? Close introspection will sharpen a company's proposal preparation skills and increase its assurance of winning the next time.

THE PURCHASING AGENT'S MOTIVES

Many purchases by government or industry are too small to warrant the issuance of a request for proposal or request for quotation, and many that are not too small for an RFP or RFQ do not use formal evaluation techniques. These purchases of goods and services are usually handled by purchasing agents or buyers, who may or may not require some form of proposal along with a price quote. With the advent of higher technology work activities and outputs, these procurements more frequently require supplemental information with the bid quote to fully describe and define the work. This supplemental information essentially constitutes a proposal and must be presented in a way that adheres to many of the principles presented in this book in order to enhance the proposer's potential for success.

It should be kept in mind, in selling to any organization, that the main objective of the buyer is profitability for the purchaser's organization. The proposal or supporting information for a price quote should emphasize how it is profitable to the buyer to purchase the item from a particular firm. Although the lowest-priced quote is usually what the purchasing agent or buyer is looking for, it is often more important that a work activity or work output be offered that exactly meets the specifications and that is delivered exactly on time to meet requirements. A buyer's profitability may depend on timely delivery more than on low cost, as delays in production may be more costly than a higher item price tag. Therefore, timing of the work performance, as well as quality and suitability of the work, can be of vital importance to a customer. These factors need to be evaluated, as mentioned earlier, in the original decision to bid. If a company perceives that it has and can supply exactly the right service or product at exactly the right time or place and that competitors are not in an equally good position, it does not have to compromise by offering a price that is significantly lower than costs plus a reasonable fee or profit. Thus, being in the right place at the right time has a value in itself.

McGraw-Hill's *Purchasing Handbook*, Third Edition, compiled by George W. Aljian, points out and describes many of the techniques used by purchasing agents and buyers to assure that their company gets the most for its money. This means buying the right quality, the right quantity, at the right time, at the right price, and from the right source. Notice the order of these criteria. Quality, quantity, and timing precede price in the list. And the source is listed last. The source will be determined by the other criteria. If all other criteria are met, the source's experience and qualifications could be a determining factor in the selection. Notice that the pricing criterion is the "right" price and not necessarily the "lowest" price.

Purchasing agents will look for much the same information that a source evaluation board will, except usually in less detail. In making a purchase they will consider the following technical or product characteristics:

Industry standard specifications
Manufacturer's specifications
Brand name
Part drawing
Part specification
Tolerances
Finishes
Material specifications
Manufacturing process specifications
Quality-control system requirements
Acceptance tests
Packaging specifications
Performance specifications
Warranty provisions
Distributor provisions
Field service requirements
Resale requirements

In addition to technical or product characteristics, purchasing agents and buyers will look for and evaluate various availability-related features such as: quantity, usage rates, lead time, long-term requirements, transportability, and chronic short supply problems.

In evaluating the bidders themselves, purchasing agents and buyers will evaluate the suppliers' management capability, technical capability, manufacturing capability, labor–management relations, past performance, financial strength, and company ethics. To obtain this information, purchasing agents and buyers will use supplier proposals, quarterly or annual financial reports, trade and industry information, field surveys, and quality surveys as inputs to their decisions.

CONTINUING EVOLUTION IN SOURCE EVALUATION

As the sophistication of work outputs and work activities increases, so also the methods of evaluating proposals increase in sophistication. The methods, techniques, and skills used in source evaluation are continually improving and expanding to include the consideration of even more factors in the selection process. Customers are becoming more aware of the techniques of profitability analysis, life-cycle costing, and detailed industrial engineering type man-hour and material-based cost estimating. As noted in Chapter 11, some private firms and government agencies have developed computer programs that not only mathematically check resource estimates down to the lowest level in the work element structure, but are able to compare an estimate on a line-for-line basis with the estimates of other bidders or with those that are independently prepared. These computer programs will provide labor-hour, material, dollar, and percentage differences between proposal line items of competitive or independent estimates. Having been provided with this information, evaluators can then dig deeper into the reasons for the larger deviations and selectively analyze a proposal in detail for completeness and credibility.

Commercial, industrial, and business customers are beginning to exercise greater care and to employ more advanced analysis techniques to determine if the work proposed will result in a higher or lower profitability to the customer than the work of competitors. It should be remembered that the customer will be doing a profitability analysis on a proposal from the customer's viewpoint just as the proposer has carefully employed profitability-optimization techniques in the proposal marketing and market analysis phase of new-business acquisition. Knowing that evaluations are beginning to include these subtleties of business acumen, analysis of customer's profitability during the marketing phase as well as a proposer's own profitability becomes even more important.

The *life-cycle cost* of products, projects, processes, and services is becoming important to commercial and government customers as well as to individuals. Costs of ownership and even the costs of disposal are becoming more important factors in the acquisition decision than initial price. Environmental issues such as biodegradability, environmental impact, and industrial safety are also becoming important factors in the selection decision. Straight-line control in proposal preparation requires attention to all of these factors if your proposal is to rank high in the proposal evaluation process.

Aside from the many factors and criteria that have been mentioned in this and previous chapters, there are still more intangible factors that must be considered in the proposal preparation process. A few of the more important of these are discussed in detail in the next chapter.

13
KEY ELEMENTS REQUIRED FOR SUCCESS

Let it be carried out with all diligence.

—Ezra 6:12

Some basic principles touched on in earlier chapters bear restatement, emphasis, and expansion. These principles are vital to winning in today's business environments. They are the keys to success; therefore, they should be reviewed carefully in light of any specific proposal.

TECHNICAL EXCELLENCE: A MUST

Being a winner in today's business environment requires technical excellence—in the company, in the work activity or work output offered, and in the means of accomplishing the work. It grows out of the desire to be the best in the field from two seemingly conflicting objectives in business: specialization and diversification. Specialization tends to narrow the scope of one's activities to those on which one can concentrate skills and abilities. Specialization of a company can result in that company carving out a niche in the industrial or commercial sector of the economy where a need exists and where techniques of fulfilling that need have been identified and can be profitably marketed. If the proposing organization has specialized in the work activity being proposed, this specialization should be described and supported by documented historical evidence or recorded actual experience. On the other hand, diversification can also be an asset to a proposer if the work being proposed can benefit through input from other disciplines or other aspects of the company's activities. Technical excellence means having the best personnel, equipment, and facilities available to do the job and having a management structure that will motivate personnel to use these resources to achieve superior results. A proposal must convey enough information to the customer to provide the assurance that technical excellence will be maintained in all of these areas. Technical excellence falls into three general categories: (1) meeting the specifications, (2) providing on-schedule performance, and (3) controlling costs to meet budgets.

Meeting the Specifications. The function of the technical portion of the proposal is to convince the buyer or evaluator that the specified performance for the activity or product will be met. In areas where specifications cannot be met, sufficient description, discussion, and evidence should be presented to demonstrate that a deviation from specified performance will not affect the overall desirability and quality of work except, perhaps, to improve it. If a stated requirement is unrealistic or unattainable, an alternative should be proposed that points out why the original requirement is unrealistic. Recognizing an unrealistic or unachievable performance, quality, or operational goal and proposing an alternative requirement is often an indication of knowledgeability and confidence, rather than of limitations, and is often graded as a plus rather than a minus in evaluations.

Providing On-schedule Performance. The overall technical, organization and management, and cost proposal volumes must emphasize and demonstrate confidence and ability to perform the work as proposed and to complete the work on schedule. Timeliness of performance can mean the difference between the customer's success or failure and can make a difference in long-term profitability. Often, the completion of the supplier's work will be keyed to the customer's overall objectives and goals. An assurance of a proposer's timely completion of intermediate as well as final milestones can be a determining factor in the choice of a particular company to do the work.

Providing Consistency in Quality. Both the technical and management volumes of the proposal should provide convincing evidence that the company intends to provide a high-quality service or product. The organization and management volume should provide evidence of independent checks and balances on quality of work. The technical volume should detail the methods, procedures, personnel, and equipment that are planned for inspecting, testing, checking, and correcting deficiencies during and after performance of the work. The experience and past performance portion of the organization and management volume should emphasize a record of continuous improvement in quality—through detailed evaluation of past mistakes and implementation of corrective measures.

INNOVATION AND CREATIVITY

Two qualities that are easily recognized and highly regarded by proposal evaluators are innovation and creativity in designing a work activity or work output that will fulfill a need. A willingness to use modern, advanced techniques and equipment for management, design, and production will improve credibility and demonstrate management and technical competence. Innovation and creativity will be evident in a company's acquisition of new skills, techniques, and equipment as they become available. Both the awareness of and use of emerging methods of improving efficiency, economy, and effectiveness will be a clear indication of a proposer's desire to provide the customer with the most for the money.

Use of High-technology Methods and Materials. High technology is affecting virtually every type of work activity and work output in our nation. Businessmen are discovering that in order to remain competitive, they must be ready and willing to take advantage of the speed, accuracy, and efficiency offered by high-technology skills, methods, and equipment. Decreasing costs, size, and improving quality of sensors, microprocessors, computers, automated office business systems, computer-aided design techniques, and automated factory manufacturing, assembly, handling, and testing equipment have caused a new industrial revolution in many fields. More businesses are adopting high-technology methods each year. Recognition that there is no such thing as status quo is an important attitude to maintain while submitting a proposal. Each new job acquired is an opportunity to update work activities and outputs to take advantage of these evolving high-technology methods.

Efficient Design and Manufacturing Techniques. Companies that are involved in design, manufacturing, and production are finding that highly organized and controlled scheduling techniques are not only useful but necessary to reduce or eliminate wasted time, effort, and materials. Any new product or service activity that has not included systems or methods for assuring precision in sequencing of hand and automated operations should analyze these systems and include them where appropriate to be competitive. The just-in-time system, applied effectively by the Japanese, is one that requires each operation to be performed and each part to be delivered just-in-time to be mated with its related operations and parts: not too soon or too late. Too early delivery of a part or performance of an operation can cause wasted storage or maintenance time for parts and too late delivery can result in wasted human and machine time.

In the field of design, computer-aided design techniques can save a significant number of labor hours by reducing or eliminating time-consuming manual production of drawings and by reducing the designer's time to performing essentially creative rather than mechanical work. Drawings and designs can be stored on magnetic tape or disk, rapidly updated, and reproduced by printer or plotter on demand in almost any size or shape desired. The use of three-dimensional and color graphics is also reducing or eliminating the many hours of technical artwork required to convert a drawing into an isometric, perspective, cutaway, or exploded view of a part, subsystem, or product. This time-conserving technology can be of considerable value to the preparer of proposals, because designs do not usually have a high degree of maturity at the time a proposal is submitted. Computer-aided design allows preliminary concepts to be converted to realistic drawings and isometrics in a very short time.

Production, logistics, and materials handling are also now using advanced electro-mechanical and computer technology to increase the speed and accuracy of these functions. Computer-aided manufacturing, advanced vision systems, sensors, robotics, and automated storage and retrieval systems are being used with increasing frequency in industries of all sizes to reduce labor hours required to accomplish industrial functions and to enhance production quality and speed. Dedication to investigating, analyzing, and adopting these systems in marketing new products or services will enhance one's competitive stature when proposing and performing new work.

The People-oriented Approach. The use of innovation and creativity in motivating and rewarding employees is becoming a recognized and effective way to build timeliness and excellence into a work activity or work output. The people-oriented approach to solving problems, when coupled with the high-technology approach, has proven to be an unbeatable combination for competing in today's business environment. Employee stock ownership, profit sharing, employee pension trusts, quality circles, and other motivational programs are demonstrating their immeasurable value in motivating employees to become interested and contributing members of the management of a company. These programs have increased team spirit, quality consciousness, and profitability in virtually every instance where they have been properly applied. An organization and management proposal should demonstrate the people-oriented approach to improving efficiency, economy, and effectiveness of operations, and the proposal should outline the ways in which the company plans to continue to use these programs to assure a continuation of high morale and high productivity. In-company proposals should also show how these programs will increase overall company and resulting customer profitability. Evidence of successful programs in other organizations, advice from consultants and motivational program experts, and reviews of the latest literature will assist a proposing company in building a convincing argument for including these programs in plans for future work acquisitions.

STABILITY AND FLEXIBILITY OF LABOR BASE

When structuring the labor base for a new work activity or work output, it is necessary to review and analyze all available options pertaining to the use of the existing company's work force and the hiring or employment of new personnel, consultants, or subcontractors. The company's management should be involved in this part of preproposal planning, because it bears not only on winning the present job but on the company's posture for the acquisition of future work. Company growth and expansion plans must be correlated with the work acquisition under consideration to determine how the skill categories, skill levels, and salary structure will be affected by performance of the new work. To be a winner in today's business environment, the work force itself must be continually updated, trained, motivated, and periodically supplemented to assure a competitive posture and a high capture potential for new contracts. Several interacting factors are important elements of this analysis: (1) a stable base of skilled personnel, (2) hiring at the low end, (3) use of consultants and subcontractors, (4) part-time and temporary employees, and (5) cross-training of the work force.

A Stable Base of Skilled Personnel. The solid foundation on which most companies build their future is a stable base of loyal, long-term, skilled, and experienced personnel. A company that has built its capabilities and reputation by continually improving and rewarding its employees is in an excellent position to assure the customer that this stable base will continue to be available to perform work in a satisfactory if not outstanding manner. Because long-term employees continue to receive merit, longevity, and quality salary increases, however, this stable work force will tend to escalate in cost as time progresses, making older companies less competitive than newer companies, unless the older companies continually supplement and add to their labor base at the lower end of the salary scale.

Hiring at the Low End. To replace personnel losses caused by retirements and attrition, a company should bring in and train personnel at the lower end of the salary scale, and proposals should reflect the use of these newer employees as soon as possible in the work cycle. Presuming that the basic skills required to perform a proposed mission have been built up over a period of years through training, development, motivation, and advancement of personnel, new skills should be acquired and developed from the ground up rather than hired in from other companies at high salary levels. Maintaining this long-term approach requires considerable discipline, because the temptation will exist to hire highly paid experts as an expediency to perform a short-term job when this job might just as well have been performed by a consultant or subcontractor. Hiring at lower levels and promotion from within are still the best practices if the appropriate skill categories and skill levels can be acquired for new work in this manner. Then the phasing in of these skills in each new job on an appropriate time-based scale will make each proposal more competitive.

The Use of Consultants and Subcontractors. When it is anticipated that specialized skills not readily available within the company will be needed to perform work on a potential contract, consultants and/or subcontractors can be highly effective in supplementing and complementing the skills and experience of the work force. These performers bring in skills that could not be developed or obtained from within the company, and they can be employed contingent upon receipt of the contract. The skillful planning and use of consultants and subcontractors, therefore, is an integral part of preproposal planning both by the proposal team and by the company's management and should be a part of the overall labor base analysis that precedes the acquisition of new work.

Part-time and Temporary Employees. An increasing number of families have more than one individual who is or can be considered as part of the labor work force. Where there is a full-time breadwinner, there is often a spouse, child, or parent that can and does engage in part-time or temporary work to assist in earning discretionary income for the family. Because there is a full-time breadwinner, these family members do not object to (and often desire) part-time or temporary employment. There is a readily available market, therefore, of part-time or temporary skilled and unskilled employees who can be brought in during peak workloads, but need not be considered part of the permanent work force.

Wives or husbands who are the second wage-earners in families may have just the skills needed for the project's duration. An expanding population of retirees exists, many of whom have retained their skills, vigor, and desire to do productive work. Companies, universities, and government organizations are finding that the experience, maturity, judgment, and seniority of many early and not-so-early retirees can be of remarkable benefit to an organization at a relatively low cost and commitment.

The increasing costs of education have caused many students to search for temporary or part-time employment to supplement their income as well as to provide them with experience in applying their newly found knowledge. Students are excellent temporary or part-time contributors to profitable business ventures. This labor market should not be overlooked when structuring a proposal for new work as it may enhance a proposal significantly from a price and skill-mix standpoint.

Cross-training of Work Force. Behavioral studies show that employee motivation increases as employees identify with a larger part of the job and as they realize and recognize the part that their work plays in the overall work activity or work output. A company policy of cross-training and broadly scoped jobs will reflect itself in a cost-effective proposal, because some flexibility exists in work assignments.

If one person has the potential to perform several types of jobs, the hiring of two or three people may be avoided. Effective cross-training and judicious use of cross-trained employees can result in significant savings in labor hours required to do a job and can avoid excessive buildup (above a company's stated growth objectives) of a work force that must subsequently be fed with new, profitable work.

Specialization of company objectives and goals does not necessitate or require overspecialization of personnel, although every company must have a certain percentage of specialists in its work force in order to maintain technical excellence in its field. Today's business environment requires proposals based on a careful analysis of the degree of skill specialization and the degree of cross-training required for and provided to each task. A common fault in proposals is the use of highly qualified and highly paid personnel for work that starts out needing their services but which can eventually be delegated to lower skill levels. These highly qualified and highly paid individuals should be transferred to the next new job as soon as possible (1) to reduce the composite labor rate as soon in the project as practicable and (2) to make these highly qualified personnel available as early as possible for new projects that require high front-end skill levels. Experience shows that, as a job progresses through time or the quantity of related experiences increases, the skill level required to do the job will decrease. The more highly paid employees that are needed to start new projects can train newly hired, lower paid personnel as the job becomes more well-defined and more routine. Then these skilled workers can be moved on to establish new jobs and to train more new employees. This time-phasing of skill mix and skill category should be used to advantage in structuring the tasks and subtasks of the proposed work and in demonstrating to the customer that skill application and utilization has been optimized.

COMPETENCE IN SCHEDULING

The successful proposal must demonstrate to the customer not only that meticulous care has been taken in scheduling the work but that the tools, techniques, and methods of scheduling are both available to and understood by the proposer for use in rescheduling the work as required as the work progresses. Development and maintenance of a detailed multifaceted work activity usually requires the use of one or more of the generally accepted methods of network analysis or bar chart scheduling. Operations analysis tools designed specifically for large, long-duration, multitask projects were brought to bear by the Special Projects Office of the U.S. Navy and published in 1958 as the PERT (program evaluation and review technique) system. Since that time, various versions of PERT, critical-path bar charting, procedure diagrams, bubble methods, and method of potentials have been developed for both manual and computerized usage. These methods interrelate jobs in sequence or time and provide the assistance a manager needs that goes beyond the limit of mental planning capacity in a multifaceted project. When a project is underway and the inevitable deviations from plans occur, the network or bar-charting techniques help the manager to determine the importance of these deviations and to take effective corrective action. Two of the many good books on critical-path analysis and bar charting are: *The Critical Path Method* by Arnold Kaufmann and G. Desbazeille (see bibliography), a book using complex mathematics adaptable to computer use; and *Analysis Bar Charting*, by J. E. Mulvaney (see bibliography), a book that can be used for manual bar-chart analysis.

Even the simplest proposal should have some form of time-oriented display of the activities that must be carried out in parallel or in sequence in order to accomplish the work on time. Presentation of this type of display in the proposal will demonstrate to the customer that intermediate milestones have been established and that detailed planning has been carried out in anticipation of the contract award. Master schedules, backed up by detailed schedules of each phase of the work, are required for more complex projects. In a complex project, the customer will want to know what type of scheduling technique will be used, how often it will be updated or adjusted, how corrective actions will be input to rescheduling exercises, and to what degree the scheduling methods will be integrated into overall project performance. An important precaution to take in the scheduling discussion is to avoid indicating a fascination with the scheduling technique itself: recognize that the scheduling and rescheduling methods are only there to help the project be completed on time.

PROPOSAL REVIEW BY MANAGEMENT AND OTHERS

Throughout the proposal preparation process, as shown in the flow diagrams accompanying appropriate chapters, critical reviews of the proposal are made by the proposal team, by special review teams, by middle management, and by disinterested experts. Independent reviews can often catch errors that are overlooked by those who are working with the proposal manuscript on a day-to-day basis and they can identify inconsistencies that may be generated by last-minute changes. After such interim and final reviews, the proposal is much more likely to be error-free and to represent the desires and policies of company management.

HONESTY AND INTEGRITY

Being a winner in today's business environment need not involve a compromise in a company's honesty and integrity. It should be recognized that there is a difference between honesty and optimism, but that the two are compatible. Optimism is a necessary element in proposal preparation. The entire proposal preparation activity must be carried out with a winning attitude in order to win. The positives that are in the proposing company's favor should be pointed out and not the negatives, which are not in its favor. But a proposer must honestly believe that the job can and will be done in the manner, on the schedule, and with the resources and costs stated in the proposal.

FINAL GUIDANCE

There is some final guidance that we must provide in order to make this book complete. This final guidance is in the form of subjective but important factors that will spell success in a proposal preparation effort.

1. *Be Complete.* A proposer should provide all required information in the proposal. A key piece of information left out could have disastrous effects on the capability of winning. Often the only thing the evaluator has to go on is the content of the proposal. Evaluators are sometimes not allowed to consider other information than that in the proposal even if this information is of general knowledge.

2. *Be Organized.* An organized flow of information in the proposal that permits the evaluator to follow the thread of rationale will result in a positive reception of the proposal. Careful consideration of the interrelationship of the technical, organization and management, and cost or price proposal volumes is an absolute necessity. Organization in the proposal will be a clear indication of an ability to properly organize the proposed work.

3. *Be Objective.* A proposer should avoid emotional comparisons, flowery words, and unfair judgments in the proposal. Frankness and objectivity are appreciated by the recipient.

4. *Be Informed.* By listening to the customer's requirements and to the marketplace, a proposer can provide an up-to-date, timely proposal. The degree to which one is informed on the work content and the customer's requirements will be evident in the proposal.

5. *Be Innovative.* A proposer should not hesitate to present new approaches and ideas as long as they are backed up by planning and analysis. Most customers do not merely want yes-men to conform unquestioningly to their requirements.

6. *Be Factual.* A proposer should not overstate or understate capabilities or costs. An accurate cost estimate should be presented and a reasonable fee requested.

7. *Be Professional.* Proposal evaluators appreciate a professional, well-organized proposal.

8. *Follow Up.* A proposer should not forget to follow up the proposal with responsive answers to requests for supplemental information. A contract is sometimes won based on this last-minute information.

Finally, a proposer should remember that compatability with the request for proposal or requirements is a measure of responsiveness. Responsiveness is a key factor in the evaluation of any proposal. Internal compatability of the proposal within volumes and between volumes is a measure of credibility. Although credibility is not a scored factor, it is an important characteristic, the lack of which can cause rejection of the proposal in its entirety. The quantities and types of resources that are proposed to apply to each aspect of the job are a measure of an understanding of the requirements, which is a scored criterion and a clear indicator to the evaluator of a proposing firm's technical competence in the work being proposed.

As aids to your proposal preparation process, a sample proposal preparation manual, which can be modified and tailored to your specific company, is included as Appendix I, and a proposal checklist is included as Appendix II.

14

CASE STUDY: A WINNING PROPOSAL

. . . it will turn out exactly as we have been told.

—Acts 27:25

Very seldom does the student or practitioner of proposal preparation have the opportunity to study a real and undisguised winning proposal. The reason for this is that companies that have won don't like to disclose how they did it, much less disclose the secrets about specific characteristics that made their product or service innovative and unique. The passage of time, however, and the desire to reveal real success stories as they emerge so that others can take advantage of the experience gained, sometimes motivates companies to permit their successful proposals to be revealed. Such is the case of the proposal by a small high-tech start-up company in Huntsville, Alabama: Rantek, Inc. The proposal is titled "High Fidelity/Precision Cell and Tissue Bioreactor." Our appreciation goes to the corporate officials of Rantek, Inc., for providing written permission to reveal the events leading up to the submittal of a $50,000 proposal to the State of Alabama's Innovation Fund along with the contents of the proposal.

A proposal in the medical field may not fit the subject matter of all small proposals contemplated by industry in the years 2000 and beyond, but the principles, techniques, procedures, and practices employed are applicable across the board to those who are seeking grant funding for ventures ranging from demonstration and commercialization of an idea to the introduction of a new high-technology product or service into the marketplace. Thus, we present the vision, goals, objectives, strategic plans, procedures, and resulting proposal of Rantek, Inc., for a high fidelity/precision cell and tissue bioreactor and we describe how straight-line control resulted in an excellent winning proposal and a decision to manufacture and market a prototype laboratory bench model device that could have far-reaching effects in the field of medical research and health care, as well as the potential for great prosperity for the founders, participants, investors, and employees of Rantek.

BACKGROUND: THE VISION AND PLANTING THE SEEDS

Rantek, Inc., is a small start-up company founded in 1984 to conduct biotechnology and related medical research that capitalizes on the latest state of the art in engineering and science. Its interests lie in exploiting high-technology aerospace and biotechnical advances to create marketable and profitable methods, tools, processes, and products in the fields of medicine and pharmaceuticals. Rantek's initial focus is on techniques, methods, and tools for processing high-quality human cells and tissue in quantity. The founders envision an organized, systematic, synergistic, and methodical growth pattern using a network of participating organizations and personnel who have had extensive experience in research, engineering, and management ranging from small experiments to large projects. Using a corporate philosophy of attracting sponsored research to fund the initial product development and start-up activities, coupled with the exchange of in-kind services with motivated participants, Rantek's goals are to grow in an orderly manner in a near-zero-indebtedness environment. The normal negative cash flow and need for extensive borrowing is being avoided by pursuing a carefully calculated growth pattern based on sponsored research that will bootstrap itself as technology advances and as sales increase. A long-range plan describing both the broad vision and the focused product areas for Rantek, Inc., was generated, and an internal company roadmap of the steps required to achieve corporate objectives was developed.

The broad vision to provide medical research tools for the betterment of mankind and a prudent and ultraconservative financial posture and growth plan were the driving forces behind Rantek's corporate goals, objectives, and implementing plans and procedures.

ASSEMBLING THE PROPOSAL MECHANISM

In late 1990, Rantek, Inc., became aware of an opportunity for funding of its initial product definition and prototyping. Rantek was contacted by a staff member of the Alabama Technology Partnership. This Small Business Development Center is not unlike those that exist in many states, regions, and municipalities throughout the United States. These centers are sponsored by a variety of organizations, such as the U.S. Small Business Administration, state governments, local universities, and state and local chambers of commerce to encourage and enhance the formation of new businesses and to facilitate their growth and expansion. The staff member explained the following ground rules for proposals that would lead to a $50,000 grant from the state's Innovation Fund:

1. Participation of a university was encouraged.
2. One or more of eight target technologies were to be addressed (*Biomedical science* and *biotechnology* were two of the eight.)
3. Economic impact on the state was to be demonstrated through provision of convincing evidence that the research would eventually result in the creation of more jobs in the state.

4. Cost-sharing and in-kind services were encouraged.
5. Both a final proposal and a business plan were required. A 25-page limit was placed on each.
6. Evidence of a patent application or patent disclosure was highly desirable.
7. Maximum grant funding was $50,000.

Rantek, Inc., officials immediately recognized that this was the opportunity they had been waiting for to get a new product idea launched. The grant effort exactly fit the company's preconceived vision, goals, objectives, and growth plans. The company set out to enlist qualified and experienced persons who could not only fill vital roles in the grant research effort, but who could also follow through as supporting participants as the product lines grew to fill the needs of the target marketplace. A university was enlisted and a principal investigator with directly applicable experience and qualifications was selected by the university. An engineering/manufacturing firm was contacted for potential participation in the product's design and fabrication with the potential for profits from production. A small technical management firm was employed to write the proposal and to manage the project once grant funds were approved and received. Qualified medical consultants were contacted and agreement was obtained for use of their résumés and potential participation.

With this highly qualified team in place, Rantek was ready to launch its efforts to win the state innovation fund grant. During the 9-month period from initial contact to submission of the proposal in late June 1991, several drafts of the proposal were developed for review by all participants. With each review, the proposal improved. Many meetings were held to clarify requirements and obtain guidance. A preliminary proposal was submitted for formal comments and responses by the state. These comments were incorporated into the proposal, and it was upgraded, modified, and amended accordingly. In the meantime, potential medical and high-tech equipment design and manufacturing firms were contacted to ascertain their interest and get their comments and approaches. The university established a 6-person market evaluation and study team to investigate the marketplace, to make recommendations to Rantek on the development of the new product, and to determine potential long-term sales and income. Results of this "market opportunity analysis," published in a 42-page bound report, were favorable for long-term Rantek profitability.

Rantek also submitted a patent disclosure for the device to the U.S. Patent Office under their "Document Disclosure" program, which permits filing of a patent within two years of the disclosure. Straight-line control consisted of meeting all of the state's requirements, responding to and addressing all comments to the proposal, internal or external, and keeping an astute eye on the political, demographic, commercial, and financial aspects of the venture.

With the above said as introduction, we note that a winning proposal speaks for itself. Therefore, we have reproduced in this chapter the 25-page proposal in its entirety. The proposal was created and produced in WordPerfect 5.1 on a personal computer and laser printer. A 25-page business plan (not included in this book) was also created with the same equipment. The proposal is reproduced on pages 278 to 302.

PROPOSAL BY RANTEK, INC. TO THE ALABAMA INNOVATION FUND FOR A PRECISION BIOREACTOR SYSTEM

Alabama Innovation Fund
Alabama Technology Assistance Partnership
PRELIMINARY PROPOSAL and ABSTRACT

Office use only: Date Rec'd ______
AIF No. ______

Please read the "General Solicitation" and the "General Instructions" before completing this Preliminary Proposal Application.

Project Title: HIGH FIDELITY/PRECISION CELL AND TISSUE BIOREACTOR

Submitted by: Rantek, Inc., Huntsville, AL
(Business or Private Entity)

Mailing Address: 810 Regal Drive, Suite A (P.O. Box 301: Huntsville, AL 35804)
City: Huntsville State: AL ZIP: 35801

(Institution of Higher Education)

Mailing Address:
City: Huntsville State: AL ZIP: 35899

Project Budget Summary:

	Source of Funds			
	AIF	Business	University	Total
Direct Costs				
Personnel	$30,000	$80,000		$110,000
Facilities				
Equipment	7,000			7,000
Materials	3,000			3,000
Travel				
Outside Services	5,000			5,000
Other				
Total Direct Costs	$45,000	$80,000	$ 0	$125,000
Indirect Costs	$ 5,000	$ 0	$ 0	$ 5,000
Total Cost of Project	$50,000	$80,000*	$ 0	$130,000

*Rantek Contribution

Current Number of Employees 4 Current Sales $100k/yr.
Projected Number of Employees 25 (1993) Projected Sales $1.472m/yr.

PROJECT TECHNOLOGY AREA:

[ ]Materials Science [ ]Advanced Manufacturing [ ]Environmental Science
[ ]Information Science [x]Biotechnology [ ]Marine Science
[x]Biomedical Science [ ]Microelectronics/Computer Science [ ]Other (explain)______

I request evaluation of this proposal by ATAP for funding by the AIF. I authorize ATAP to furnish relevant information contained in this proposal to appropriate evaluation activities, however, I expect that information to be held in strict confidence by them. In consideration for this evaluation, I waive all claims against ATAP and other appropriate evaluation activities.

Endorsements by Authorized Officials:

Business/Private Entity	Institute of Higher Education
Name:	Name:
Title: Executive Vice President	Title: Assoc. Provost & V.P. for Research
Signature:	Signature:
Date:	Date:

Alabama Innovation Fund
Alabama Technology Assistance Partnership
Preliminary Proposal/ Abstract

Project Title: Precision Bioreactor System
Submitted by: Rantek, Inc.

Abstract:

Please describe the proposed project. Include identification of the problem, objectives, methodology, anticipated results and end product expected from AIF funded efforts. No system presently exists that adequately meets the needs of the medical research and health care fields for cell and tissue replacement. Trauma victims, persons suffering from degenerative diseases, arthritis, and autoimmune diseases require high quality live cells for transplantation. Our objective is to develop a precision cell and tissue processing system for the medical research and health care industries which offers features not present in today's cell/tissue culture systems. The Rantek Precision Bioreactor System will produce multidimensional (versus "flat Surface") cell growth and "bulk thickness" tissue. Our design methodology allows accessibility of active cell populations to controlled quantities of oxygen, organic sustaining materials, proteins, hormones, and enzyme activators. This targeted delivery permits controlled matrix growth which will provide tissue segments suitable for transplantation.

This system will serve as: (1) a tool for major medical research which focuses on disease treatment involving cell differentiation and regeneration technology; and (2) a cell and tissue processing system for the health care industry. The Rantek system supports both basic and applied research, thus expanding potential markets for the system. The AIF funding will provide for: (1) concept expansion, preliminary design, fabrication and functional verification of an engineering model; and (2) initial market penetration.

Project Status:

Please describe any prior, current and/or competing research and any future efforts necessary for commercialization. Include commitments for manufacturing and follow-on funding. Rantek, Inc., and the University of Alabama in Huntsville have performed detailed assessments of the current state of the art in cell and tissue culturing and have investigated the market for availability, accuracy, fidelity, and utility of existing laboratory equipment for cell and tissue processing. There is no current equipment available for precision processing in quantity: as a result, we are in final negotiations with Teledyne Brown Engineering of Huntsville, AL, for engineering model development. An engineering model is needed to serve as a suitable system for laboratory demonstrations. This model will precede full scale production and marketing and will aid in verifying the functions of the final system.

Economic Impact:

Please describe the anticipated economic impact on the State of Alabama accruing from completion of this project. Include local, regional, statewide and national impacts where possible.

Prompt funding and early market outreach of the High Fidelity cell/tissue Processing system could result in employment of up to 14 persons in Alabama in support of this program by the end of calendar year 1992. Continued expansion should result in the creation of 300 plus jobs in the state of Alabama. Regional, national, and international impact on the pharmaceutical, medical, and health care industries could be significant as a result of the medical products made possible by the existence of the High Fidelity cell/tissue processing system.

Facilities and Key Personnel:

Please describe the physical facilities, including equipment, to be used in the conduct of this proposal. Identify the essential personnel who will be conducting this project.

Life sciences, environmental, and biotechnical facilities at the University of Alabama in Huntsville will be used for research to support breadboard testing and preliminary design. The development of the prototype model will be performed at the Teledyne Brown Engineering shop facilities. Requirements definition, model and subsystems testing will be conducted at UAH.

TABLE OF CONTENTS

1.0 GENERAL

1.1 INTRODUCTION

Literature surveys by Rantek, Inc. (1) and market surveys by the University of Alabama in Huntsville (2) indicate that there is currently no device on the market or planned that is capable of being scaled up to produce human cells and tissue of the quantity, quality, and configuration required for widespread general medical use. Current devices produce only small quantities of cells and tissue; growth is inhibited by absence of full protection from environmental factors; and growth configuration flexibility and control is inadequate. Rantek, Inc., has filed a patent disclosure with the U.S. Patent Office for a device that will be able to process production quantities of cells and tissues of tailored configurations in a controlled environment. The inherent design will permit its scaling up to larger sizes. This proposal is for the design and development of a Precision Cell and Tissue Processing System, a device incorporating this design concept.

We will design, fabricate, and test an engineering model of a Precision Cell and Tissue Processing System, and we have identified and enlisted the required engineering, scientific, and management skills from within the State of Alabama to design, develop, test, analyze, and perfect the device. The Precision Cell and Tissue Processing System will: (1) enable significant advances in a number of medical and biomedical arenas; (2) create a base of research and scientific activities in the state in the field of bio-technology; and (3) stimulate high-technology research and economic growth in Alabama. Using aerospace-derived technology, engineering, and management principles, accompanied by talented in-kind services from participating organizations, a cell and tissue processing system engineering model will be built and tested with the support of the innovation funds to be provided by the Alabama Technology Assistance Partnership. The high fidelity cell and tissue processing system will permit both research and routine production of living cells and tissues that have been tailored to meet specific medical and biotechnical needs.

1.2 RATIONALE AND DESCRIPTION OF THE NEED

In precisely the right kind of carefully controlled environment, human cells and tissues can be reproduced, modified, separated, mixed, and transformed in ways that promise to offer dramatic new methods of curing, repairing, or replacing organs, tissues, and body fluids. In addition to being a cell/tissue processing system, the Precision Cell and Tissue Processing System is a diagnostic tool for research for cures in a number of critical diseases. It is a device that will permit a wider range of independent cell culture studies of high quality that will identify and confirm methods of establishing immune system cells and cells for bone marrow and organ transplants with greatly reduced danger of rejection. Table 1.1 lists some medical conditions that can be addressed by the research made possible through the use of the Precision Cell and Tissue Culturing System.

TARGET DISEASES AND CONDITIONS	
Anemia	Kidney transplant rejection
Blood clots	Multiple sclerosis
Diabetes	Osteoporosis
Hairy-cell leukemia	Pituitary deficiency
Heart attacks and vein transplantation	Rheumatoid arthritis
Hemophilia	Severe burns
Hip surgery	Severe wounds
	Viral infections

Table 1.1. Target Diseases and Conditions.

There are a number of important biomedical applications that could be improved by enhancements in cell and tissue processing methods. For example; isolation of beta pancreatic cells to determine how their production of insulin is regulated leading to the possibility of cell implantation for gene therapy; purification of whole islets of Langerhans for transplants as a cure for juvenile onset diabetes; isolation and purification of hematopoietic stem cells for treatment of certain types of leukemia; isolation of cells from organs that produce various hormones and enzymes such as urokinase, erythropoietin, and human growth factor; separation of bull sperm for selective breeding; high-resolution separation of various proteins such as subclasses of immunoglobulins and cell surface antigens on an analytical scale to determine molecular structure and function; and of special interest to Rantek, purification of proteins and cells such as those present in bone marrow for research and as a potential for pharmaceutical products.

A variety of separation techniques have been developed over the years to attack these problems: density gradient centrifugation, various types of chromatographies such as affinity chromatography and high-pressure liquid phase chromatography, florescence-activated cell sorting, various types of electrophoresis such as gel electrophoresis and continuous flow electrophoresis, isoelectric focusing, and liquid phase partitioning (sometimes referred to as counter-current distribution). Each of the available techniques has its own problems that currently limit the range of application. For cell separation and selection, a number of these techniques, particularly electrophoresis and isoelectric focusing, are limited by convection and sedimentation.

The purpose of the research made possible through use of our precision system is to identify target cell functions, and to develop processing approaches to eliminate selected subgroups such as T-lymphocytes, which react with major histocompatibility (MHC) antigens to cause immune rejection of human tissue transplants. Higher specific antigen-compatibility donor-to-host is expected to allow an increased acceptability rate of bone marrow transplants and thus further reduce or eliminate host resistance response to grafts. It has been demonstrated that continuous flow electrophoresis can separate certain subgroups of lymphocytes. Our research will involve use of a model lymphocyte system to test new methods of altering the surface charge of the target cell so that undesirable cells can be eliminated from the population to be transplanted. The use of microgravity research is planned to gain a better understanding of processing methods for identification and control of these materials. This hopefully will lead to enhancements in the overall cell selection process and to a more successful approach in transplantation.

There are more than one hundred different types of diseases affecting the joints and connective tissues. Collectively, these diseases are known as arthritis. In addition to joint symptoms, patients may have other pathological changes such as nodules or lumps under the skin, inflammation of connective tissue of the lungs, skin, blood vessels, muscles, heart and even eyes. The Arthritis Foundation and National Institute of Arthritis, Metabolism and Digestive Diseases estimate that thirty-one million Americans alone were suffering from arthritis as far back as 1980. Today we see little progress in alleviating many of these conditions. The most severely crippling form of this disease is rheumatoid arthritis. This chronic disease does not respect age, sex or financial status and victimizes over 6,500,000 people in the United States alone. An estimated 250,000 children under the age of 18 and forty-one percent of the population over the age of 65 suffer from rheumatoid arthritis - and surprisingly, more women than men are usually affected. Although various treatments may alleviate symptoms in most cases, there is no treatment known today that will cure or help in all cases. If the right tools and equipment can be made available to researchers and suppliers of biomedical products, revolutionary new substances and treatment methods can be investigated, refined, and produced. The Precision Cell and Tissue Processing System is expected to provide a tool that will help identify the causes and provide potential treatments to victims of arthritis.

Bone marrow transplants are currently used to treat a number of diseases; however, immune rejection (graft vs. host) is still a major problem. Some partial success has been realized in reducing the immune rejection by various methods of removing all T cells from bone marrow cells prepared for transplant. Not all of T cell subsets are responsible for the immune rejection. The object of this research is to develop methods to remove only those immune cells that can cause graft vs. host rejection, thus allowing transplantation of specific marrow cell populations. To date, no one has used antigen-specific treatments plus free-flow or continuous flow electrophoretic methods to isolate and remove certain subpopulations of T-Lymphocyte cells which are suspected to be involved in immune rejection.

Two main approaches involving the major human histocompatibility complex (MHC) are employed in making marrow transplantation available to a broad range of patients. One involves searching for donors with close antigenic similarity to the patient and then using the same regimens employed for transplantation between genotypically identical siblings. The other involves accepting a larger degree of antigenic disparity and devising special conditioning regimens to make such incompatibility acceptable.

The objective of our research is to gain better understanding and control of the T-Lymphocytes subpopulations present in bone marrow and to remove those which are responsible for or stimulate transplant rejection.

Likewise, research into cell culturing, mixing, diffusion, and controlled growth in this device could provide new hope for AIDS victims by potentially restoring immune capabilities; provide insights into cancer causes and cures; and address disease control and human organ function restoration.

1.3 FEATURES OF THE SYSTEM

As described in more detail in Section 2.0, Technical Plan, a proprietary and patentable system has been conceptually designed that will provide environmental control for cell and tissue growth, mixing, separation, and process control. The device will also control nutrient concentration in a protected environment. The design permits a wide range of cell and tissue studies and processing activities but promises to allow marketing at a price affordable to most research laboratories, institutions, and hospitals. Materials used in the processing system will be non-toxic and autoclavable for sterilization purposes. Simple controls and sample access provisions are included.

1.4 USERS AND MARKET POTENTIAL

Initial users of the system will be those involved principally in research and limited processing. Immunology laboratories, hospitals, research institutes and foundations, and pharmaceutical houses will be the first to acquire and use the processing system. Eventually, as processes become standardized and as the use for modified cellular and tissue material becomes more widespread, health clinics and individual physicians may have a need for versions of the system (see Figure 1.1). The market survey results are included in detail in the accompanying business plan.

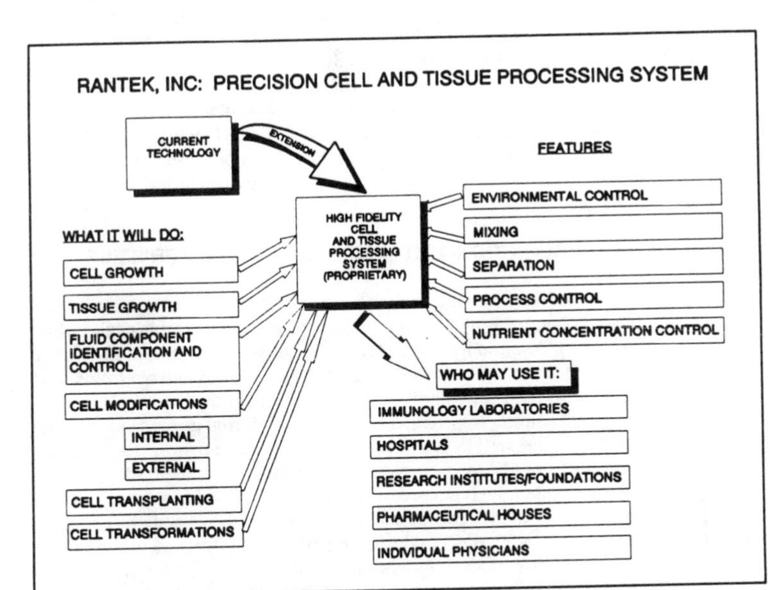

Figure 1.1 Features and Uses of High Fidelity Cell and Tissue Processing System

2.0 TECHNICAL PLAN

2.1 DESCRIPTION OF END ITEM AND USE

This technical plan is for the proposed design, manufacture, assembly, and demonstration of an engineering model of a unique multi-dimensional cell/tissue differentiation and growth control pre-production device that promises, in production, to have high market potential as a research and diagnostic tool in the medical and biotechnical industries.

2.1.1 GENERAL DESCRIPTION AND PROPOSED CHARACTERISTICS

We propose to design, manufacture, and test the first bench scale portable and uniquely designed biofactory system which will meet several research needs while offering the health care community and pharmaceutical suppliers a unique tool for continuous process cell separation, cell culturing, cell regeneration and alteration, and structured tissue growth. This capability is not yet available in the marketplace. The completely automated culture environment system will permit processing to take place at less than 1-g. The system is a substantial improvement over other similar systems that offer little of the process controls the Rantek reactor offers. The Rantek system will provide a high degree of fidelity in fluid motion and constituent control, providing a powerful process control capability with a high system reliability. The system will not only offer the standard temperature controls of other similar systems but will allow for unique temperature needs as dictated by specific cell requirements during processing.

The engineering model will be patterned after, and incorporate, the basic features of the production design shown schematically in Figure 2.1. The device is unique and innovative in that it incorporates precision high-technology aerospace developed materials, drive systems, support mechanisms, electrostatic and electromagnetic interference shields, surface hardened materials, image analysis systems, and environmental control to permit uninhibited and customized natural growth of human cells and tissue based on optimization of each specific cell type. For example, a precision computer-controlled monitoring system will be capable of maintaining a nominal chamber temperature of 37°± 2°C; pH at 7.2 ± 0.2; and CO_2 and O_2 levels at 40 ± 10 mm Hg and 80 ± 20 mm Hg respectively. The glucose level will be maintained in the range of 200 to 40 mg per dl. Vibrations of drive and support systems will be masked through the use of a fluid drive, an air suspension system, and acoustic dampening. Air/gas environmental control will keep the growth chamber in an inert, temperature controlled environment. Anti-static materials and electromagnetic shielding will protect the delicate cell and tissue process from outside forces that adversely affect cell or tissue growth and processing. Atomic surface

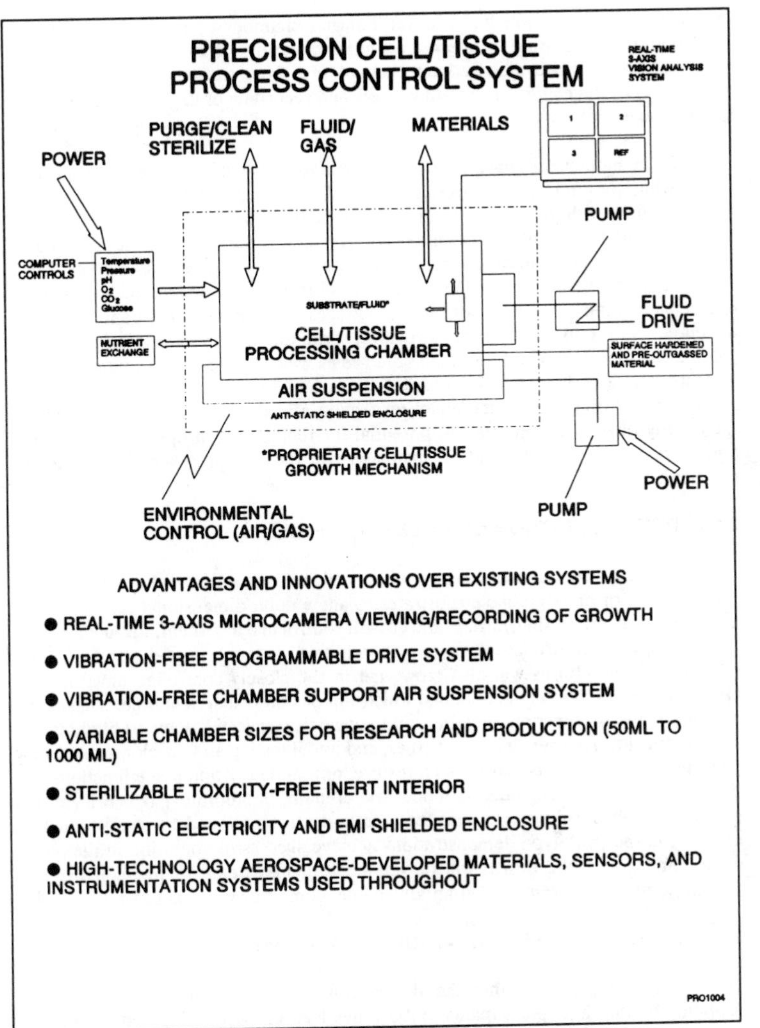

Figure 2.1. Schematic of Precision Cell/Tissue Process Control System.

hardening techniques will be used for internal chamber components to prevent ionization and outgassing of internal chamber materials. Minicomputer-controlled video cameras will be mounted on each of the three axes to provide real-time observation of the cell/tissue growth process and recording of the growth process for future image analysis.

Since human cells have an electrical component (a membrane potential) that affects what goes into and comes out of the cell, nontoxic and nonionizing materials of high purity will be used to remove all factors that influence research and processing fidelity. Multiple precision controls and safeguards will be included to protect against perturbation of natural processes and cell-to-cell association. High purity cells and tissue, once exposed to this controlled processing and selected for reduced antigenic contamination, which reduces the potential for donor/recipient rejection, will retain their original structure, form, and integrity. Variable chamber volume will permit research on small volumes as well as processing of large volumes of material. The high purity biological fluids and tissues made possible through control, monitoring, and contaminant protection afforded by this device will exceed the quality of present materials in potential for reassimilation into the body of the human recipient. Tailor-matched materials can thus be provided from donor to recipient.

2.1.2 PRE-PRODUCTION ENGINEERING MODEL

The engineering model design will permit cells to attach and grow on micro-carrier beads or on degradable substrates to allow multi-dimensional cell and tissue differentiation, processing, mixing, and growth control in a low-shear, turbulence-free, environmentally controlled volume. In addition to anchorage-dependent cells, suspension cell cultures will also grow well in the closely controlled environment. High density cultures can be obtained which will permit research in areas such as the immune response function, metabolism, toxicological investigations, transplantation of synthetically produced cells and tissues, and initial testing and development of new pharmaceutical products that protect, restore, or modify cell and tissue functions and characteristics. The engineering model will contain, in laboratory bench form, the principal systems, subsystems, and features that are envisioned for the production model. As the prototype demonstrations achieve successful operation in the hands of potential users, the design will be refined to achieve ease of manufacture and assembly, cost effectiveness, and an affordable, competitive market price.

2.1.3 INTENTION TO MANUFACTURE IN ALABAMA

It is our intention that the device will be manufactured in the State of Alabama by the same participants that are involved in the development effort.

2.2 INNOVATIVE TECHNOLOGY EMPLOYED

The Precision Cell and Tissue Processing System uses a novel, proprietary design based on proven technologies. It is an outgrowth of and new combination of previously proven concepts. These concepts have been synthesized into a workable cell production system.

2.2.1 FEASIBILITY

The prototype device design and the design of its more refined and sophisticated production counterpart draw upon proven and conservative scientific and engineering principles and demonstrated aerospace and biochemical technologies. The work is based on a consistent on-going effort by NASA and other agencies to advance the state-of-the-art, and we are capitalizing on past and current government-conducted and sponsored research.

2.2.2 INNOVATIVENESS

The unique and innovative elements of the Precision Cell and Tissue Processing System are: (1) the accommodation of a multi-dimensional cell/tissue growth process; (2) capability to permit maximum consistency and homogeneity of cell surface tension; and (3) allowance for the processing of many different cell systems and types. The design integrates all of the features desirable in an optimum cell/tissue processing system into a single, compact design that enhances accuracy of results, ease of operation, and flexibility of application.

2.2.3 ORIGINALITY

The design is an original concept which has been disclosed to the U.S. Patent Office by Rantek, Inc., under U.S. patent law and represents a unique combination of existing concepts and new approaches. Patent rights and applications will be negotiated and exercised by the various participating organizations as new concepts, subsystems, components, and approaches evolve.

2.2.4 PATHWAY TO IMPROVED TECHNOLOGY

The engineering model design is sufficiently modular and flexible to permit the addition of new technology advances in materials, manufacturing processes, and system integration as they emerge during the demonstration and operational use of the device. The engineering model will prove and demonstrate the concept, while the production model may take on both generic and special forms and requirements identified during initial market penetration phases.

2.2.5 PRODUCTIVITY POTENTIAL

The Precision Cell and Tissue Processing System's inherent design permits more rapid cell growth, larger cell and tissue masses, tailoring of tissue shapes, and more efficient culture processing than the systems that currently exist in the marketplace. Rapid change-out of sample processing systems, real-time control of environment, ease of removal of metabolic wastes, and automated control are features that can allow a productivity potential increase of up to five times greater than that of existing systems. In the long term, one can envision cell/tissue processing systems of this type in many medical treatment or medical research institutions in this country and in many foreign countries. Potentially, a production model could be an adjunct to the standard medical equipment of immunologists, dermatologists, oncologists, hematologists, nephrologists, orthopedic physicians, and physicians dealing with cosmetic surgery, arthritis, burn patients, trauma victims, and transplant donors and recipients. We envision an expanding commercial marketplace as the usefulness and application of cell and tissue processing multiplies. Pharmaceutical companies, in the meantime, such as those shown on Table 2-1, are candidates for purchase of systems for research and product development.

PHARMACEUTICAL COMPANIES THAT ARE POTENTIAL USERS OF THE PRECISION CELL AND TISSUE PROCESSING SYSTEM

Abbott Laboratories--Abbott Pharmaceuticals, Inc.
Adria Laboratories
Alza Corporation
Ayerst Laboratories
Berlex Laboratories, Inc.
Borneman, John A., and Sons
Braintree Laboratories, Inc
Bristol Laboratories
Bristol-Meyers Oncology Division
Brown Pharmaceutical Co., Inc.
Burroughs Wellcome Company
CIBA Pharmaceutical Company
Dista Products Company
Du Pont Pharmaceuticals
Flint
Geigy Pharmaceuticals
Hoechst-Roussel Pharmaceuticals Inc.
Janssen Pharmaceutica Inc.
Key Pharmaceuticals, Inc.
Lactaid, Inc.
Eli Lilly and Company
Marion Laboratories, Inc.
McNeil Consumer Products Co.
McNeil Pharmaceutical
Ortho Pharmaceutical Corporation
Parke-Davis
Pharmacia Laboratories
A.H. Robins Company
Ross Laboratories
Savage Laboratories
Searle Pharmaceuticals, Inc.
Smith Kline & French Laboratories
Stuart Pharmaceuticals
Syntex Laboratories, Inc.
Wallace Laboratories
Winthrop-Breon Laboratories

Table 2.1. Pharmaceutical Companies That Are Potential Users.

3.0 SCHEDULE FOR THE WORK

This work is expected to be completed within nine (9) months of the grant award date. The work will be performed in three phases, resulting in a engineering model that will be ready for commercialization after conducting production planning and making final manufacturing arrangements in keeping with its verified market capture potential. A user demonstration activity will provide processing verification leading to the acquisition of initial production orders. The program schedule is shown below in Figure 3.1.

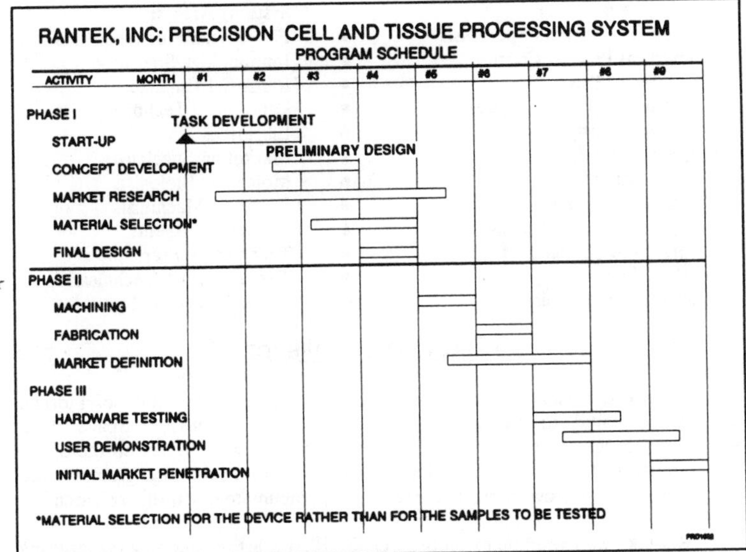

Figure 3.1. Program Schedule for Precision Cell and Tissue Processing System

4.0 SKILLS REQUIRED

Initial skills required and which will be employed under this grant will be a microbiologist, junior level mechanical engineer, tissue culture technician, and a manager. These will be part-time personnel who are employed in other synergistic ventures which will permit the use of in-kind services, facilities, tools, and equipment.

Farther into the future, the growth of Rantek, Inc., and its associates will require a mix of high-technology skills to successfully transition the product and its associated products and services into the marketplace. As initial sponsored research is funded, and as product orders increase, a full range of scientific, technical, management, and support functions will be required. The following are example skills and professions that will be needed to successfully carry out the growth to that envisioned in the five-year plan:

- Administrative Specialists
- Analytical Biochemists
- Bio-Engineers
- Biomedical Graphics Specialists
- Biomedical Process Control Specialists
- Biomedical Support Services Specialists
- Biophysics Analysts
- Cell Biologists
- Computer Analysts
- Computer Science Technicians
- Configuration Specialists
- Detailed Designers
- Electron Microscopists
- Electronics Technicians
- Enzyme-Linked Instrument Systems Analyst
- Flow Cytometer Operators
- Immunology Specialists
- Industrial Engineers
- Laboratory Technicians
- Librarians
- Marketing Specialists
- Molecular Biologists
- Precision Machinists
- Purchasing Agents
- Technical Writers
- Tissue Culture Technicians
- Word Processing Specialists

5.0 RELEVANT EXPERIENCE

Rantek is a small business, incorporated in 1985 in the state of Florida, currently dedicated to biotechnology and aerospace research. Space flight experiments in cell/tissue growth are currently underway with NASA on the Space Shuttle, Joust, and Consort rockets. Plans are being laid for space and ground-based research in bone marrow transplant research and rheumatoid arthritis research.

Rantek flew experiments on the CONSORT I Starfire sounding rocket using the ITA (Instrumentation Technology Associates of Exton, PA) Materials Dispersion Apparatus (MDA), sponsored by the Consortium for Materials Development in Space, a NASA-supported Center for Commercial Development of Space based at the University of Alabama in Huntsville. The CONSORT I Rocket flight took place on March 29, 1989. A second series of similar experiments were undertaken on the CONSORT II rocket flight on November 15, 1989, but was errant. These experiments were then rescheduled and flown on the CONSORT III flight in May 1990.

Following are some details of the experiments flown on the CONSORT series and results from the research thus far.

Using the ITA Materials Dispersion Apparatus (MDA), Rantek experiments on the CONSORT I sounding rocket studied the diffusion of two biological materials: (1) urokinase, which is an enzyme that dissolves fibrin blood clots after they are formed and (2) a monoclonal antibody (MAb). Concentrated solutions of the smaller urokinase molecule were placed into two diffusion chambers, and solutions of the larger antibody were placed into two other corresponding chambers. Urokinase diffused across the interface into the urokinase chambers when the opposing chambers were exposed to each other during the micro-g portion of the flight. The amount of diffused molecules which are measured after the flight can be used to calculate the effective diffusion constants for both proteins. Diffusion in opposite directions in the same chamber will allow a double check of the accuracy of the diffusion constant observed for that particular chamber. This also is important for determining the effective mixing which occurs as the two chambers are brought together before diffusion begins.

The flight samples were recovered from the MDA at the launch site, frozen and transported back to the National Institutes of Standards and Technology in Boulder, CO. The samples were then taken to NASA's Johnson Space Center for analysis. Data indicated that diffusion between the chambers did occur as expected.

A Rantek experiment flew on the most recent Shuttle Atlantis flight (STS-37) in April 1991. This experiment investigated interferon production by human T-lymphocytes under microgravity conditions. A magnetic mixing principle and cell surface antigen expression in low gravity, and antigen-antibody binding are being evaluated on upcoming Joust and Consort rocket flights.

6.0 RESUMES OF KEY PERSONNEL

COL. RICHARD L. RANDOLPH: PRESIDENT, RANTEK, INC.

Col. Randolph holds an AB degree in Physics (Univ. of Southern California), an MA degree (George Washington Univ.), and is a graduate of Air War College and Air Command and Staff College. He began his career with the US Air Force as a flying officer holding several increasingly responsible management assignments including the area of research and development of nuclear weapons. He was Chief, Aircraft Requirements, Headquarters Far East Air Forces, and Commander of a major USAF technical training school. After retirement from the Air Force in 1966, Col. Randolph launched an active career in industry. His research and patent contributions involve research and experimentation in unique methods of communication through the earth. This work led to the development of prototype electronic apparatus with two patents issued. He has pursued further research associated with the development of medical diagnostic apparatus and specialty tools. Since 1967 Col. Randolph served as President and Director of The World Security Fund, and in 1979 he became President and a Director of Microgravity Research Associates, Inc. Seeing a critical need to utilize the increasing advancements in technology, Col. Randolph and several of his associates founded Rantek, Inc. to conduct biotechnology and related medical research capitalizing on the latest state-of-the-art in engineering and science.

MR. WILLIAM K. VARDAMAN: EXECUTIVE VICE PRESIDENT, RANTEK, INC.

Mr. Vardaman holds a B.S. in Industrial Design and Science from Stamford University in Birmingham, AL, and an M.S. degree in Administrative Science from the University of Alabama. He began his professional career with Hayes International in 1955 with increasing responsibility in development of engineering specifications, hardware design, and development of scientific and technical publications involving contracts with the Air Force, Army, Navy, Army Ballistic Missile Agency, and the National Aeronautics and Space Administration. Mr. Vardaman joined the National Aeronautics and Space Administration in 1961 and held several key roles in staff and management positions related to materials research and scientific flight payload development. He played a key role in the initial development of NASA's Materials Processing in Space (MPS) program which led to the creation of the Agency's Commercialization Office. After his retirement from NASA in 1984, Mr. Vardaman joined Wyle Laboratories and supported its thrust in MPS and Scientific Hardware Development. Mr. Vardaman later joined the Center for Space and Advanced Technology (CSAT) in Washington, DC, in 1987 and currently serves that company in an advisory and program development and implementation role.

DR. MARIAN L. LEWIS: UNIVERSITY OF ALABAMA IN HUNTSVILLE: PRINCIPAL INVESTIGATOR

Dr. Lewis has more than thirty years of experience in cell culture technology including metabolic assays, biochemistry and evaluation of cellular natural product secretion. She received her PhD in Biophysical Sciences from the University of Houston in 1979. She has more than eight years experience as a virologist and was trained at the Baylor University College of Medicine and the M.D. Anderson Hospital and Tumor Institute (Texas Medical Center) and the Communicable Disease Center in Atlanta and Phoenix. She served as virologist and laboratory supervisor at Northrop Services at NASA/JSC during the Apollo Missions. Dr. Lewis' areas of specialty include all aspects of cell biology, microgravity cell biology including bioreactor development and process verification testing, diagnostic and experimental virology, cellular immunology, and diagnostic bacteriology. She has expertise in laboratory supervision and management, operation of analytical instrumentation, biochemical and immunochemical analyses and project management. Dr. Lewis is a published scientist with over 80 publications and technical presentations. She has received awards for her outstanding work in cell culture technology among which was a Lyndon B. Johnson Space Center Group Achievement Award for her work on the Space Bioreactor Program Bioreactor Engineering and Cell Science Research Team in 1989. She is a co-applicant for a patent on three dimensional cell to tissue assembly and recipient of a NASA Tech Brief Award (Three Dimensional Cell to Tissue Assembly) 1990 and 1991. Dr. Lewis holds a research faculty appointment in the Department of Biological Sciences at UAH and is a PI on biological experiments on the Shuttle and sounding rockets.

MR. RODNEY D. STEWART: PRESIDENT, MOBILE DATA SERVICES: PROJECT MANAGER

Mr. Stewart is a project manager and marketing specialist for high technology, aerospace, defense, and commercial ventures. He specializes in cost estimating, proposal preparation, acquisition management, project engineering, and the multiple-disciplined facets of emerging complex projects. He holds a Bachelor of Science Degree in Civil Engineering from Virginia Polytechnic Institute and State University (1951); was a Guided Missile and Special Weapons Officer in the US Army Ordnance Corps for three years; served as Senior Design Engineer for Thiokol Chemical Corporation for seven years at its Redstone Division near Huntsville, Alabama; worked as an Aerospace Project Engineer, Project Manager, and Cost Analysis Manager at NASA/MSFC for twenty years; and has managed his own firm, Mobile Data Services, for ten years. He has written or co-authored eight books on cost estimating, proposal preparation, and acquisition management; served as Vice Chairman of the Major Systems Acquisition Study Group of the Commission on Government Procurement in 1970 and 1971; and is Series Editor for the John Wiley & Sons' "New Dimensions in Engineering" book series.

DR. TOMMY A. BROCK: ASSOCIATE PROFESSOR OF MEDICINE; UNIVERSITY OF ALABAMA AT BIRMINGHAM: CONSULTANT

Dr. Brock is a cell biologist who is currently Associate Professor of Medicine at the University of Alabama at Birmingham. He specializes in vascular cell biology and vascular pathology with a heavy emphasis on cell culture techniques. He has developed new methods for culturing vascular endothelial and smooth muscle cells. Currently he is collaborating with Dr. Suzanne Oparil (see resume below) in studies of the physiology and pathophysiology of pulmonary vascular smooth muscle cells under hypoxic conditions. Dr. Brock has made three major contributions which have changed the cell biology of hypertension research. He was the first to demonstrate that phorbol esters, presumably acting through protein kinase C, inhibit the early phase of angiotensin II induced stimulation of phospholipase C in vascular smooth muscle cells. This mechanism, which is presumably mediated endogenously by diacylglycerol, is very likely an important control mechanisms for turning off the early signaling events. Second, Dr. Brock is a pioneer in the new technology for measuring intracellular calcium concentrations to define kinetics of hormone stimulated calcium translocation in vascular smooth muscle cells. Dr. Brock has further advanced the field by developing technology for imaging the concentrations of calcium and other cations in single cells. Third, most recently, Dr. Brock has made important contributions in defining hormonal signaling mechanisms in cultured endothelial cells. He is clearly one of the leaders in this field. His recent demonstration that loading of cultured endothelial cells with guanine nucleotides stimulates phospholipase C and uncouples ATP receptors is particularly noteworthy.

DR. SUZANNE OPARIL: PROFESSOR OF MEDICINE; UNIVERSITY OF ALABAMA AT BIRMINGHAM: CONSULTANT

Dr. Oparil is an established scientist who has an outstanding track record in medical research. She has over 250 full length publications and an equal number of shorter contributions in highly respected peer reviewed journals. Her investigation has focused on biochemistry, physiology and regulation of angiotensin converting enzyme and the role of central neural mechanisms in the pathogenesis of hypertension. In recent years, her research has been on the cellular and molecular level. She and her colleagues have demonstrated that expression of the gene for angiotensin converting enzyme (ACE), an important enzyme in the pathogenesis of many forms of hypertension, is increased in cultured vascular smooth muscle cells and neurons after exposure to inhibitors of the enzyme and hypoxic atmospheres.

7.0 FACILITIES AND EQUIPMENT

The principal two laboratories that will be used for testing and development of the high fidelity cell and tissue processing system are shown on Figure 7.1. These laboratories consist of approximately 400 square feet in the Science Building at the University of Alabama in Huntsville. Three other laboratories in the science building and clinical science building are also available to support this work. These laboratories and their equipment are shown on Figures 7.2, 7.3, and 7.4. New equipment to be purchased or leased with the $7,000 equipment allocation under this grant include a fluorescence phase microscope, a vapor pressure osmometer, an Eliza microplate reader, and a flow cytometer.

BIOTECHNOLOGY LABORATORY

DESCRIPTION

400 Square Feet, Room 351-A and 351-B, Science Building

EQUIPMENT

1) Incubator
2) Autoclave (Market Forge)
3) Fume Hood
4) Analytical Balance (Sartorlus)
5) Shaking Incubator (Damon/IEC)
7) Shaking Incubator (New Brunswick Scientific)
8) Orbital Shaker (Lab Line)
9) Ball Mill (Norton)
10) Freezer
11) Sample Cooler
12) Water Bath (Blue M)
14) Flammable Chemical Storage
15) Baker Class II Biological Safety Hood
16) Controlled Environment Tissue Culture Incubator (Forma)
17) Invented Microscope (Nikon)
18) Laminar Flow Bench
19) Cryogenic Freezer
20) Coulter counter

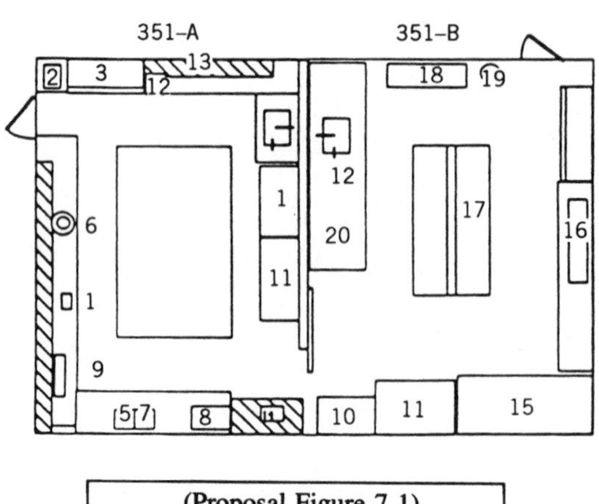

(Proposal Figure 7.1)

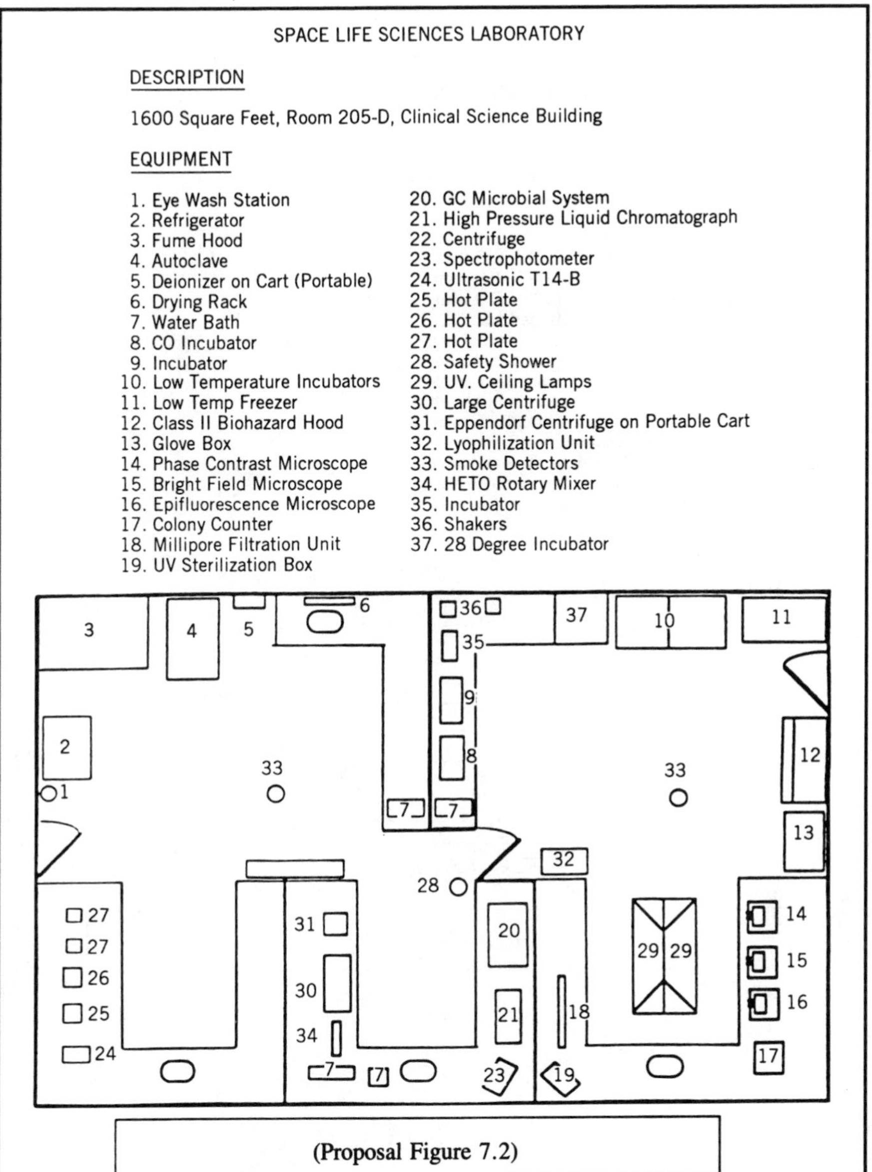

(Proposal Figure 7.2)

INSTRUMENTAL ANALYSIS LABORATORY

DESCRIPTION

800 Square Feet, Room 349, Science Building

EQUIPMENT

1) Gas Chromatograph - TCD & PID (Hewlett Packard)
2) Gas Chromatograph ECD/FID (Hewlett Packard)
3) Gas Chromatograph - FID/NPD (Hewlett Packard)
5) Ion Chromatograph (Shimadzu)
6) High Performance Liquid Chromatograph - UV. RI & Fluorescence (Shimadzu)
7) Total Organic Carbon Analyzer (Astro)
8) Inductively Coupled Spectrophotometer (IL 200)
9) Mercury Analyzer System (Perkin Elmer)
10) Atomic Absorption Spectrophotometer (Perkin Elmer)
11) Flame Auto - Sampler (Varian)
12) Atomic Absorption Spectrophotometer (Varian)
13) Graphite Furnace Auto - Sampler (Varian)
14) Analytical Balances (Mettier, Sartorius)

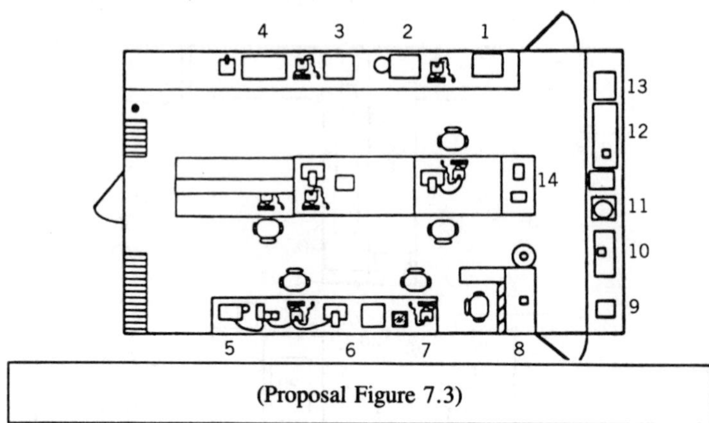

(Proposal Figure 7.3)

ENVIRONMENTAL ANALYSIS LABORATORY

DESCRIPTION

1000 Square Feet, Room 347, Science Building

EQUIPMENT

1) Solvent Storage Cabinet
2) Sample Drying Oven (FREAS)
3) Muffle Furnace-(Thermodyne)
4) Solution Column Rack
5) Gas Cylinder Serving Area
6) Sample Processing Area
7) Sample Receiving Area
8) Conductivity Meter (Yellow Springs Instruments)
9) Specific Ion Meter (Orion)
10) pH Meter (Fisher)
11) Water Deionizer System (Continental)
12) R.O. Water Filter (Millipore)
13) Fume Hoods
14) Auto Glassware Washer
15) Sample Cooler
16) Chemical Storage Area

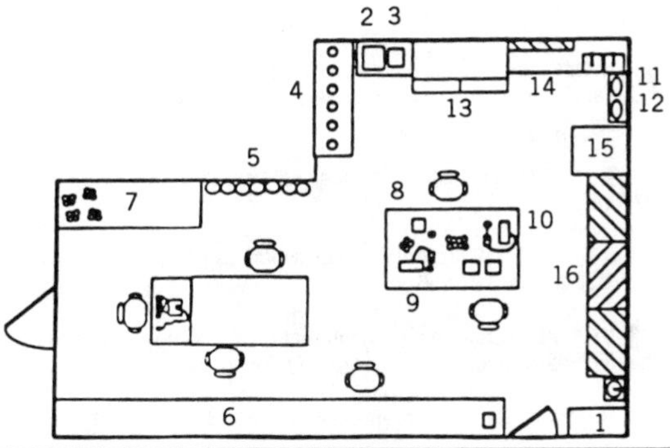

(Proposal Figure 7.4)

8.0 BIBLIOGRAPHY

(1) Bibliography of documentation on selected research and tools associated with or related to "Precision Bioreactor."

A. Rotating-wall, Vaned Vessel: Designed and developed at NASA's Johnson Space, Houston, Texas. System is being designed for use in the environment of low gravity. i.e., for use in space research. NASA Tech Briefs; Volume 14, Number 12; Dec. 1990.

B. Burridge K (1986) Substrate adhesions in normal and transformed fibroblasts: Organization and regulation of cytoskeletal, membrane and extracellular matrix components at focal contacts. Cancer Rev 4:18-78.

C. De Brabander M, Nuydens R, Greerts H, Huyens R, Leunissen J. Jacobson (1990) Using nanovid microscopy to analyze the movement of cell membrane components in living cells. In: Herman B. Jacobson K (eds) New Developments in Optical Microscopy. Alan R. Liss, New York. In press.

D. Armstrong PB (1989) Cell sorting out: The self-assembly of tissues in vitro. CRC Critical Reviews in Biochemistry and Molecular Biology 24:119-149.

E. De Brabander M, Nuydens R, De May J (1986) The use of submicroscopic gold particles combined with video contrast enhancement as a simple molecular probe for the living cell. Cell Motility and the Cytoskeleton 6:105-113.

F. DePasquale A (1975a) Locomotory activity of epithelial cells in culture. Exp Cell Res 94: 191-215.

G. Gabbiani G, Chaponnier C, Zumbe A, Vassalli P (1977) Actin and tubulin co-cap with surface immunoglobins in mouse B lymphocytes. Nature 269:62D (1989)9-689.

H. Harris AK (1973a) Location of cellular adhesions to solid substrata. Devel Biol 35:97-114.

I. Soranno T, Bell, E (1982) Cytoskeletal dynamics of spreading and translocating cells. J Cell Biol 95:127-136.

J. Cheng LY (1987b) Deformation analysis in cell and developmental biology. Part II: Mechanical experiments on cells. J. Biomech 193d:18-24.

K. Honda H. (1989) Geometrical models for cells in tissues. Int Rev Cytol 81:191-246.

(2) Jay Clayton, Henry Chin, Jim Englert, Mark Gann, and Brenda Thrasher; HIGH PRECISION BIOREACTOR MARKET OPPORTUNITY ANALYSIS; University of Alabama in Huntsville, Huntsville, AL; June 6, 1991.

FINAL WORDS

Despite the fact that there are principles, rules, guidelines, equipment, methods, and techniques that help produce a winning proposal, every proposal situation is different and each proposal must be addressed with its own particular requirements in mind. Straight-line control is more of a mind-set than it is a guaranteed formula for success. Winning proposals are essentially documented evidence of in-depth creative work by the proposal preparer(s) in meeting or exceeding all requirements.

A convincing and credible proposal can only be created in an atmosphere of willingness to do almost anything (providing it is legal and moral, of course) to win. The convincing document that results from this process, coupled with continued astute and prudent marketing and follow-up, will produce proposals that have the inherent capability to win new work for your firm.

APPENDIX 1
SAMPLE PROPOSAL PREPARATION MANUAL

This sample proposal preparation manual assumes a hypothetical company, which is a division of a larger company. The division is headed by a general manager, who reports to the corporate management, along with general managers of other divisions. A director of advanced planning is responsible for the division's acquisition of new work. Other directors manage various portions of the division's activities.

SECTION 1. PURPOSE

This manual establishes uniform guidelines for the preparation and processing of proposals within the company. Its purpose is to provide a general guide for proposal preparation; it is not intended to limit or constrain any procedures made necessary by unique requirements or special circumstances not addressed in this manual. Proposals in all cases should be prepared to meet specific customer requirements.

SECTION 2. SCOPE

This manual is applicable to all proposals for which a proposal task authorization is issued, and to engineering change proposals for which a task authorization has been issued, regardless of cost/price basis or cost magnitude, and regardless of whether the proposer is responding to a formal request for proposal or request for quotation or is making an unsolicited offer.

SECTION 3. APPROVALS

Proposal Task Authorization

Proposal task authorizations for expenditure of $ __________ or less require approval of the cognizant director. Those in excess of $ __________ require the general manager's approval. Supplementary proposal task authorizations increasing the cumulative authorization for a proposal to more than $ ________ also require the general manager's approval.

Proposal Signature Authority

Policy No. ________ specifies proposal signature authority and dollar amounts for the various levels of management.

SECTION 4. PROPOSAL CYCLE

General

The proposal is a primary mechanism for the acquisition of new business. It is the company's committed offer to undertake certain work for a firm dollar amount and to complete that work within a specified time. It is essential, therefore, that the proposal be:

- Technically sound
- Accurately cost-estimated
- Structured for proper program control
- Responsive to the customer's requirements.

There are three distinct but interrelated parts of all proposals: technical, organization and management, and cost. The form in which they appear in the final published proposal—as separately bound volumes, separate sections in a single volume, or merely separate paragraphs—is incidental to the preparation process, which in all cases must be thorough, orderly, and complete.

Because of the short time normally available for the preparation of proposals, the technical, organization and management, and cost sections are prepared concurrently—many times by separate groups. Because of the interrelationship of each section to the others, continuous liaison between groups is an absolute necessity.

The controlling authority throughout the proposal preparation process is the proposal manager. The proposal manager is responsible for the total proposal. He or she is designated by the director to whom the proposal has been assigned. Proposal team members normally include an assigned engineering staff; a contract administrator; an administrator from program management; a marketing representative; a cost analyst; a technical documentation/data management representative; and support personnel from such groups as program management, product assurance, manufacturing, reliability, design and drafting, procurement, and test. Support participation is solicited by the proposal manager and varies with the nature and scope of each proposal. Participation is required by all groups that will be committed in the proposal to do work on the job. A general checklist of proposal responsibilities is shown later in this manual.

Classified proposals are handled in accordance with standard security practices (Policy No. __________) or special requirements of the RFP/RFQ. The proposal manager is responsible for the overall proposal security under the guidance of the company security officer. The manager of technical documentation supervises the marking, handling, reproduction, and storage of sensitive proposals and related materials processed through technical documentation. (A typical proposal flow chart is shown in Figure A1.1.)

Proposal Meetings

Bid and Proposal Meeting. The contracts department, upon receipt of any request for proposal (RFP) or request for quotation (RFQ), or if organizing an unsolicited proposal, advises and forwards data to the director of advanced planning for the division. The director of advanced planning reviews the RFP/RFQ or in-house proposition to determine its business potential and notifies the director of corporate advanced planning that a proposal cycle is being initiated.

If the RFP or RFQ is determined to be a joint division opportunity or is found to apply to another division, the director of division of advanced planning, forwards it and the data package to the director of corporate advanced planning.

When notification from the corporation offices to proceed is received, a "bid/no-bid" decision meeting is scheduled in coordination with the appropriate director. At this meeting, if a preliminary "bid" decision is made, a proposal manager is then selected.

If a decision is made not to respond to the RFP or RFQ, the director of advanced planning informs the customer and corporate management. If it is to be accepted, the proposal manager, in coordination with the director of advanced planning, prepares a preliminary proposal plan, which includes the following information:

1. Estimate of resources required to implement proposal
2. Identity of the customer
3. Proposal preparation time
4. Preliminary proposal team
5. Expected contract work scope (work statement)
6. Special requirements
7. Funding available to the customer
8. Type of contract anticipated
9. Past relations with the customer and future prospects, including company's obligation to the customer
10. Competitive chances to win the contract, including both the company's and the customer's competitive position
11. Estimate of contract value and possible future business potential
12. Preliminary proposal strategy
13. Facilities required

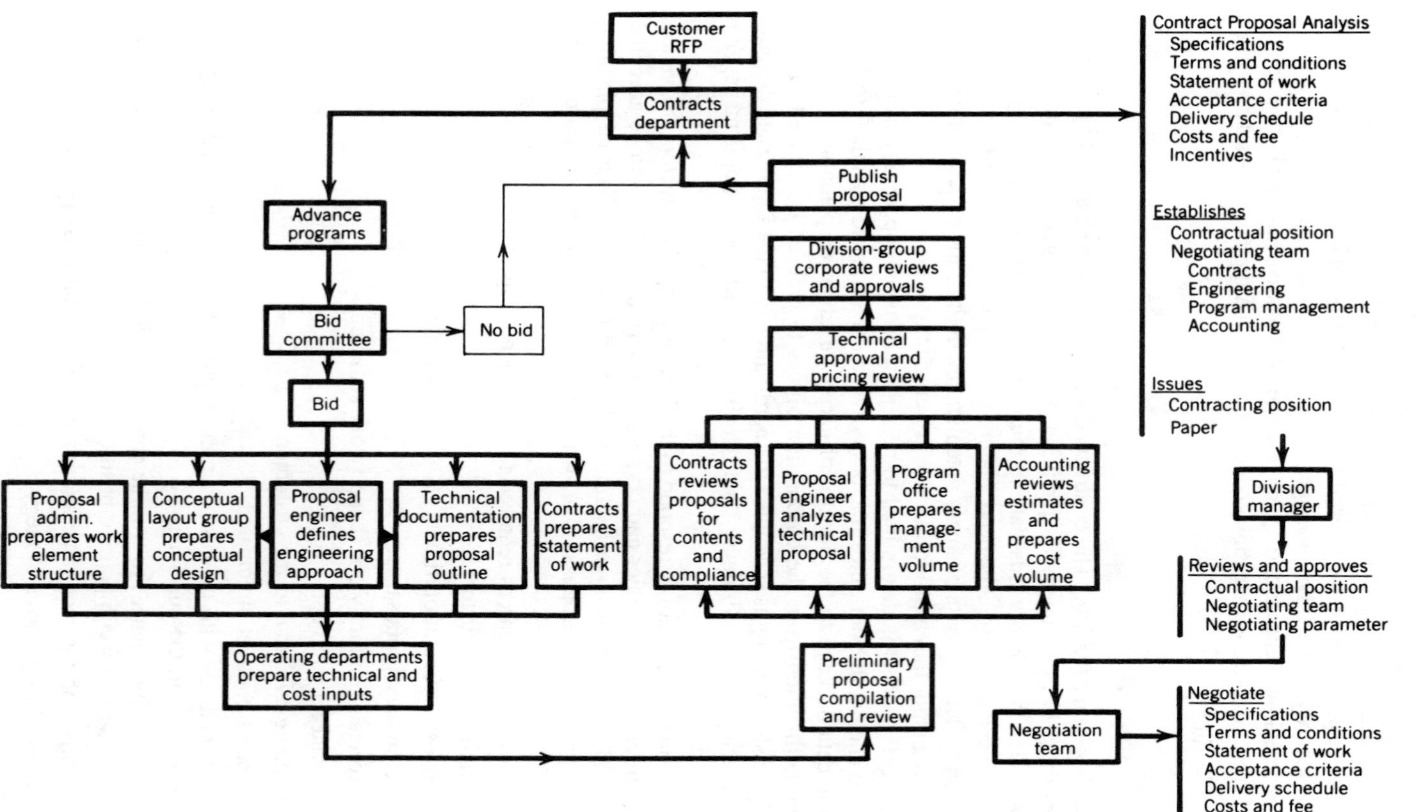

FIGURE A1.1 Typical proposal flow chart.

14. Equipment and/or property required
15. Special terms and conditions.

The above information is summarized in a proposal task authorization (Figure A1.2). Also a cost estimate checklist (Figure A1.3) is attached to all proposal task authorizations in excess of $ ________ .

The bid package, which includes a copy of the request for proposal, the proposal plan, and the proposal task authorization, is presented at the division bid committee meeting within ________ working hours of RFP/RFQ receipt. The bid committee usually consists of the following management personnel: (1) general manager; (2) technical director; (3) cognizant activity director; (4) division controller; (5) director of administration and contracts; (6) director of advanced planning; (7) director of engineering; (8) director of advanced technology; and (9) manager of contracts. In addition, the following personnel may be members of the committee, as appropriate: director of corporate advanced programs; director of corporate advanced planning; director of manufacturing; and director of quality assurance.

The division bid committee reviews the RFP/RFQ and:

1. If a decision is made not to bid, the data package is returned to the division director of advanced planning to inform customer/requestor accordingly.
2. If the RFP or RFQ is accepted, the committee approves the following and forwards data package to the director of corporate advanced planning and the division controller:
   a. Proposal manager
   b. Proposal schedule
   c. Proposal task authorization.

The division controller reviews the information in the proposal task authorization for compliance with bid and proposal guidelines, assigns a proposal charge number, and returns the proposal task authorization to the proposal manager.

If an engineering change proposal (ECP) is being processed, the contracts department will assign an appropriate ECP number, indicating that the change has been approved. A task authorization is then issued by the program office, defining the ECP requirements. The task authorization goes through the same approval cycle as a proposal task authorization.

Proposal Kickoff Meeting. On receipt of proposal task authorization approval, the proposal manager and data management department prepare detailed proposal outlines (with writing assignments) for each proposal section: (1) technical; (2) organization and management; and (3) cost. The proposal manager, within 24 working hours following proposal task authorization approval, convenes a proposal kickoff meeting. Attendees include representatives from all contributing, operating,

PROPOSAL TASK AUTHORIZATION (PTA)

Work Order	Task	Rev. No.

Title		Prime Division
Log No.	Customer	Business Area

Rep. No. or Cust. Msg. no.	Security Classification	Department having prime responsibility

PROPOSAL WORK STATEMENT

Business Potential, (Estimated Contract Price)—is requirement budgeted? Yes ☐ No ☐ Unknown ☐

Marketing Strategy
Contract Possibility
Past Relationship with Customer
Prime Competitors
Conversion Date
Contract Type
☐ CPFF ☐ CPIF ☐ FFP ☐ T&M ☐ FPI ☐ Other

Proposal Task Budget

Previous Authorization ________
This Authorization ________

Total Authorization ________
(Include this PTA)

End Item User	DOD	NASA	Other Gov't	Comm'l	Comm'l Purchaser Supplying Gov't

Proposal Budget														TOTAL	
Department		Jan	Feb	Mar	Apr	May	Jun	Jul	Aug	Sep	Oct	Nov	Dec	MM	$
	Lab $														
	N/L $														
	Lab $														
	N/L $														

	Lab $														
	N/L $														
	Lab $														
	N/L $														
	Lab $														
	N/L $														
Total Lab	$														
Total N/L	$														
**Totals*	$													THIS AUTH.	

Does this contract require significant additional capital equipment or facilities? (describe)	$ Amount Capital Equipment

Identify key contributors required for this proposal.	Proposal Manager	Submission Date:

Marketing (1)	Date	Dir. Adv. Plan (2)	Date	Dir. Adv. Prog. (3)	Date
Director (1)	Date	Gen. Mgr. (2)	Date	Controller (4)	Date
Div. Controller (1)	Date	Asst. Controller (3)	Date	Sr. V. Pres. (4)	Date

* Totals in excess of $1,000 require bid comm. approvals—total authorization determines approval level: (1) $1,000 or less ☐ (2) over $1,000 ☐ (3) over $10,000 ☐ (4) over $25,000 ☐ N/L = Non-labor

FIGURE A1.2 Proposal task authorization form. CPFF: cost plus fixed fee; CPIF: cost plus incentive fee; FFP: firm fixed price; T&M: time and materials; FPI: fixed price incentive. BID COMM: bid committee.

Type of Proposal	Contract Effort			Proposal Volumes
___ Cost-Type ___ ______	___ Engineering Study	___ Production	___ Field Support	___ Letter ___ ______
___ Fixed Price ___ ______	___ Breadboard	___ Spares	___ ___________	___ Technical ___ ______
___ ROM ___ ______	___ Prototype	___ Data	___ ___________	___ Management ___ ______
				___ Cost ___ ______

REQUIRED EFFORT*	Electrical	Mechanical	Test	Systems Integration	Reliability	Data	Installation	Advanced Programs	Quality Assurance	Manufacturing	Tooling	Project Management	Computer	Technical Staff										
Bidders conference (include travel)																								
Detailed proposal schedule																								
Kick off meeting																								
Vendor/sub survey (include travel)																								
Teaming arrangements (include travel)																								
Detailed program schedule																								

Preliminary design																								
Validate customer specifications																								
Engineering costing																								
Hardware costing																								
Detailed bill of materials																								
Special tooling/test equipment																								
Proposal preparation (rough, final)																								
Customer presentation (include travel)																								
Negotiations (include travel)																								

* Explain all items checked but not included in budget request. ROM = rough order of magnitude

FIGURE A1.3 Proposal cost estimate checklist. ROM: rough order of magnitude (costs).

and support departments, as determined by the proposal manager. At this meeting, all pertinent information is distributed and discussed. The graphic representation of work (GROW) process is explained (if it is going to be used) as well as the resulting preparation techniques for storyboards. Storyboard blank forms are distributed. As a minimum, the following additional information is supplied to the attendees:

1. Proposal control form (see Figure A1.4)
2. Copies of the RFP/RFQ
3. Master storyboard, proposal theme and proposal outline
4. Proposal preparation schedule and budgets
5. Marketing intelligence and proposal strategy
6. Technical approach and costing guidelines

PROPOSAL TITLE: ______

RFP TITLE and NUMBER: ______

CUSTOMER NAME and ADDRESS: ______

PROPOSAL CHARGE NUMBER: ______

PROPOSAL MANAGER: ______

PROPOSAL ADMINISTRATOR: ______

CONTRACT ADMINISTRATOR: ______

FINANCIAL REPRESENTATIVE: ______

TECHNICAL WRITER: ______

COURIER EXPEDITER: ______

CRITICAL DATES:

TECHNICAL SECTION to TECH WRITING: ______

COST SECTION to ACCOUNTING: ______

SIGN-OFF MEETING: ______

DELIVERY to PUBLICATION ACTIVITY: ______

MAILING DATE: ______

FIGURE A1.4 Proposal control form.

7. Proposal preparation schedule of all proposal contributors' work
8. Assignment of tasks, as appropriate
9. Coverage of any other proposal effort items, as applicable

The proposal manager may appoint, if required, a technical team to evolve the technical aspects of the proposal and to determine conceptual feasibility and design, including risk areas and program scope.

Responsibilities of the Team

Following the kickoff meeting, the customer's requirements are reviewed and the program manager and members of the proposal team prepare a work breakdown structure (WBS), a preliminary program schedule, and a program milestone chart.

Storyboards are prepared. Upon completion of storyboards, the proposal team conducts a baseline concept and storyboard review prior to initiation of proposal writing. The object of this review is to summarize the design concept, confirm proposal strategy, identify any further engineering or design to be done, and verify responsiveness to known customer requirements. The team evaluates the adequacy of available drawings, sketches, and technical data for cost-estimating purposes. After this meeting (and additional review team assessments, as appropriate) proposal writing begins.

The proposal manager calls a management/cost meeting with representatives of management, accounting, and contracts, as appropriate. This meeting is the kickoff for the cost-estimating activity required to support proposal preparation. The meeting covers management and cost factors, such as:

Organization and personnel availability
Program management techniques
Company capability and experience
Facilities and equipment availability
Program plan
Task descriptions and task responsibility matrix
Schedules
Work breakdown structure
Cost estimating guidelines and instructions
Drawings/sketches/specifications
RFP/RFQ terms and conditions

At this point of the procedure, the three major proposal activities are under way concurrently: (1) preparation of technical proposal drafts; (2) preparation of organization and management proposal drafts; and (3) preparation of cost volume draft. As a guide in estimating the various tasks required, the definitions and procedures given in Sections 6, 7, and 8 below should be utilized by assigned personnel.

SECTION 5. RATES

Direct Labor

Direct labor rates (bid rates) are preestablished for the pricing of all proposals and are reviewed periodically to determine whether current conditions support their continued use. These rates are revised at least once a year, employing actual average rate data by skill category and skill level. An escalation factor that includes the effect of merit increase budgets, labor turnover, promotions, and reasonable anticipated changes in wage patterns is applied to these data.

Direct labor rates are applied in proposals by fiscal year to the estimated number of actual productive hours required to perform the required tasks.

Overhead Rates

Overhead rates are calculated as follows for each overhead pool:

$$\frac{\text{Projected overhead expense dollars}}{\text{Projected direct labor dollars}} = \text{Overhead \%}$$

Overhead and direct-dollar projections are based on the latest division workload and business projection and on related historical indices.

General and Administrative (G&A) Rates

G&A rates are calculated by adding the division expenses to corporate and independent research and development allocations that are based on cost input. The makeup of these expenses is as follows:

1. *Local Division Expenses*. These are based on the latest division cost projections for such functions as purchasing, contract administration, accounting, the general manager's office, and other support functions.
2. *Independent Research and Development*. This amount, if applicable, is based on an annual advanced negotiated agreement and is allocated on projections of cost input, by division.
3. *Corporate Expenses* These are allocated to the division on the basis of cost input. Corporate expenses include bid and proposal costs that are fully burdened and covered by an advanced negotiated agreement.
4. *Division cost input* includes all direct labor, overhead, and nonlabor dollars in the month of expenditure.

It is the responsibility of the accounting department to insure that all cost exhibits, including computerized labor/cost estimates include the proper labor and burden rates.

SECTION 6. DIRECT LABOR

Definition

Direct labor (ref. Policy No. ______) is the cost of productive time of personnel classified as direct labor employees, based on the hourly rate paid. Productive time is not limited to labor input that is incorporated in the end product, but is limited to the effort that can be identified with a particular cost objective (e.g., that of engineers, designers, project administrators, mechanical assemblers). Personnel who are classified as *direct* are those employees who can properly account for a majority of their direct productive effort and can properly charge a majority of their direct productive effort to cost objectives. Direct labor employees are outlined by job classifications in Table A1.1.

Estimating Direct Labor

The estimating of direct labor embodies a variety of techniques and methods. In a specific instance, an estimate may represent a combination of many methods and techniques and, of course, judgment. Figure A1.5 graphically depicts methods of

TABLE A1.1. Direct Labor Categories

Job	Labor Code
Senior Staff Engineer	1001
Senior Engineer	1002
Engineer	1003
Associate Engineer	1004
Junior Engineer	1005
Designer	1006
Mechanical Assembly Technician	1007
Electronic Technician	1008
Technical Writer	1009
Junior Production Engineer	1010
Field Service Representative	1011
Project Administrator	1012
Material Technician	1013
Production Engineer	1014
Metrology	1015
Machinist	1016
Toolmaker	1017
Product Electronics	1018
Calibration Technician	1019
Mechanical Shop Inspection	1020
Electrical Inspection	1021

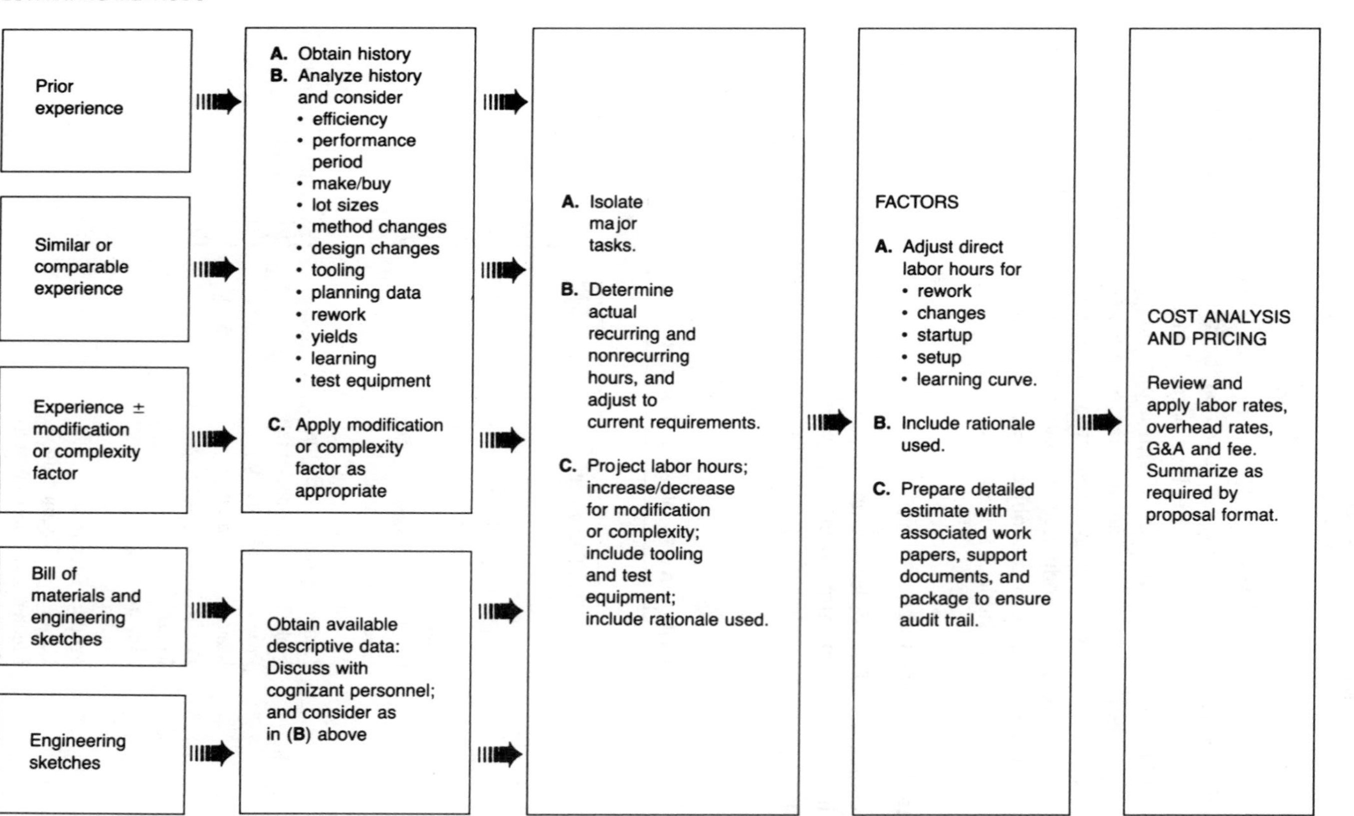

FIGURE A1.5 Flow chart for estimating direct labor charges.

estimating these labor requirements in addition to emphasizing specific aspects of each method. Whichever estimating technique is used, it must be for a specific task as defined by the work breakdown structure or statement of work. It is the responsibility of the estimator to prepare a concise audit trail through the data and be able to demonstrate that prudent business judgment was used in making the decisions on the labor values selected.

After the basis for estimate has been determined, the estimator will complete the detailed estimating sheets using the number of actual productive hours estimated to be required to perform the task, or block of tasks, for which the particular estimator is responsible.

SECTION 7. OTHER DIRECT COSTS

Materials

Definition of Materials. Materials classified as *direct* are purchased and/or requisitioned from stores for a sales order, work order, or divisional production order, and they will be incorporated into a contract end item or will be consumed or expended in the performance of a contract. Further, materials classified as direct should be easily and economically identifiable (and measurable) with a specific cost objective, which objective is necessary to meet the requirements of the contract. (Ref. Policy No. ______ , "Direct Versus Indirect Charging.")

Estimating Materials Costs. The estimating of materials requirements can encompass a variety and combination of methods. These are summarized in flow-chart form in Figure A1.6. It is impossible to formulate an approach that will cover every condition, and prudent business judgment must always be exercised by the estimator. All purchased items identified as "buy" on the bill of materials will be substantiated by vendor quotations, a catalog reference, or a copy of a previous purchase order for an identical or like item. If a previous purchase order price needs adjustment for quantity difference or anticipated increase due to inflationary factors, such adjustments must be accompanied by a reasonable explanation. Any overages for loss, shrinkage, breakage, waste, anticipated changes, or any other reason must be substantiated. Sources of all purchased item costs, if known, should be noted on a "Purchased Parts Justification Data Sheet." All procurements must conform to Policy No. ______ , "Make/Buy."

Travel

Travel expenses are those expenses incurred by an employee that are necessary while the employee is engaged in duties away from the normal place of business. The classification of travel expenses should be consistent with the classification of the associated labor charges, i.e., if an employee charges labor directly while away from the normal place of business, the related travel expenses should be charged

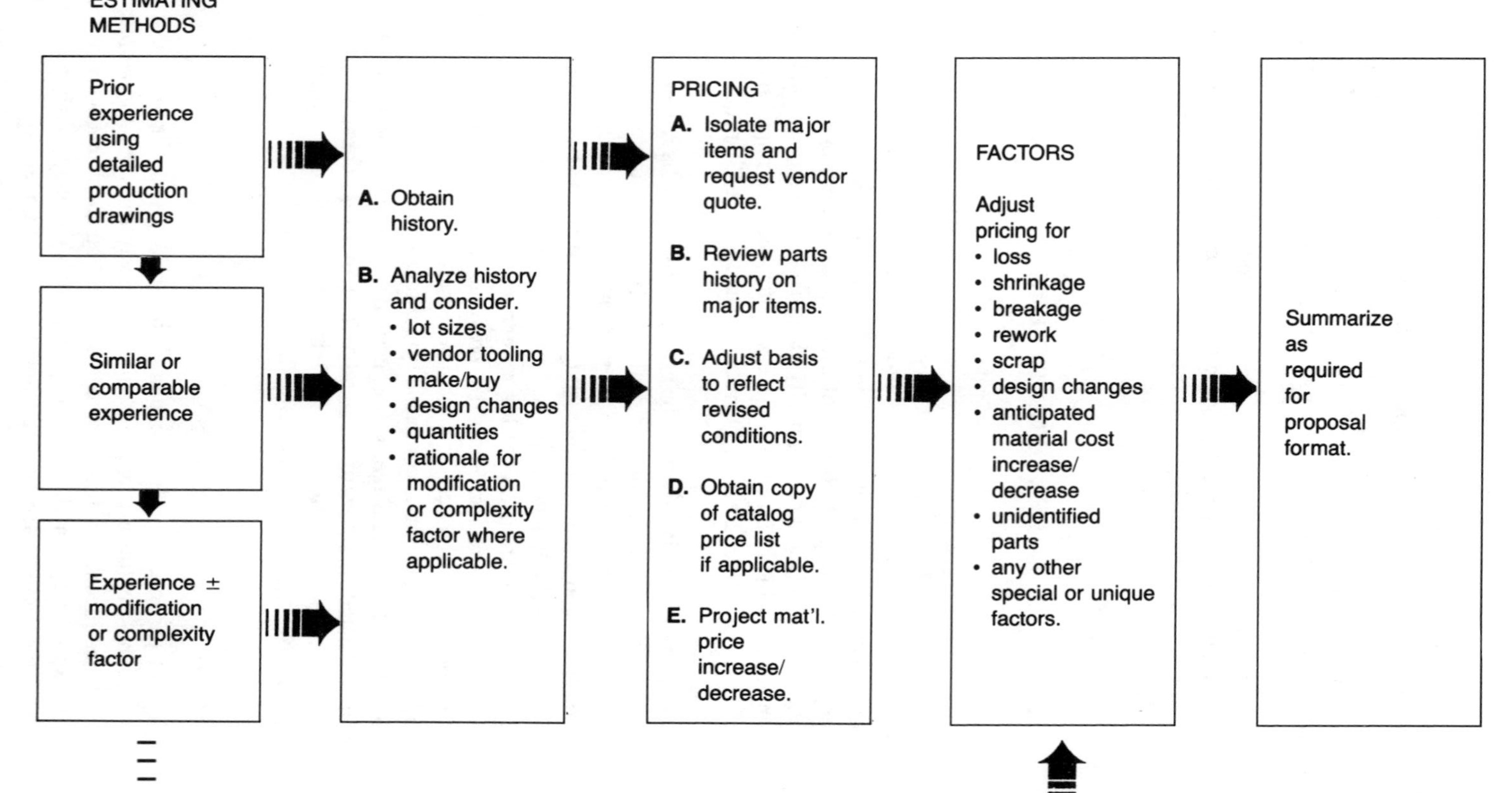
ESTIMATING
METHODS
Prior
experience
using
detailed
production
drawings
Similar or
comparable
experience
Experience ±
modification
or complexity
factor
A. Obtain
history.
B. Analyze history
and consider.
• lot sizes
• vendor tooling
• make/buy
• design changes
• quantities
• rationale for
modification
or complexity
factor where
applicable.
PRICING
A. Isolate major
items and
request vendor
quote.
B. Review parts
history on
major items.
C. Adjust basis
to reflect
revised
conditions.
D. Obtain copy
of catalog
price list
if applicable.
E. Project mat'l.
price
increase/
decrease.
FACTORS
Adjust
pricing for
• loss
• shrinkage
• breakage
• rework
• scrap
• design changes
• anticipated
material cost
increase/
decrease
• unidentified
parts
• any other
special or unique
factors.
Summarize
as
required
for
proposal
format.

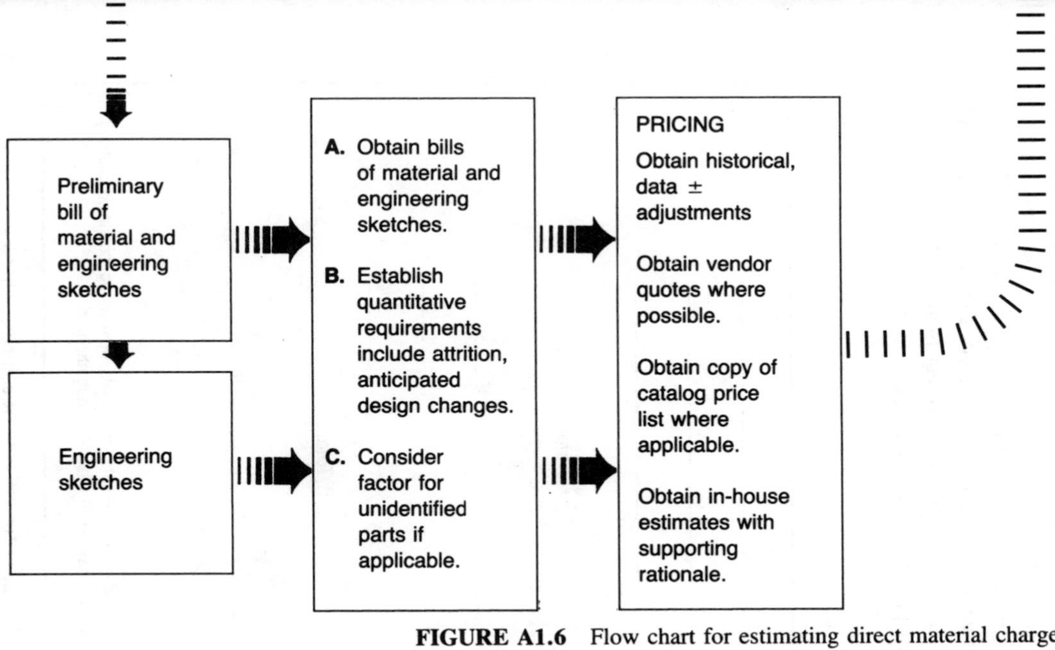

FIGURE A1.6 Flow chart for estimating direct material charges.

directly; if there is a proration of direct and indirect charges of an employee's labor charges while he or she is away from the normal place of business, the related travel expenses should be prorated in the same manner (Ref. Policy No. ______).

Travel requirements will be specified by each participating functional department in terms of destination, duration, number of personnel, and the reason for each trip on a travel estimate form (shown in Section 9). Travel and subsistence rates are supplied by accounting, utilizing the latest published airline fares and approved auto rental and daily subsistence rates.

Computer Costs

Computer costs are estimated for each task required to perform the contract.

In order to comply with company regulations and cost-accounting standards concerning consistency in estimating, it is required to quote corporate computer facility costs in the manner in which they are accumulated. Figures A1.7 and A1.8 are forms to be used for the backup of computer cost estimates. These forms must be submitted as backup for all proposals and "estimates to complete."

Proposal Number: ______________________

Project Name: ______________________

Task Number: ______________________

		Rate		
Computer Time	(minutes) ________ ×		= $ ________	
Programming Time	(hours) ________ ×		= $ ________	
Data Storage	(weeks) ________ ×		= $ ________	
Miscellaneous*			= $ ________	
		Total		$ ________

Estimate Prepared by ______________________ Date: ______________

*Miscellaneous charges result from the usage of dedicated terminals, networking, scanning, plotting, etc. A separate form should be used to break down miscellaneous charges.

FIGURE A1.7 Worksheet for estimate of computer charges.

Proposal Number: ______________________________

Project Name: ______________________________

Task Number: ______________________________

		Rates		
Dedicated Communications Terminal	(months)	________	×	$ ________
Dedicated Computer Terminal	(months)	________	×	$ ________
Network Usage	(hours)	________	×	$ ________
Optical Disk Drives	(hours)	________	×	$ ________
Keyboarding	(hours)	________	×	$ ________
Laser Printers	(hours)	________	×	$ ________
Scanners	(hours)	________	×	$ ________
Plotters	(hours)	________	×	$ ________
		Total		$ ________

Estimate Prepared by ______________________ Date: ______________________

FIGURE A1.8 Form for miscellaneous estimate of computer charges.

Consultants

In the event that the services of consultants are required, the proposal manager will inform the contracts department of the particular type of specialist required. A subcontracts administrator will be assigned to solicit quotations.

The support for a consultant cost should consist of an identification of the consultant by a brief biography of experience in his or her specialized field, the number of days the consultant is required, the fee rate per day, and travel expenses.

Subcontracts

Major Subcontracts. A major subcontract is distinguished from purchase orders and routine subcontracts by the following criteria:

1. A major subcontract is the procurement of a subsystem or major component, where the supplier has a systems management responsibility, or where the

supplier performs to a work statement that includes development and/or performance specifications, or where the supplier performs major engineering and design effort.

2. A "major" subcontract excludes:
   a. Supplier catalog items
   b. Supplier-published specification items
   c. Supplier items built to detail or assembly drawings.
3. It is not possible to define a major subcontract in terms of dollar value alone. However, experience indicates that all individual procurements in excess of $ ________ or which represent of the program's prime cost normally qualify as major under the definition outlined above.
4. In unusual circumstances, where a particular procurement does not meet the normal criteria established above, the potential procurement may be considered a major subcontract only after the circumstances are specifically documented in the procurement files and approved by the senior vice-president.
5. All subcontracts that qualify as major within the definition outlined above will be reviewed and approved in writing by the controller prior to incorporation in a proposal or quotation that commits the corporation.

Routine Subcontracts. Subcontracts are designated as either *routine* or *major* for the purpose of determining whether the major subcontract handling rate should be applied or, if designated routine, the normal general and administrative rate.

Whether the subcontract qualifies as major, as defined previously, or is considered routine, the procurement and management of the subcontract becomes the responsibility of the subcontract administration section of the contracts department. Policy No. ______ specifically defines the processing responsibilities involved in subcontract procurement.

Where it is necessary to subcontract major subsystems, subassemblies, critical components, or black boxes, detailed information should be provided on subcontractors. When a subcontract is defined as major, a special section of the proposal should be set aside to integrate the proposed subcontract work into the proposal. The subcontractor's proposal should be presented in the same format as that prescribed for the prime.

Request of Data from Subcontractor. In addition to complying with the description, work statements and specifications supplied to the bidders in conjunction with the RFP/RFQ requirements, the subcontractors' proposal should include the following:

1. Type of contract
2. Cost breakdowns for each task/item and a summary cost breakdown for all tasks showing material, hours by labor classifications wage rates, burden rates, and extended amounts in all areas
3. Unit cost breakdown by element of cost for each type of equipment furnished

4. Priced bills/list of major materials
5. Exhibit of approved wage, burden, and other applicable rates. Specify approving agency and approval period
6. Summary and separate man-loading tabulation for each task and subtask specified in work statement
7. Special tools and special test equipment, priced in the same format as above but each as a separate item
8. Profit/fee. If either weighted profit guidelines or profit incentives are applicable, the subcontractors should submit substantiating justification in support thereof
9. If factors or contingencies are used, they must be explained and justified
10. Details of any deviations from the terms, conditions, schedule, or technical requirements of the statement of work and specifications supplied by the buyer.

Interdivisional Work

Request for interdivisional cost estimates will be initiated by the subcontract administrator assigned to the proposal. If the request is for other than a price-listed catalog item, the interdivisional proposal should be requested in the same format and detail as that prescribed for prime procurements.

SECTION 8. ESTIMATING

Engineering labor is estimated on the basis of the defined tasks in the work breakdown structure and/or the statement of work. The engineering directorate is responsible for estimating tasks such as studies and analyses, design, reliability requirements and analyses, technical documentation, manufacturing liaison, subcontracting, interdivisional work and vendor liaison, data/report manuscript preparation, customer reviews and liaison, and field support.

Design and Drafting

Design and drafting estimates are based on the drawing count and sizes established from the preliminary engineering design concepts, drawing tree, indentured parts list, customer drawing form requirements and estimated drawing complexities, and other available information. Standard drawing estimating factors are then applied against the drawing count.

Technical Documentation (Data Management)

Technical documentation estimates are based on customer data requirements and related internal requirements supplemented by page counts, illustration counts/complexities and overall scope of the writing, editing, and production effort as defined by the contract data requirements list, as well as the release and control of such documents (document control, data retrieval/retention/ microfilming). Estimates also consider labor involved in managing subcontractor data requirements utilizing routine surveillance/ audit techniques and communications.

Reliability

Reliability estimates are based on the labor involved in preparing a reliability program plan that includes a detailed listing of specific tasks and procedures to implement and control the reliability function. The magnitude of the reliability plan is determined by the operational use of the end item produced as well as the specific standards required by the request for proposal. Reliability labor estimates include but are not limited to the effort involved in the identification, assessment, and solution of problems that impact the specified contract requirements. Estimates include liaison efforts to disseminate and implement reliability requirements by all cognizant engineering, manufacturing, quality control and test personnel, as well as vendors and subcontractors.

Manufacturing and Testing

Manufacturing labor is used when the primary function is the production of end-item hardware. This labor effort encompasses production planning; design and fabrication of special and production tools; special test equipment and handling equipment; parts fabrication and assembly; and in-process, final, and environmental testing.

Manufacturing labor and material estimates will be based on the drawing tree and/or the bill of materials as drawn up from the drawings supplied as cost-estimating input, i.e., drawings, sketches and specifications from design and drafting, engineering, and the customer. Assembly and major subassembly conceptual layouts with overall dimensions, general tolerances, materials, similarity identifications, process specifications, and major special equipment identifications are considered the minimum for adequate manufacturing estimating.

Functional, acceptance, and environmental tests, as well as plans and equipment for the tests are the responsibility of this area, as are any modifications to facilities.

Product Assurance

Product assurance estimates are based on the hardware, specifications, drawings, and assembly/test activities previously defined by the request for proposal or developed during the estimating activity. These estimates include the following areas:

1. *Quality Control.* Source and receiving inspection, fabrication and assembly in-process inspection, test witness, discrepancy resolution, special process control, and subassembly build support

2. *Quality Engineering.* Design review, quality planning, malfunction analysis, subcontractor control, system test witness, discrepancy resolution, and field technical support
3. *Configuration Management.* Drawing and change control, as-built vs. design comparisons, configuration management audits, subcontractor configuration management control.

SECTION 9. COST COLLECTION AND SUPPORTING DATA

The project administrator for each contributing department or section collects all the pertinent data from the estimators assigned the various tasks or subtasks defined by the statement of work and work breakdown structure. These data will consist of time-phased input sheets, which detail, by labor classification, the hours per task per month or year for the duration of the program and the estimated nonlabor dollars shown in the time period of commitment. Each input sheet must be accompanied by a task element substantiation sheet (Figure A1.9), properly describing the task and the estimator's breakdown of the hours assigned each portion of the task, with rationale for the estimate. When material costs are included, they are to be described and listed on a purchased parts justification sheet (Figure A1.10), showing quantity, unit cost, total cost, and the source or rationale for the estimated costs. If travel is anticipated in the task, a travel and subsistence cost substantiation form (Figure A1.11) must be submitted showing destination, number of people, number of days, and the reason for the trip.

Computer cost backup will be furnished as described in Section 7. Interdivisional and/or subcontract estimates, if required, will be supported as described in Section 7. The responsible project administrator reviews the cost data for proper format and backup material, and if satisfied, submits it to the department manager or section supervisor for review and signature approval. The various cost packages are then submitted to the proposal program management administrator for review.

If the data meet the approval of the proposal administrator, they are then transferred to the cost analysis office. The cost analyst assigned to the proposal reviews the cost packages for errors, omissions, format, and the overall reasonableness of the material cost justification. Travel forms and computer estimates are checked for proper rates and all calculations verified. The input forms for engineering, manufacturing, and, if applicable, field support are reviewed and corrected, if necessary, for titling, hour count, and period of performance. Interdivisional and subcontract costs are compared with the latest interdivisional and subcontract proposals for proper costs and period of expenditure. The information on the input data sheets is then entered into the computer and a time-phased, dollar, and manpower computerized run is produced. The cost analyst checks the tab run for proper rates and keyboard input errors. If all is in order, the analyst prepares the cost volume or cost proposal (including any updates or changes) and the cost backup packages are placed in the cost analysis master proposal file.

SECTION ________ TITLE ______________________________ DATE ____________

The labor and nonlabor elements of cost summarized below are required for performance of the following functions:

Element and Amount	Description of Performance

SPO NO.	TASK NO.	TITLE	SHEET
			OF

FIGURE A1.9 Task element substantiation form.

PARTS—TOOLS—TEST EQUIPMENT

REPORTING SECTION: ________ ATTACHMENT: ________

Part Number	Vendor	Estimate Source			Qty.	Price		Lot Charge	Total Price	Notes:
		PO	Quote	Date		Unit	Total			

Proposal Number:	Title:	SPO:	Task:

FIGURE A1.10 Purchased parts justification sheet.

Destination	1	2	3	4	5	6	7	8	9	Reason for Trip
	No. Pers.	No. Days	Total Person Days	Subsistance	Automobile Rental	Air Fare	Total Trip Costs	No. of Trips	TOTAL	
			(col. 1 × col. 2)	(col. 3 × $34)			(cols. 4 + 5 + 6)		(col. 7 × col. 8)	

FIGURE A1.11 Travel and subsistence expense form.

The cost proposal, together with the technical and management proposals (considered preliminary at this time), are then reviewed by the proposal management team for technical and pricing approval. If no revisions or updates are required, the proposal package is then submitted for division and corporate review and approval, as required by company policy.

The proposal cost analyst, with the cooperation of all departments that input estimates to the proposal, will inform the proposal manager and the division controller of any revisions or additions to previously provided cost information that affect the proposal. Any significant changes are brought to the attention of the appropriate management team and the contracts department notifies the customer. In the event that the division succeeds in winning the contract, this updating process continues up to the time of contract negotiations.

SECTION 10. MANAGEMENT REVIEWS

Order of Review Activities

The review cycle for proposals is as follows:

1. Department managers review technical and cost estimates for text, format, supporting data, and compliance with statement of work and work breakdown structure requirements.
2. Proposal administrator performs preliminary proposal compilation and review.
3. Proposal manager reviews technical and cost estimates.
4. Contracts manager reviews the proposal for contents and compliance with the request for proposal.
5. Proposal engineer/manager reviews the technical proposal.
6. Program office reviews the organization and management volume.
7. Cost analysis team reviews estimates and prepares the cost volume.
8. At this point in the review cycle, any revisions necessary are completed and the proposal package is again reviewed by the proposal team and approved for submission to management.
9. The proposal package is submitted for division management review and sign-off.
10. The proposal package is submitted for corporate review and sign-off (if required by corporate policy).

The final review will be conducted by the following personnel in addition to the division general managers:

1. Director of the company segment that will be responsible for the work (Directorate)
2. Accounting: division controller

3. Contracts: manager of contract administration
4. Technical director
5. Marketing: director of advanced planning
6. Manager of the following support groups if their input is significant:
   a. Purchasing
   b. Product assurance
   c. Reliability
   d. Design and drafting
   e. Data management
   f. Manufacturing and testing
   g. Engineering.

These people are notified after the responsible directorate has established a specific time and location for the general manager's review.

Proposal Responsibilities

A list of participants and their proposal responsibilities follows:

A. Proposal Manager
   1. Overall proposal responsibility
   2. Prepares proposal task authorization and identifies participants, services required, and special facilities/equipment
   3. Establishes proposal schedule, calls key meetings
   4. Develops master storyboard and proposal theme
   5. Furnishes copies of request for proposal to all participants as required
   6. Establishes proposal budgets and assignments
   7. Supervises preparation and disseminates technical data, work breakdown structure, and detailed program schedule for cost estimating
   8. Supervises/reviews technical and management proposal
   9. Keeps director informed of proposal progress/problems
   10. Establishes program plan
   11. Establishes program organization
   12. Participates in preparation of, and approves, bill of material with incorporated make–buy decisions
   13. Establishes technical, management, and pricing strategies
   14. Approves technical proposal management proposal, and cost package
   15. Writes statement of work
   16. Approves program schedule

17. Serves as chairperson of proposal kickoff meeting, baseline concept review, all director reviews
18. Establishes proposal review cycle and resolves all reviewer comments

B. Director of Responsible Company Segment
1. Assigns proposal manager
2. Approves proposal task authorization and general assignments
3. Approves technical, management, cost strategies
4. Approves proposal theme and approach
5. Approves technical, management, cost proposals
6. Recommends fee/profit structure
7. Approves statement of work
8. Recommends fact-finding, negotiation actions
9. Serves as chairperson of division review

C. Technical Documentation/Data Management Group
1. Writes assigned sections
2. Critically reviews, edits, and/or rewrites input material
3. Formats, illustrates, produces all volumes
4. Supervises security aspects, including marking, storage, and handling
5. Prepares estimates for documentation/data management support services
6. Recommends special artwork and presentation techniques
7. Prepares concept illustrations and engineering drawings

D. Product Assurance Group
1. Prepares quality-assurance program plans
2. Prepares quality-assurance estimates
3. Prepares bill of materials addendum for special inspection equipment
4. Writes product assurance section of proposal

E. Reliability Group
1. Conducts reliability analysis
2. Prepares reliability program plan
3. Prepares reliability estimates
4. Participates as an advisor in bill of materials preparation
5. Writes reliability section of proposal

F. Manufacturing, Assembly, and Test Group
1. Prepares preliminary and final bills of materials
2. Prepares manufacturing, assembly, and test support estimates

3. Prepares bill of materials addenda for special test equipment
4. Writes manufacturing and test sections of proposal

G. Contract Administration Group

1. Reviews request for proposal terms and conditions, identifies risks, recommends exceptions and/or qualifications
2. Establishes, via consultation with engineering and technical documentation, a position on "rights in data" (legal position of the firm in the possession of patents, copyrights, and trademarks)
3. Establishes proposal restrictive legends (notations that the information contained in the proposal is the sole property of the proposer and that it should not be distributed by the customer to other bidders or competitors) and/or other proprietary markings
4. Reviews statement of work
5. Prepares incentive or award fee plans
6. Obtains subcontract, interdivisional, and consultant quotations
7. Makes recommendations regarding contractual aspects
8. Furnishes and executes required certifications
9. Prepares transmittal letter in concert with proposal manager and advanced planning group
10. Recommends fact-finding, negotiation plans

H. Procurement Group

1. Provides purchased parts quotations and delivery schedules
2. Identifies, with engineering, long-lead items (for both proposal quotations and program delivery)
3. Participates as an advisor in bill of materials preparation and make–buy decisions
4. Prepares procurement/manager plans

I. Security Group

1. Provides overall guidance on security aspects
2. Handles all special security requirements

J. Cost Analysis Group

1. Provides any special cost-estimating forms to all estimating groups
2. Prepares special cost volume supporting text as required
3. Verifies final cost elements with contributors
4. Prepares all required cost exhibits, including computerized manpower/cost spreads if required
5. Provides all labor/burden rates

6. Prepares computer use forms
7. Maintains proposal cost backup files

K. Advanced Planning Group

1. Furnishes marketing intelligence, including recommended technical, management, and pricing strategies
2. Recommends proposal theme and approach
3. Updates intelligence throughout proposal preparation with emphasis at proposal meeting, baseline concept review, cost/management kickoff meeting, and division review
4. Recommends proposal support tactics
5. Prepares proposal follow-up plan
6. Drafts corporate management interest letter and proposal submittal letter
7. Establishes proposal delivery method and date
8. Responsible for coordinating and controlling all contacts with customer during proposal preparation

APPENDIX 2
CHECKLIST FOR PROPOSAL REVIEW

Technical Approach

1. Are the technical requirements as seen by the customer clearly interpreted and not simply parroted from the request for proposal?
2. Does the proposal reply to the basic minimum requirements as given by the customer?
3. In the event of deviations or alternatives, are detailed reasons for these recommendations given, especially in terms of the desirable results that the customer will receive? (These results might include such items as improved performance, lower costs, greater producibility, earlier delivery, and simpler maintenance.)
4. In the event that certain program objectives are incompatible with other program goals (e.g., range versus speed, payload versus speed) does the proposal show that the optimum solution, all factors considered, has been chosen?
5. Does the approach consider such matters as logistics, maintenance, retrofitting, and problems of the using organization?
6. If originality has been identified as a requirement, does the proposal present a unique, imaginative approach?
7. Does the approach avoid overengineering and oversophistication?

Technical Competence

1. Does the proposal provide convincing assurance of specific technical competence for this project?
2. Does the proposal give specific examples of similar projects successfully completed by the company?

3. Do the résumés of key personnel who have been designated to execute the program relate their experience to the specific needs of this project? Has extraneous biographic information been eliminated?
4. Is the availability of qualified manpower clearly detailed in terms of man-hours for both full- and part-time people?
5. Does the proposal clearly indicate that the company has adequate space and facilities, both general and special, to perform the work efficiently and on schedule?
6. Are special facilities (such as dust-free laboratories, data-processing equipment, construction tools, special laboratory equipment, etc.) to be provided for the project clearly enumerated?
7. It is clearly indicated that all required facilities will be available when needed for this project?
8. Where arrangements with subcontractors are proposed, is specific evidence given of the subcontractor's commitment to make people and facilities available when required?

Schedule

1. Does the proposal provide convincing assurance that the customer's delivery and due dates will be met?
2. Is sufficient detail regarding master scheduling, programming, follow-up, and other like functions given to reinforce this assurance?
3. When subcontractors and major suppliers are involved, are sufficient safeguards built into the proposed scheduling system to ensure that their schedules will comply with the master program schedule?

Reliability and Quality Control

1. Does the proposal spell out reliability and quality-assurance provisions in such a manner that the customer can have no doubt that quality standards will be met or surpassed?
2. Is a clear understanding of reliability and quality control requirements reflected in the proposal? Are deviations, if any, satisfactorily explained? Are testing procedures outlined in sufficient detail?
3. Is evidence given that the system of inspection guarantees the same high level of quality and reliability from subcontractors as from in-house production?

Program Direction and Management

1. Does the proposal clearly demonstrate an understanding of the customer's concern with the management of this project?

2. Is evidence given that top level management effectively communicates with and inspires its personnel?
3. Does the proposal specify the required number of the right types of management people?
4. Is evidence given that supports the selection of subcontractors, not only from the standpoint of their technical and manufacturing capabilities but also their management philosophy and talent?
5. When subcontractors are involved, has it been clearly stated how their management will participate in the program?

Manufacturing Competence

1. Does the proposal clearly indicate that the company has adequate manufacturing space and facilities, both general and special, to perform the work efficiently and on schedule?
2. Is the specialized equipment and the processes required for the project given sufficient prominence in the proposal through photographs and descriptive information?
3. Does the proposal call attention to the high standards of production and test procedures?
4. Does the proposal specifically state that all required facilities are available for the project at the times required in the project?

Field Support

1. Does the proposal provide documented assurance that field service and support activity, as required by this project, are of a high caliber at reasonable cost?
2. Does the proposal highlight the magnitude and scope of the field service and support area?
3. Does the proposal provide specific examples of accomplishment in the field service and support area?

Price

1. Is this the lowest price considering (a) long range versus immediate return and (b) probable competitive price range?
2. Is there adequate evidence that subcontractors and suppliers have submitted their lowest realistic cost estimates?
3. Have man-hour, space, facility, and other cost factors been estimated properly?
4. Are overhead, burden rates, and proposed profit or fee reasonable for this type of project?
5. Is the extent of pricing detail given consistent with the importance of these details?

6. Does the proposal provide convincing evidence that the company is properly oriented and organizationally structured to meet the specific management demands of this project, especially in terms of providing the necessary communication functions, both internal and external; of traceability of decision-making; and of integration of all project bits and pieces?
7. Is evidence given of management's understanding of how the specific project fits into the customer's overall requirements? Does the proposal show how this project relates to long-range business objectives of the company?

APPENDIX 3
GSA'S AUTOMATED PRODUCT LISTING SERVICE (APLS)

Overview

The APLS is a computer database system which has been developed as a tool to reduce procurement research time and to provide basic assistance to General Services Administration (GSA) customers in the product selection phase of their procurement efforts. Suppliers are requested to submit their price lists in a specific format for entry into the database. This service is available to any U.S. government customer with access to a computer that has telephone communication capability. Customers will initially be able to review product pricing information for the contracts on the Federal Supply Schedule for copiers. Later, many other office machine items will be added.

Milestones in the Development of the APLS

April 1990: The APLS was introduced to suppliers of electrostatic copying equipment at an open conference held in Crystal City, Virginia. Federal Supply Schedule 36 IV was selected as the test vehicle for this exciting new program, which uses an electronic data acquisition system to facilitate the item selection process.

June 1990: The feedback that GSA received after the April conference was favorable, with more than 95% of prospective vendors expressing interest. Therefore, the Engineering and Commodity Management Division of GSA furnished vendors with instruction manuals and data-entry program disks to allow the vendors the opportunity to familiarize their staff with the data collection format.

August 1990: At the Office Equipment '90 Expo held in Crystal City, Virginia, GSA conducted two workshop sessions designed to acquaint customers with the APLS database system.

September 1990: As awards were being made by GSA for the copier products, the Engineering and Commodity Management Division was providing the final production versions of the APLS data entry disks and instruction manuals to the contractors. The participating contractors use these program data entry disks to furnish GSA with the relevant data on their awarded products for incorporation into the APLS database.

October 1990: The first data entry disks were completed by participating contractors and the APLS went on line for customer access.

January 1991: An online bulletin board system (BBS) was added to update APLS users on important announcements (contract modifications, promotional sales, discounts, etc.) which may apply to current GSA contracts. An integral E-Mail system (electronic mail) is implemented to facilitate communication between GSA, contractors, and customers.

April 1991: The APLS database is expanded to include award information for typewriter contracts and required consumable supply items for copiers and laser printers (toner, developer, etc.). An indexed contractor address file for these items is placed on line.

June 1991: Despite some late awards, vendor acceptance of the system for the initial year was at the 70% level, with additional participation anticipated.

APLS Features

The APLS provides customers with 24-hour access to current information on contracts. The menu selections and search techniques (based on the selection of required system features) enable the customer to rapidly identify suitable products to match specific configuration requirements. Within the APLS, both detailed and summary report formats are available.

The APLS partitions the products by their acquisition plan (i.e., rental, lease to purchase, purchase, renewal). Within a specific acquisition plan, products are arranged based on their monthly volume band. As a final step, within each of the monthly volume subgroups, model configurations are ordered by their calculated recurring costs, as derived from life cycle cost (LCC) formulas. This process allows customers to readily identify the most affordable system to meets their needs.

A Look at the Future of the APLS

The next few years will see continued expansion of the APLS online database to encompass additional items. Current plans include support for higher speed communication modems (9600 baud), network connectivity of the APLS with other government computer systems, and the evolution of the APLS to include multiline fax and voice-processing systems to handle user requests for copies of federal supply schedules and vendor price lists.

BIBLIOGRAPHY

Aljian, George W., ed. *Purchasing Handbook*, 3rd ed. New York, McGraw-Hill, 1984.

Blanchard, Benjamin S., and Wolter J. Fabrycky. *Systems Engineering and Analysis*, 2nd ed. Englewood Cliffs, N.J.: Prentice-Hall, 1990.

Blanchard, Benjamin S. *System Engineering Management*. New York: John Wiley & Sons, 1991.

Burt, David N. *Proactive Procurement*. Englewood Cliffs, N.J.: Prentice-Hall, 1984.

Commerce Business Daily, published daily by U.S. Department of Commerce, U.S. Government Printing Office, Superintendent of Documents, Washington, D.C.

Dyson, Robert G. *Strategic Planning: Models and Analytical Techniques*. John Wiley & Sons, Ltd. Chichester, England: 1990.

Hartley, John R. *Concurrent Engineering*. Productivity Press, Cambridge, MA, 1991.

Heinritz, S., and F. Farrell. *Purchasing Principles and Applications*. Englewood Cliffs, N.J.: Prentice-Hall, 1981.

Helgeson, Donald V. *Handbook for Writing Technical Proposals That Win Contracts*. Englewood Cliffs, N.J.: Prentice-Hall, 1985.

Hirano, Hiroyuki. *JIT Factory Revolution*. Productivity Press, Cambridge, MA, 1989.

Holtz, Herman. *The Consultant's Guide to Proposal Writing*. New York: John Wiley & Sons, 1986.

Holtz, H., and T. Schmidt. *The Winning Proposal: How to Write It*. New York: McGraw-Hill, 1981.

Kaplan, Marshall H. *Acquiring Major Systems Contracts*. New York: John Wiley & Sons, 1988.

Kaufmann, Arnold, and G. Desbazeille. *The Critical Path Method*. Gordon & Breach Science Publications. New York, 1969.

Lefferts, Robert. *Getting a Grant*. Englewood Cliffs, N.J.: Prentice-Hall, 1982.

Loring, Roy, and Harold Kerzner. *Proposal Preparation and Management Handbook*. New York: Van Nostrand, 1982.

Marakon Associates. *The Interaction of Growth and Profitability*. (A presentation to the Institute of Management Sciences) November 11, 1980.

McCready, Gerald B. *Marketing Tactics Master Guide*. Englewood Cliffs, N.J.: Prentice-Hall, 1982.

Moder, Joseph J. et al. *Project Management with CPM, PERT and Precedence Diagramming*, 3rd ed. New York: Van Nostrand Reinhold, 1983.

Mulvaney, J. E. *Analysis Bar Charting*. ILIFFE Books, Ltd, London, 1969.

Ostwald, Phillip F. *Cost Estimating*, 2nd ed. Englewood Cliffs, N.J.: Prentice-Hall, 1984.

Peters, Thomas J., and Robert H. Waterman, Jr. *In Search of Excellence*. New York: Harper & Row, 1982.

Stewart, Rodney D. *Cost Estimating*, 2nd ed. New York: John Wiley & Sons, 1991.

Stewart, Rodney D., and Ann L. Stewart. *Managing Millions, An Inside Look at High-Tech Government Spending*. New York: John Wiley & Sons, 1988.

Tracy, John A. *How to Read a Financial Report*. New York: John Wiley & Sons, 1980.

U.S. Office of the Navy, Program Evaluation and Review Technique, Washington, DC, 1958.

QUICK LOOK INDEX

SIXTEEN KEYS TO A COMPETITIVE PROPOSAL

INDEX